D0321808
al drift and the
OC: S
D

New views on an old planet

NEW VIEWS ON AN OLD PLANET

Continental drift and the history of earth

TJEERD H. VAN ANDEL
Stanford University

Cambridge University Press

Cambridge
London New York New Rochelle
Melbourne Sydney

To the memory of J. H. F. Umbgrove, whose *The Pulse of the Earth* gave me my first taste for the subjects discussed in this book

Published by the Press Syndicate of the University of Cambridge
The Pitt Building, Trumpington Street, Cambridge CB2 1RP
32 East 57th Street, New York, NY 10022, USA
10 Stamford Road, Oakleigh, Melbourne 3166, Australia

First published 1985
Reprinted 1986, 1987, 1989

Printed in the United States of America

Library of Congress Cataloging in Publication Data
Van Andel, Tjeerd H. (Tjeerd Hendrik),1923–
New views on an old planet.
Includes index.
1. Geology. 2. Continental drift. I. Title.
QE26.2.V36 1985 551.1'36 84 – 14251
ISBN 0 521 30084 3

Contents

Contents

Preface

The study of geology was a science which suited
idle minds as well as though it were history.
Henry Adams, *The Education of Henry Adams*

This is a book about change. Change is much in evidence in our society, where it appears to accelerate continuously, but others have dealt with that theme. Here I shall deal with our continuously changing earth.

We are inclined to view nature as constant, the earth solid, the sea eternal, the climate dependable, on any timescale that matters to us, though sometimes perturbed by landslides, storms, or earthquakes. Geologists, too, have long regarded the earth as well-behaved, its evolution slow and driven by the same forces, with the same intensities and in the same combinations, that drive it still. Rather a dull history, were it not for the excitement provided by the evolution of life.

This view seems no longer valid today. More probably, the history of the earth was a history of many brief intervals of dramatic change between longer times of relative quiescence. Mountains rose, then disappeared; ice ages came and went; even the positions of the continents changed continuously.

There is more: To discuss earth history today is to do it against the backdrop of a rapidly changing earth science. In the 1960s, geology was altered profoundly by the adoption of a new ruling theory that included the acceptance of continental drift. This revolution has left nothing untouched; few of our views of the earth remain as they were 20 years ago, and more revision is yet to come.

A dynamic earth, seen through the eyes of a dynamic science – that is the theme of this book. To deal with this in concepts rather than through sequences of events, and to deal with it on behalf of anyone who considers the subject worthy – that is its mode.

In books and lectures, knowledge of the earth is usually purveyed in terms of its processes, the work of wind and water, the deformations of the crust that make mountains, the origin of rocks. Yet the earth cannot be fully understood by such processes alone. There is more to the Grand Canyon than the physics of flowing water. It, like every other feature of the earth, is the product of a chain of events reaching far into the past, each event dependent on those that went before, each influencing what happens next. Without question, geology is a historical science, and geologists are often at their best when they think and speak as historians. To read the history of the earth from the fragmented and incomplete archives of the rocks requires more than an appropriate dose of physics, chemistry, or biology, more than ingenuity, perseverance, and a critical eye. It also demands a subtle quality that is difficult to define but whose absence is quite obvious: a historical sense. And courage, as well, because the canvas is large, and the brush strokes must be bold.

The following pages present what I regard as the main features of the history of the earth. The format is unusual, because we begin at the end rather than at the beginning, the end being more familiar to us than that remote planet where everything began 4 billion years ago. This is not a step-by-step account of the sequence of events, and it is not in these pages that the specific history of any given time or region can be found. Rather, I have woven together the major strands of change: the drifting of continents, the fluctuations of climate, the procession of life, illustrated when necessary with brief examples here and there. Not always has my knowledge sufficed to keep each strand separate from the others and in sharp focus. Frequently I have had to raise questions rather than provide answers. Sometimes things that seemed good to know did not fit well into the narrative; a few such stories I have told anyway, because I thought them fascinating, but my account of the history of the earth does not strive for even modest completeness.

Another perspective of this book is time. Change occurs on many scales of time and space. The weather changes by the hour and by the day, whereas climate does so by decades and centuries. Life may suffer drastic setbacks in a very short time, then take millions of years to recover and rise to greater complexity than before. Continents drift in tens and hundreds of millions of years. Inevitably, long times and vast spaces, rather than brief events and local details, dominate this book.

Change can be steady and progressive or episodic and oscillating, even cyclic if events recur with predictable regularity. The history of the earth embraces all these modes. Geologists once were sharply divided into two camps: the gradualists, emphasizing the steady march, and the catastrophists, dreaming of sudden change. The catastrophists lost out more than 150 years ago, but these days we appear to be converging on a middle way.

This book, written for anyone who has an interest in the history of the earth around us, requires no prior knowledge of geology, and I have held formal terminology to a minimum and have kept mathematics out of it. I have also, in the interest of a smooth flow, ignored some standard practices, such as adding the term "period," "epoch," or "era" to every stratigraphic name. The book was born from an undergraduate course at Stanford University taught mainly to students who did not intend to become geologists. It was my impression that they found the approach challenging and rewarding, and I hope that my readers will do so as well. If it should serve other teachers as background material, I shall be pleased, but I certainly did not intend it just as a textbook.

Unavoidably, the route I have taken is personal, and the choice of subject matter and the level of its treatment have been influenced by my experiences and preferences. A balance must be struck, of course, but that may have to be accomplished with further reading, for which I have made suggestions at several levels of difficulty. On the other hand, the customary apparatus of references and documentation seems out of place here.

Collectively, I thank those who wrote or said what, over the years, my mind has selectively absorbed, and I owe much to many, not the least to those who bore with patience the effects these labors had on my moods and my accessibility. Gratitude is especially due to the students in my Earth History course who helped me, not all unknowingly, to find out whether or not a discourse would work. The figures have been drawn especially for this occasion; where appropriate, sources have been listed in the back of the book. I thank my wife Marjorie for doing those drawings in Chapters 17 and 18 that were beyond my competence. James D. Valentine, Thomas J. M. Schopf, Michael O. McWilliams, and Alfred Kroner, as well as some anonymous reviewers, have saved me from many mistakes and from sowing a good deal of confusion. David Tranah, by his confidence, kept me going at a few critical times. Several nongeologists patiently read successive versions of the text; their comments were probably the most important of all. It is customary to acknowledge that mistakes and aberrant views are entirely one's own; I believe that the reader will find this self-evident.

Stanford T.H.v.A.

Acknowledgments

The following publishers have given permission to reprint material from copyrighted works: From "The Waste Land" in *Collected Poems 1909–1962* by T. S. Eliot, © 1936 by Harcourt Brace Jovanovich, Inc., © 1963, 1964 by T. S. Eliot (reprinted by permission of Harcourt Brace Jovanovich, Inc., and Faber & Faber); from "Burnt Norton" in *Four Quartets* by T. S. Eliot, © 1943 by T. S. Eliot, renewed 1971 by Esme Valerie Eliot (reprinted by permission of Harcourt Brace Jovanovich, Inc., and Faber & Faber); from *The Abyss of Time* by Claude Albritton, © 1980 by Freeman, Cooper & Company; from *The Pulse of the Earth* by J. H. F. Umbgrove, © 1947 by Martinius Nijhoff, Publishers; from *The Creative Mind* by Henri Bergson, © 1946 by the Wisdom Library, a division of the Philosophical Library; from *Ode to the Sea and Other Poems* by Howard Baker, © 1966 by Swallow Press (reprinted by permission of the Ohio University Press, Athens, for the United States, its dependencies and territories, and Canada, and by permission of Howard Baker); from "Bubbles Upon the River of Time" by M. M. Waldrop in *Science* 215: 1082–1083 (February 26), © 1982 by the American Association for the Advancement of Science.

Itinerary

What are the roots that clutch, what branches
grow
Out of this stony rubbish? Son of man,
You cannot say, or guess, for you know only
A heap of broken images.

T. S. Eliot, *The Waste Land*

Rocks have been important to us for millions of years. They are plentiful and have many uses, but not all kinds serve us equally well. To find the best flint and obsidian for tools, or the finest soapstone for carving, requires skill and experience. Quarries, and therefore also quarrymen, go back at least 20,000 years and surely more. More recently, the search for metals has called for other and much more sophisticated expertise, to which medieval books on mining bear eloquent witness.

This does not, however, imply a medieval science of geology. Until the arrival of a fresh spirit of inquiry with the Renaissance, scholarly issues tended to be considered in a philosophical or religious context, and logic rather than experiment or observation seemed to hold the key to truth. This spirit prevailed longer in our attitude toward the earth than in any other discipline, and it has not fully vanished even today, as the continuing debate over evolution versus creation demonstrates. Not until we reach the late eighteenth or early nineteenth century, when geologists learned to construct hypotheses from observations and test them by more observations or by experiment, can we speak of a real science of the earth.

Even so, a few pioneers, such as Leonardo da Vinci, quite early asked modern questions about the rocks and obtained modern answers. Fossils were generally regarded as quirks of nature, formed from inanimate matter as minerals are, although Theophrastus already knew better in the sixth century B.C. The existence of an ancient life, different from that of the present world, would, if taken seriously, have raised troubling questions of a theological nature. Recognizing that those shells and bones were indeed the remains of animals, da Vinci concluded that the sea had been where the land is now. It also seemed to him that the Great Flood of the Book of Genesis would not suffice to explain this. Its traditional forty days would not be adequate for such sedentary animals as clams and corals to migrate and set down so far from the initial shore. Moreover, the alternation of fossil-bearing and barren beds appeared to him to exclude a priori a single flood.

Over the next two centuries it became the common view, fed by the desire to reconcile the geological evidence with the Scriptures, that there had been a series of floods, each exterminating life, each followed by a fresh creation. Indeed, the record of the rocks does reveal a number of great extinctions accompanied by mountain building and invasions of the sea, each followed by the expansion of new life forms. In the late eighteenth and early nineteenth centuries, practical-minded geologists in Britain, France, and Germany used these discontinuities to erect a framework of earth history that, unchanged in its essentials, still stands today. Gradually stripped of supernatural elements and calling on less metaphysical forms of natural upheaval than acts of God, the many-floods concept evolved into the theory of catastrophism, which was accepted by most paleontologists around the turn of the eighteenth century. It remained for Charles Darwin to explain the progressive changes in life forms with time, but most scholars had long since disconnected this from any concept of repeated creation.

In the meantime, a much different view of the history of the earth had quietly emerged in the late years of the eighteenth century, propounded most firmly by James Hutton, Scottish physician and gentleman farmer. After much experience with the geology of his native land and a great deal of travel abroad, Hutton became convinced that no more was required to explain the past than the processes anyone can observe every day. His dictum "the present is the key to the past," which sums up what is now called uniformitarianism, still dominates geological thinking. In his view, past states of the earth were not fundamentally different from the present one, catastrophes were not required, and the history of the earth was one of gradual and steady change. Inevitably, for the earth to have accomplished with the slow processes seen around us all that the rocks record, much more time has been needed than was generally imagined. Just how much would have surprised even Hutton.

It fell to a Huttonian follower, Charles Lyell, to convert the world in its entirety to uniformitarianism. In his book *The Principles of Geology,* first published in 1830, Lyell put his stamp on geological thought for more than a century. His uniformitarianism was considerably more rigid than that of Hutton – impossibly rigid, even by today's standards. In its simplest form, Hutton's principle said no more than that the laws of physics and chemistry, independent as they are of space and time, operated in the past as they do today. It is a wise assumption, restraining us from invoking unprovable or even supernatural events as explanations. Lyell, however, went well beyond that, postulating that the processes not only were the same but also operated in the same combinations, at the same rates, and with the same intensities. His was a static world, and we know now that the history of the earth has been far more lively than he envisaged. Although the constancy of the laws of nature is not in doubt, we must admit the possibility that processes may have shaped the earth in the past that are not in evidence in the brief instant of eternity that is ours to observe.

Lyell was quite successful, and catastrophism is now officially dead. How close it came to describing reality is another matter, however, and we shall return to that question at the end of the book. To achieve the necessary vantage point, we shall trace the history of the earth in terms of three themes. These are (1) the evolution of the solid earth, an engine driven by its internal heat and slowed by gradual cooling, (2) the history of oceans and atmosphere, another engine, this one fueled by the heat of the sun, and (3) the evolution of life, which, being dependent on the first two, greatly modifies the external earth and, in doing so, reshapes itself as well. Earth and sun must die as their fuel sources become exhausted, and life, although itself capable of infinite continuation, must die with them. As we follow these three paths through the long history of the earth, we shall find them intricately woven together, forcing us to many detours and some retracing of our steps.

We shall begin with the latest installment in an unfinished story: the Great Ice Age and the impact of its climate on the earth. This world of only yesterday is easily envisaged by the minds of all, and by the eyes of those who may have seen Greenland or Antarctica, but the early earth, as alien as another planet, is not accessible. Moreover, the record contained in very young rocks is fresh and still reasonably complete, so that we can hone our interpretive skills on them. What causes an ice age? We do not know in full, but understand parts of the answer, one of which is a large change in the behavior of the ocean. What might cause the ocean to change? Here we are on firmer ground; rearrange the continents, close or open seaways, and the ocean will be diverted from its present habits. We must then look at the concept of continental drift and at the scientific revolution that made it, a quarter century ago, the ruling theory. Driven by the earth's internal heat engine, continental drift is behind many events in the history of the earth, such as geographic change, mountain building, and sealevel changes. These events, in turn, influence ocean, climate, and life.

To a large extent, the history of the earth is a history of the advances and retreats of the sea across the edges of the continents. At times, ours has been a planet mostly covered with water; at other times, like today, the continents rise high above the sea. The record is laid down in the sediments of the continental margins, a record rich in information, but one that fails to tell us the causes of the endless inundations and withdrawals, though we look once again to the drift of continents for an explanation.

Far more alien is the earth's childhood, the first 4 billion years of a 4.5-billion-year history. The end of that childhood came rather suddenly when, a little more than 500 million years ago, life proliferated in forms that are well preserved and the world took on a rather familiar aspect. The early rocks are so old, and so much has been lost, altered, or deformed, and everything was so different from what we see around us, that for the time being logical reasoning rather than the rock record guides us, although diligent research is beginning to change that. That period, the Precambrian, was the time when everything began: ocean basins and continents, rivers and seas, the atmosphere and life, the earth itself.

Last, we shall look at life. Its history can be discussed from many points of view, but in the context of this book the interaction between evolving life in its ever greater complexity and the changing external environment of the earth is the most appropriate perspective. Life and environment are mutually dependent and mutually modify each other, but it is the environment, through natural selection, that drives evolution. Tracing this forward in time from largely surmised beginnings, we shall return to the present and come to a new contemplation of the whole of history, to ponder whether a gradualistic, a uniformitarian, or a catastrophic view best suits what we have seen.

Foundations

It has not been easy for man to face time.
Some, in recoiling from the fearsome prospect
of time's abyss, have toppled backward into
the abyss of ignorance.

Claude Albritton, *The Abyss of Time*

ROCKS, EVENTS, AND TIME

To reconstruct the history of the earth one must know what happened and when it happened. We have to assume that time and events have been recorded in some way in the rocks and that we shall be able to read this record.

To decipher the archives of the earth I shall draw on the uniformitarian principle that the present is the key to the past. This allows me to infer a volcanic eruption when a deposit resembles the ash that fell from Mount St. Helens, and to deduce inundation by the sea because another deposit looks like a beach sand on top of a dune. With knowledge of modern environments and their deposits I shall attempt to reconstruct past events, provided the record has not been too much ravaged by time. I must, of course, remind myself from time to time that past events may have occurred for which I know of no modern counterpart, but it would be dangerous to resort to that assumption too readily. The deus ex machina of ancient tragedy has no place in science.

If rocks hold the record, we must bring some order among their infinite variety. Some rocks crystallize from molten magma and are called *igneous.** If they cool in the depths of the earth, they are *intrusive,* and because cooling there is slow, they will be coarse-grained. Granite is an intrusive igneous rock. If the magma pours out at the surface as lava, fast cooling yields a fine-grained or even glassy *extrusive* igneous rock, such as basalt or obsidian, the latter also known as volcanic glass.

At the surface of the earth, all rocks are attacked by wind and water, by frost and heat, and by the activities of organisms. They weather, break down into small fragments, or dissolve altogether. Carried away by water or wind, the gravel, sand, silt, and clay are deposited elsewhere to form sediments, which eventually harden into sedimentary rocks, sand becoming sandstones, clay converting to shale. Dissolved substances precipitate and form evaporites, such as rock salt. Organisms build biogenic deposits and so form reef or shell limestones, bone beds, and coals.

Rocks of any kind may be returned to the depths of the earth by mountain building or by burial under a thick layer of later deposits. High temperature or pressure can transform them into *metamorphic* rocks. Granite is metamorphosed to gneiss, clay turns into slate, limestone into marble.

* A glossary of terms can be found at the end of the book.

Rocks thus contain a record of their own history. The events that brought them into being can be deciphered and assigned to their proper relative positions in the history of the earth. Did this limestone form before or after that granite? Did fish precede dinosaurs? Arranging events in their correct sequence is the most substantial of the many tasks of the geologist engaged in stratigraphy, in reconstructing events and time.

Having accomplished this reconstruction for a piece of history in a given spot, we must next establish its position in the whole of earth history so that we may know, for example, what happened in England while the sea inundated California. For this correlation between distant fragments of the rock record, we use mainly fossils. Beyond that, we should like to know when a certain event took place, how long it lasted, and at what rate it proceeded. Sequence is not enough; there is no history without time. We must measure time in units: seconds, hours, years.

I

Reading the record of the rocks

Unpromising as a barren mountainside may seem to the historian, it may in reality contain a record rich in information about an astonishing array of subjects. An example is the fine banding exhibited by many fossil clams and corals. This banding, a result of intermittent growth, can be seen in modern representatives as well. Clams of the intertidal zone grow when the tide is high, then stop while the water falls. Each growth layer represents one tidal cycle, two tides per day. The longer spring tide produces a thicker layer, and so we can mark off the lunar month. Where the winter is cold, the growth of clams slows or stops, allowing us to identify the seasons and count the days of the year. In deeper water, corals respond to the sun: a layer for each day, but little growth at night. The same layers are seen in fossil molluscs and corals and can be interpreted in the same way (Figure 1.1).

Why should one wish to know the number of days or of lunar months for years so long ago? Mostly because the rotation of the earth has gradually slowed as a result of tidal friction; we find more days in the year and fewer hours in the day as we go farther into the past. There is also reason to believe that the moon has not maintained a constant distance from the earth; if that is so, it ought to be reflected in the length of the lunar month. Of course, the first flush of discovery of these things in the 1960s made it all seem rather more simple than it has turned out to be. Nonetheless, it appears that 400 million years ago the earth had about 400 days, the lunar month was 30 rather than 28 days long, and there were 22 hours in each day.

Not all information contained in the rocks bears on such esoteric questions. Much is even less certain, more subject to conflicting interpretations. Even so, over two centuries geologists have extracted from the rocks a remarkably consistent and very rich history. How, precisely, was this done?

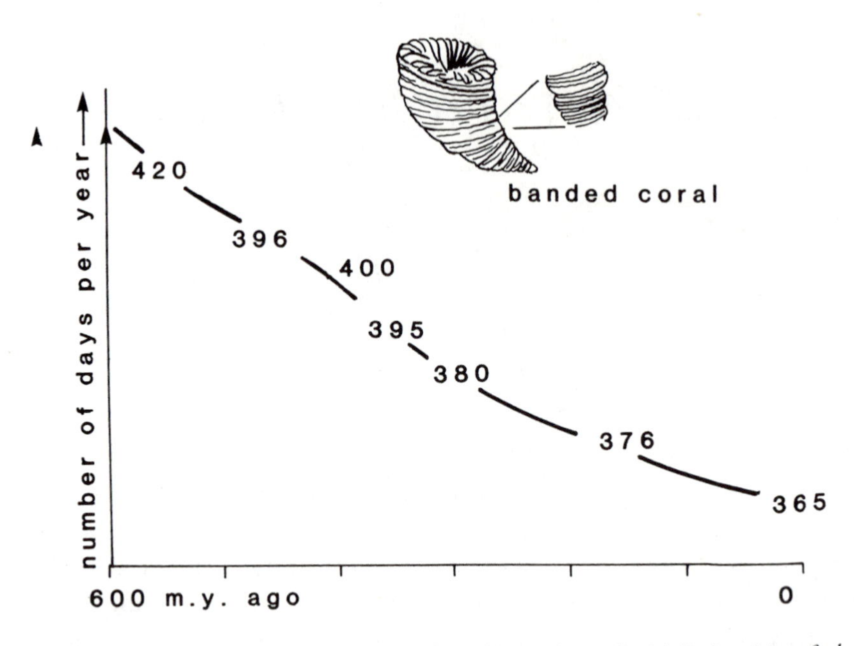

Figure 1.1. Many corals and molluscs show fine ribbing, the result of daily deposition of a layer of calcium carbonate. Because the thickness of the layer varies with the lunar month and seasons, we can count the annual number of days, finding more days per year the further back we go.

ENVIRONMENT AND FACIES

The earth expresses itself in myriad ways. Running water, helped by sand and gravel, has cut river channels and the Grand Canyon, eroded cliffs, and excavated potholes as well as the Niagara Falls. A huge literature exists relating earth processes to the landforms and rocks they produce, and this enables us to infer from a sedimentary rock the environment in which it was originally formed. The reflection of the environment of origin in a rock is called its *facies;* a sandstone may appear in a beach or a river facies. A swamp facies can be recognized because it is rich in plant remains contained in a dark, fine clay. And so forth.

A detailed discussion of all facies and their meanings would fill the rest of this book, but we can achieve some understanding by looking at broad facies implied by some very common sedimentary rock types. Limestones composed of fossil debris indicate a marine facies, usually laid down in clear shallow seas. Coals are the products of thick layers of dead plants. Such materials are best preserved under water, and coals therefore indicate a swamp facies. Salt and other minerals precipitate from seawater. When we see these evaporites, we think of a lagoon facies: shallow, sheltered bays in a dry, sunny climate. Redbeds, orange, brown, and red gravels, sandstones, and shales present a most striking facies that generally indicates a desert or a semi-arid environment.

Careful study of the physical properties of sediments is important. Fossils can help even more, for plants and animals respond sensitively to their environments, although if the species are extinct, we must infer what these responses might have been. With patience and good outcrops, the interpretation can often be quite specific. We might, for instance, try our hand at an interpretation of the example shown in Figure 1.2.

The unfossiliferous limestone at the base says "marine," but not much else. The absence of any land-derived material, even clay, suggests that the land was distant or that no mud-carrying streams entered this sea. A shore without streams causes us to think of a dry climate. The next limestone contains oysters and, toward the top, a little clay and sand, suggesting that the land was near. Oysters do not live in deep water or in the open sea; therefore the sea must have become shallower. Thin-bedded sandstones followed by coal indicate first a beach, perhaps here and there with a ridge of dunes, then a swampy coastal plain. Evidently the land was not so dry after all, or the climate changed. Dunes cover the coal, and then comes a surface of erosion. Quite a bit of history must have been removed, because the next deposit is suddenly and unmistakably marine. Did the land sink or the sea rise?

It is common for parts of a rock sequence to be missing, because some events destroy rather than create deposits, or simply leave no trace. Limestones and evaporites may dissolve and disappear. Deep burial, intense folding, or metamorphism may so distort the record that we can no longer decipher it. Sometimes, nothing happens for a while; no rock record is being formed today outside my window, nor is any being removed. Most often, however, erosion is the culprit (Figure 1.3). Although its work is at times difficult to spot, more often it leaves tracks: a residue of cobbles, a ragged surface. It may cut across folded beds, indicating that much history has been lost forever: first the deposition of sediments, then their subsequent folding into mountains, and finally removal of those mountains by erosion.

Eventually a new deposit will bury the eroded surface and preserve the break. The erosion plane is called an *unconformity*, and a *hiatus* is the interval of time for which we no longer possess a record. This interval includes the duration of the erosion itself and the time that was recorded in the layers that have been eroded away. Thus, vigorous

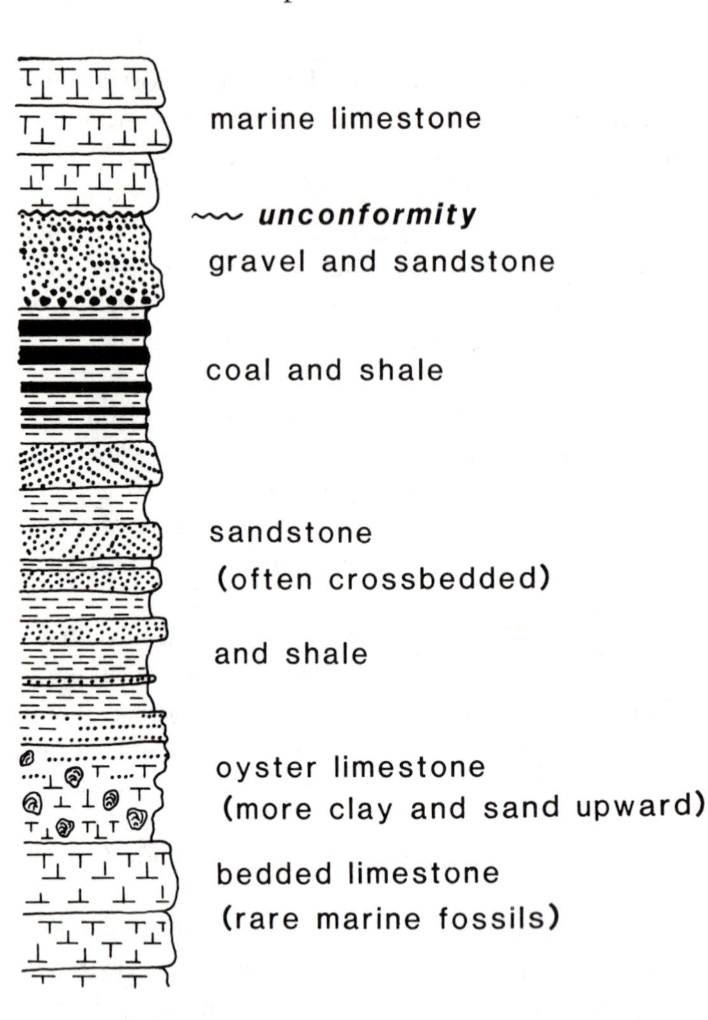

Figure 1.2. This rock record begins with a marine limestone deposited far from land but in shallow depth, proceeds to near-shore sands, a coal swamp, and then a phase of erosion (unconformity). On the unconformity rests another marine limestone. What we see is a record of the fall and subsequent rise of the sea.

erosion may leave a hiatus longer than the period of erosion itself. On the deepsea floor, where sedimentation is very slow and the beds therefore thin, a single pulse of erosion may remove layers spanning tens of millions of years.

EVENTS IN TIME AND SPACE

Having deciphered the events recorded in Figure 1.2, how shall we determine whether the presumed sealevel change was purely local or

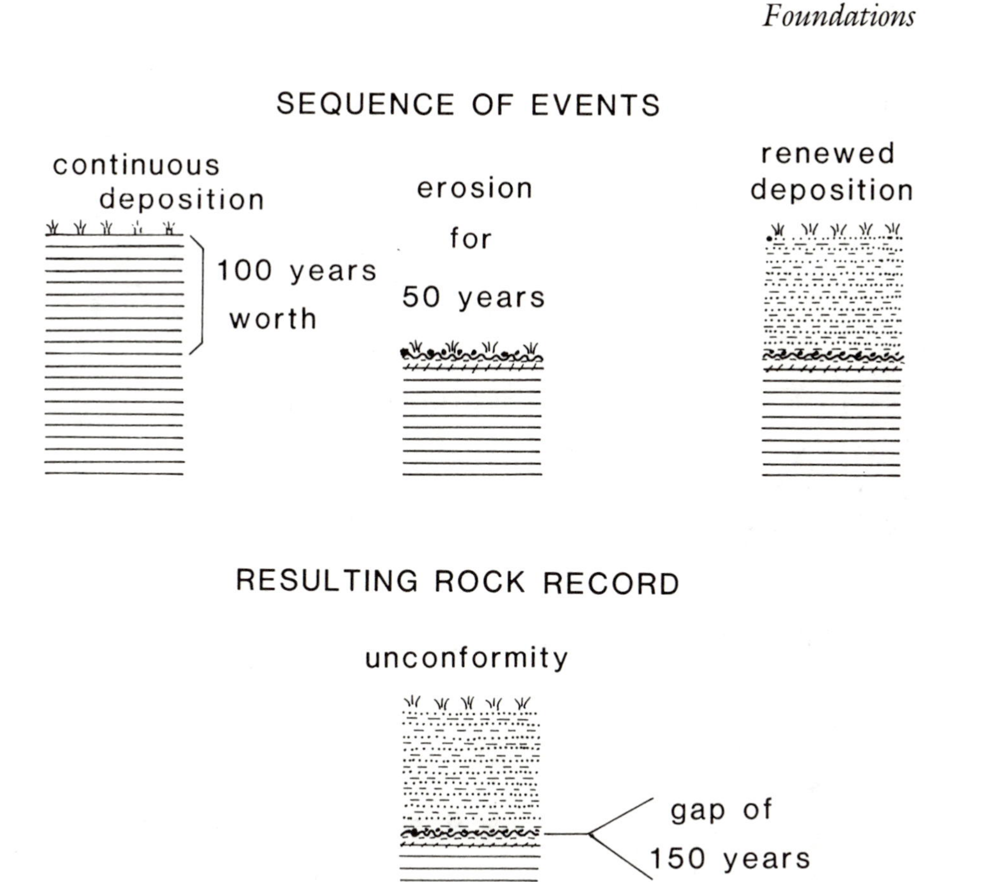

Figure 1.3. Erosion interrupts the deposition of the rock record and may remove part of what already existed. The hiatus, the gap in time represented by the unconformity, almost always includes more than the number of years that the erosion itself lasted.

something more, or find out what happened before or followed after our small segment of history? To answer such questions we must fit this fragment of local history into a broader context of time by correlating it with sequences of rocks found elsewhere. Four simple principles or laws help us do so (Figure 1.4). Three are of venerable antiquity, having been formulated in 1669 by Nicolaus Steno, a Danish expatriate at the court of Florence who in this way laid the foundation of modern stratigraphy.

The first principle, known as the law of superposition, says that in a sequence of sedimentary rocks or lava flows the age increases downward, each bed being younger than the one underneath and older than the one above. We have already applied this law in our discussion of

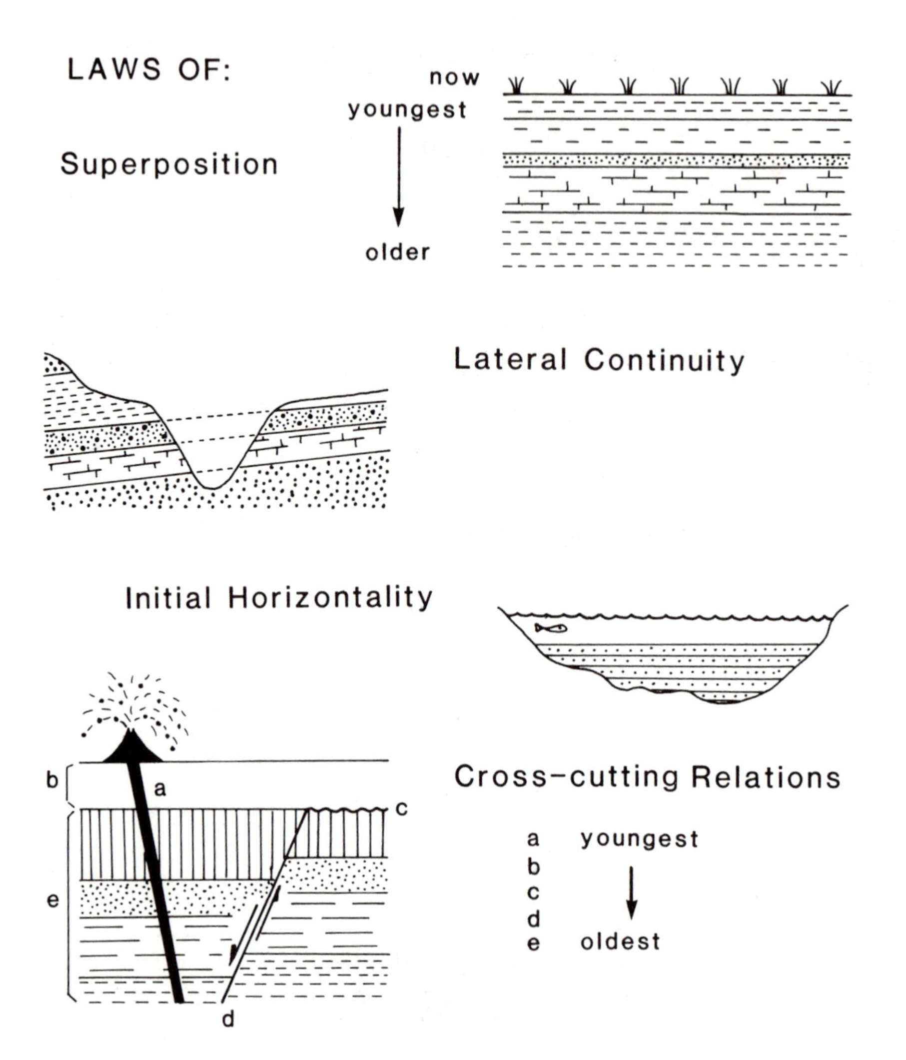

Figure 1.4. To connect events deduced from the rock record in one place with others observed elsewhere, we use a process called correlation. Correlation is facilitated by four principles or "laws" that, simple as they are, constitute a powerful tool for reconstruction of the history of geological events, the stratigraphy, of a region.

the facies of Figure 1.2. It has wide application, but there are exceptions, such as intrusion of lava between two sedimentary beds.

The second principle is that of initial horizontality. Most sediments, having been deposited in seas or lakes, were originally laid down as horizontal beds or, at most, with very low inclinations. If we now observe them to be tilted, we must conclude that they have been deformed after deposition.

The principle of lateral continuity states that sediments initially form continuous strata, changing their character only when the environment changes. If we find an abrupt lateral termination to a bed, we must suspect that something intervened after deposition: dislocation by a fault or erosion on a shore or in a stream valley. With this principle we can reconstruct strata across gaps that were once continuous. The last principle is that if a bed is traversed by another at some angle, for example by an intrusive dike or by a fault, the bed must be older than the cross-cutting event. Thus, the volcano and fault in Figure 1.4 are younger than the rocks they cut across.

In practice we can extend the scale of our observations by walking along the outcrop (Figure 1.5). If the beds are horizontal, we remain in the same time interval but range farther in space, possibly noting facies changes. If the beds are tilted, walking one way will carry us forward in time, the other back. Nevertheless, every outcrop ends eventually, in a landslide, a clump of trees, a stream cut, and we lose track of our sequence.

If, however, we have described our set of strata quite precisely and found them distinct from those above and below, we may recognize a similar sequence elsewhere in some other outcrop and make use of the principle of lateral continuity to postulate that the two were once connected. Such a distinct set of strata, different from those above and below, and traceable ("mappable" the geologist says) across the countryside, is called a formation. If we assign the limestones of two different outcrops to the same formation, we mean that they are, or at least were, part of the same stratum. Placed in the proper temporal order, formations form a stratigraphic sequence.

Sooner or later, however, every formation must end, usually because it is replaced with rocks that are different because they were formed in a different environment. A famous case of such a facies change was debated at length in Britain in the early nineteenth century. In the southwest of that country, two major sediment series can be seen, both about 350 million years old. One is a group of redbeds found in Wales, the other a suite of fossil-bearing sandstones, shales, and limestones in Devon. The two sequences are separated by the Bristol Channel, which conceals the boundary between them. An argument long raged whether one was older than the other (and if so, which one), or whether the clearly continental redbeds gradually changed into the more marine limey sequence. Finally, careful fieldwork turned up a few outcrops in which one facies was interstratified with the other, and the issue was resolved. Different as they look, the two formations represent separate facies laid down at the same time.

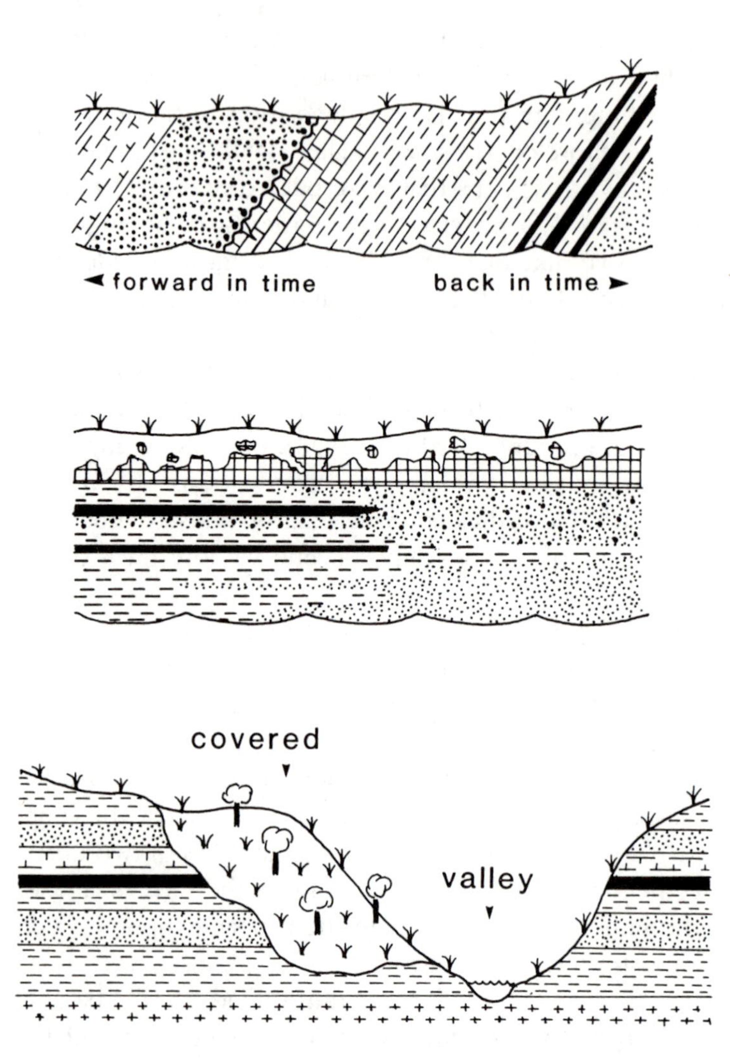

Figure 1.5. Walking along a sequence of tilted beds is like taking a walk in time (top), while walking along a set of horizontal ones may take us from one facies to the next. Plants and erosion (valleys), however, may make life difficult for the geologist (bottom).

Suppose we have established that the beds in two outcrops belong to the same formation. Can we also assume that they were deposited at the same time? Alas, no law requires that all parts of a formation must be of the same age, and quite often they are not. Consider the seashore during a rise in sealevel (Figure 1.6). Soon, a beach will form where dry land existed before, and deep-water clays will be laid down on top of the sand in the former surf zone. As the sea rises, the shore moves

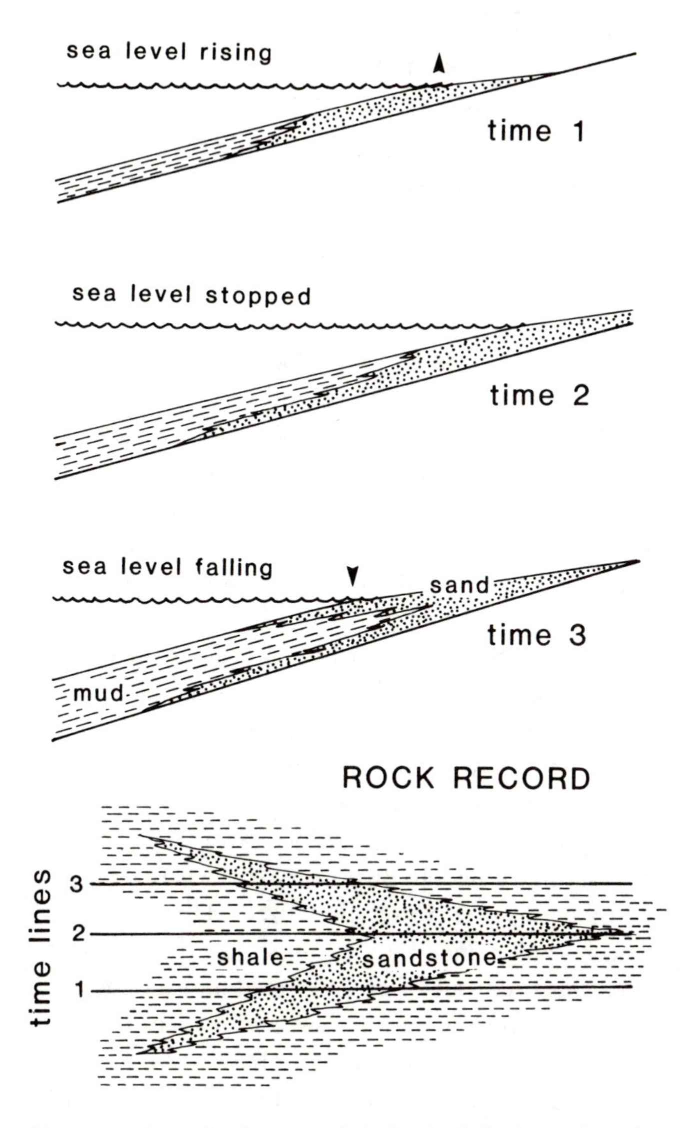

Figure 1.6. A set of sedimentary beds that is distinctive and can be traced over a large area is called a formation. Although a useful concept, the various parts of a formation need not be of the same age everywhere. In this example, as the sea rises and the shore migrates inland, the beds of sand and mud migrate with it, landward and forward in time. Consequently, the boundaries between the sandstone and mudstone beds are at an angle to time lines, and their landward portions are not of the same age as the seaward ones.

inland and also forward in time. A series of beds forms, each continuous and throughout of the same distinctive nature, but progressively younger in the landward direction. It is obvious that if we wish to decide whether or not events in separate places are contemporaneous, we shall need tools other than the physical properties of formations.

FOSSILS AND CORRELATIONS

The time-honored solution to the problem of correlation over long distances is a stratigraphy based on fossils (Figure 1.7). Long before Darwin explained why life changes over time, geologists knew that older beds contain fossil assemblages different from those in younger beds. Species become extinct, new ones appear, and whole assemblages change with time. With fossils we can erect zones to subdivide the record of the history of the earth. Fossils can thus be used to assign a formation or an outcrop to its proper place in the historical sequence, the geological timescale, assuming, of course, that adequate fossils can be found, something that is possible only for the last 600 million years of earth history.

There are, as always, difficulties and impediments. Fossils, too, are subject to facies changes. One only need think of the differences between the dwellers in a tidepool and the animals that live on an open sandy beach to realize the problem. Some species depend on a limited, narrowly defined environment; others are more tolerant and therefore are widely distributed or, as in the open ocean, occupy an environment that is vast. Obviously the more widely distributed species make better stratigraphic markers than those that are specialized. There are other prerequisites for a good fossil zone. The fossils should be abundant, they must not have been washed out of an older bed and redeposited, and they must be in good condition so that they can be identified with confidence. To have many species is better than to have a few: A single coral tells us less than 10 different species of molluscs. Consequently, microscopic fossils found by the hundreds or thousands in a single sample are better than mammoth bones. It is also helpful if the species evolve rapidly, because then there will be many extinctions and new appearances to define our zones.

How well do fossils serve the task of correlation and of subdividing earth history? Quite well indeed; numerous fossil zones mark the last 600 million years, and many of them are valid for large areas and are not unduly influenced by facies differences. For example, a few groups of calcareous oceanic plankton, microscopic organisms with calcareous

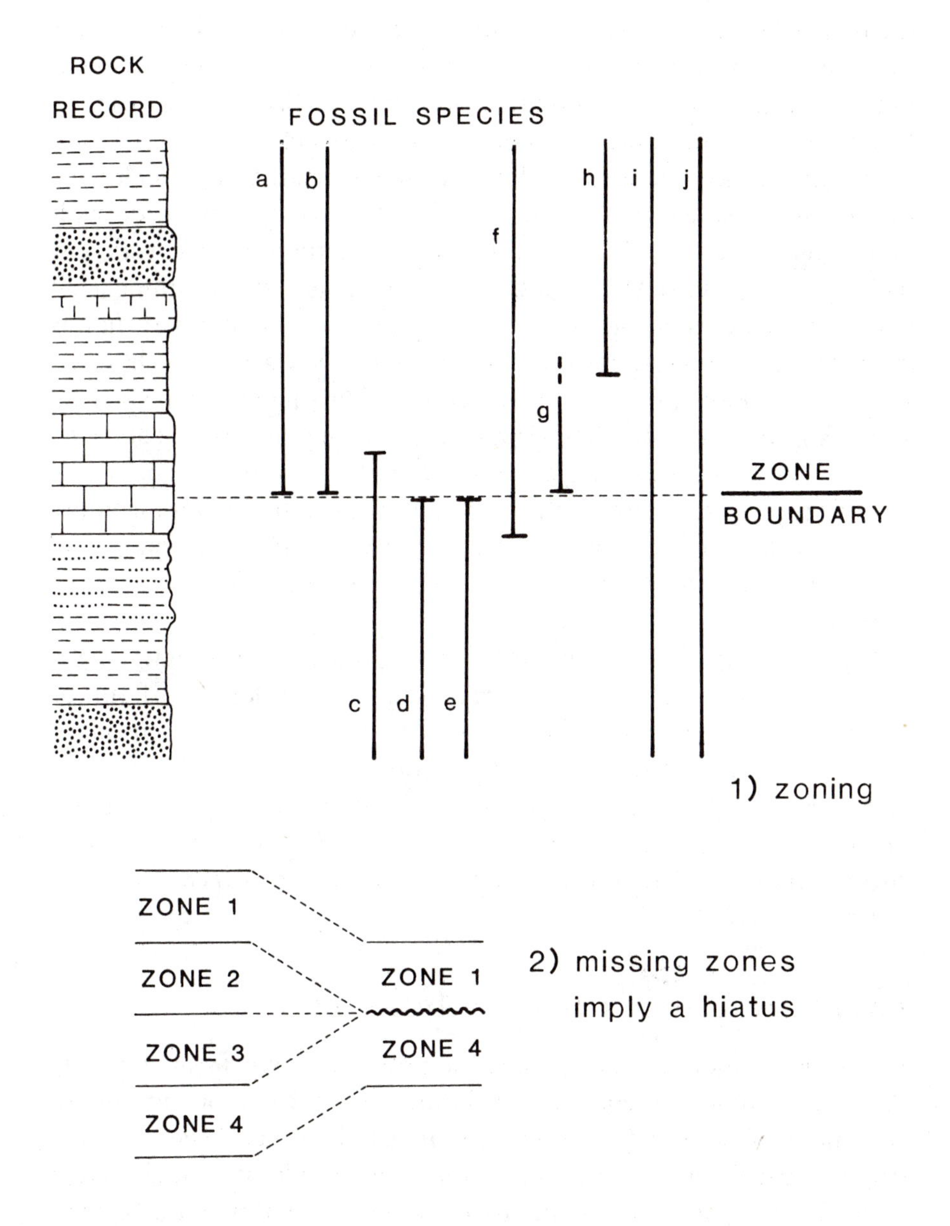

Figure 1.7. To determine whether a rock outcrop is younger or older or of the same age as another, we can use fossils. Fossil zones, based on the extinction of some species and the first appearance of others, often extend over large areas and are a key tool in stratigraphy, The absence of one or more fossil zones not only confirms the existence of an unconformity but also provides an estimate of its duration.

shells that live near the surface of the sea, allow us to subdivide the last 65 million years of history into no fewer than 50 or 60 zones. With so fine a timescale we can pinpoint events with a precision of better than half a million years. Further refinement appears possible.

It is much more difficult to deal with sediments that have no fossils, or with igneous and metamorphic rocks. And what is one to do with events that make no rocks, such as folding or faulting? Sediments deposited on land rarely contain fossils, because the chance that anything at all will be buried is slim, and the opportunity for permanent preservation is even slimmer. Some terrestrial fossil beds are spectacular, such as the frozen mammoths of the Siberian tundra or the dinosaur bone beds of Wyoming, but those are the exceptions. The dating of barren sediments, of other rocks without fossils, and of events without rocks rests on the law of superposition and on the cross-cutting rule. It requires ingenuity worthy of a Sherlock Holmes. The best way to understand the procedure is to study some examples of the kind geologists encounter all the time (Figure 1.8).

It is clear that the framework of earth history, the geological timescale, rests mainly on fossils. The timescale was developed early in the nineteenth century, and although much fine detail has been added since, it has withstood almost two centuries of use very well. The largest advance has been the addition of a calibration in years by means of radioactive age dating. Before we turn to that subject in the next chapter, however, let us first examine the speed with which geological events proceed.

RATES OF GEOLOGICAL PROCESSES

It was long thought that geological processes were invariably very slow, a natural consequence of Hutton's emphasis on an infinite amount of time and Lyell's insistence that there had been no catastrophes. Small forces can produce impressive results if enough time is available. Still, it is important to know how fast things do happen. Whether a large ancient delta was built by a river the size of the Mississippi in a few thousand years or by a much smaller one in a few million is not a trivial matter. The suddenness of some faunal extinctions may be only apparent, because from the vantage point of a hundred million years later, a single million seems like an instant. Many so-called catastrophes actually proceeded in a rather stately manner.

PLACING STRUCTURES AND IGNEOUS ROCKS IN THE TIME SCALE

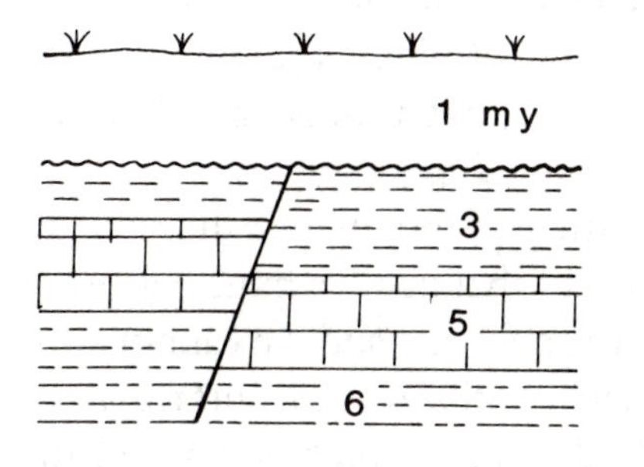

HOW OLD IS THE FAULT?

YOUNGER THAN THE THREE LAYERS IT CUTS ACROSS AND OLDER THAN THE TOP BED WHICH IS NOT AFFECTED. THEREFORE BETWEEN 1 AND 3 MILLION YEARS.

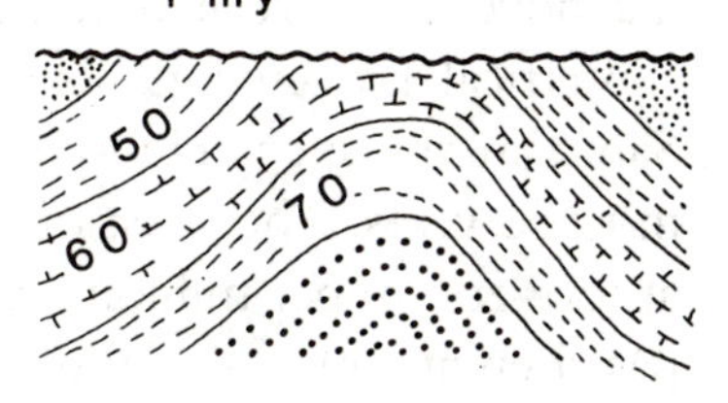

HOW OLD IS THE FOLD?

YOUNGER THAN 50 AND OLDER THAN 1 MILLION YEARS IS ALL ONE CAN SAY.

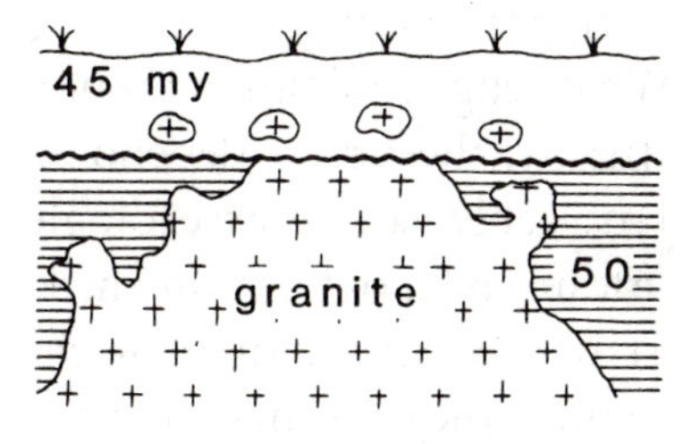

HOW OLD IS THE GRANITE?

IT MUST BE YOUNGER THAN THE SEDIMENTS IT HAS INTRUDED. GRANITE PEBBLES ARE FOUND IN THE YOUNGER SEDIMENT. THUS THE GRANITE MUST BE BETWEEN 50 AND 45 MILLION YEARS OLD.

HOW OLD IS THE BASALT?

IF IT FORMED AS A LAVAFLOW ON THE SURFACE, IT MUST BE OLDER THAN BED A AND YOUNGER THAN BED B. IF IT WAS INTRUDED BETWEEN THE TWO BEDS, IT IS YOUNGER THAN EITHER AND THAT IS ALL ONE CAN SAY.

HOW CAN ONE TELL? LAVAS HAVE TYPICAL FLOW FORMS, WHEREAS INTRUSIONS SHOW EVIDENCE OF SLOW COOLING (LARGER CRYSTALS) EXCEPT AT THEIR CHILLED EDGES.

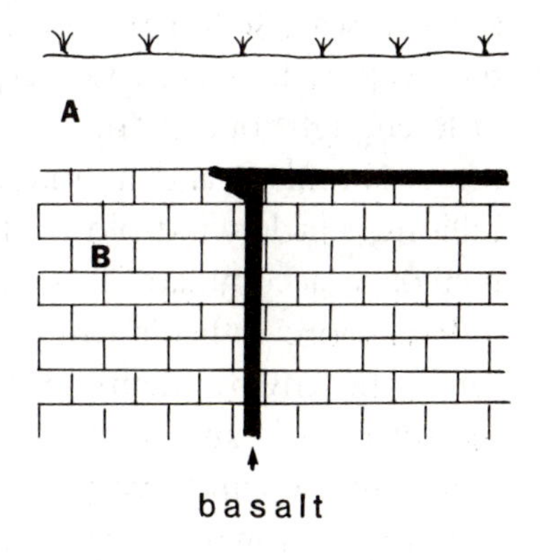

Figure 1.8. Many strata do not contain fossils, and most of our "laws" do not apply to igneous and metamorphic rocks, nor to the traces left by such events as folding or faulting. The four cases shown here depict some of the ways in which geologists deal with these situations.

Study of the present world has taught us that the processes of the earth are not always imperceptible, some of them are surprisingly rapid. The steady drop may hollow the stone, but a single landslide can remove a million stones in an instant.

Tectonic events, the folding and uplifting of mountains, the rise and fall of the land, are momentous occurrences that involve huge masses of rock or water. They appear to require large forces, and we cannot conceive of them as rapid. Yet whole continents drift across the surface of the earth at many centimeters per year, and since Columbus reached the Americas, the Atlantic has widened by more than 20 m. Western California moves north relative to the rest of North America at 30 m per century.

Vertical movements also can be impressive. The island of Espiritu Santo in the New Hebrides is emerging from the sea at a rate of 5.5 m/1,000 years. The coast of Oregon has been rising for millions of years at 100–200 m per million years. A submarine ridge off the coast of Peru has risen at a rate of 6 m per century, and the Netherlands sinks into the North Sea at a significant fraction of a meter per century.

How fast do sediments accumulate? What length of time is represented by this limestone or that shale? Naturally, these rates depend on the environment, and they range from remarkably fast to exceedingly slow. The Mississippi delta near New Orleans is about 100 m thick. Around 1,500 A.D., when the Gulf Coast was being explored by the Spaniards, this was still a shallow sea. Since then, the coast has moved 25 km south, adding 50 m of land per year to the state of Louisiana. In shallow seas, sedimentation is not as fast, centimeters per year, and in the deepsea it is even slower, centimeters per millennium. This seems little enough, but in "oceanic deserts" far from the places where plankton is abundant and its microscopic shells sink to the seafloor by the billions, a little wind-blown dust, a few fish teeth, and an occasional tiny meteorite accumulate at a rate of a mere millimeter per 1,000 years.

It is more difficult to give typical sedimentation rates on land, because the environments are so unstable and ephemeral there. Rivers meander back and forth, eroding their own deposits; dunes bury a road, but the wind sweeps them away again. Ultimately, erosion planes everything down to near sealevel, the final stage in a process that began with mountains, unless and until another uplift intervenes. Sedimentation is, in this history, but a trivial interlude. In Australia, nothing has intervened for a hundred million years, and much of that continent is now close to the final, planar stage. Erosion is most rapid in mountain country, where, even over large areas, an average of 1–3 m of rock may be removed every 1,000 years. This implies that the

Rocky Mountains, unless they keep rising, will be gone in a few million years, and so will the Alps and the Himalayas. The whole of North America is being lowered at a rate of 5–10 cm per millennium. Ultimately, the debris all finds its final resting place in the sea.

2

Perspective on time

It is a habit peculiar to geologists to speak of millions of years as casually as politicians dispose of billions of dollars. For the latter, it does not seem like real money, and I suspect that, to us, it is not really time. To be comfortable with great lengths of time is a habit which it took a century to develop.

There is, indeed, a need for much time. James Hutton had already perceived that the earth was old, with "no vestige of a beginning, no prospect of an end," but the record of the rocks is ambiguous about time. The building of a delta, even a large one, may take but a few centuries, but the raising of the Alps or the Sierra Nevada is obviously time-consuming. Early in the nineteenth century, a clear outline of the history of the earth had been established by the methods we have just discussed, but the true time dimension remained elusive. Darwin's theory of evolution heightened the sense that much time had been involved in making man out of a single-celled alga, but no one could say just how much. Over the decades, the estimates of the age of the earth went up and up, but it remained for the discovery of radioactivity in the last years of the nineteenth century to provide firm underpinnings for the time perspective of the geological past.

HOW OLD IS THE EARTH?

In 1654, James Ussher, Archbishop of Armagh in Ireland, computed, on the basis of the Scriptures, and with guidance from his imagination, that the Creation took place in the year 4,004 B.C., an estimate later refined by others to October 26 at 9 a.m., a sensible hour. For some time this was regarded as a reasonable computation, but gradually it became obvious that 6,000 years would not suffice for all that had been recorded in the rocks. In 1749, the French naturalist G. L. de

Buffon estimated that at least 75,000 years had been needed to produce all of the fossil-bearing strata he knew of. Late in the nineteenth century, an Irish chemist, J. Joly, reasoned that rivers continuously deliver salt to the sea, while nothing goes out, and he calculated that 90 million years were required to produce the present salinity. Today we have better data and would make this 260 million years if we use common sea salt, or 45 million years with another abundant element, magnesium. Silica yields a mere 8,000 years. The method is obviously flawed, and the flaw lies in the assumption that nothing leaves the ocean. We shall return to this in Chapter 12.

The 90 million years of Joly were deemed far too many by some of his contemporaries. One of them, the great physicist William Thomson, Lord Kelvin, was provoked by Lyell's strict uniformitarianism, which, because it claimed that the processes of the earth have neither slowed nor speeded up with time, smacked to him of a perpetual-motion machine. Perpetual-motion machines, of course, violate basic laws of physics. The earth is an engine driven by heat, and when the earth cools, as it must, the engine will slow and eventually stop. Kelvin could think of only one fuel, the initial heat of the planet when it formed as a hot, molten ball. The cooling rate can be estimated, and after a few attempts he settled on an age for the earth of 20 or 30 million years.

Geologists must deal with a complex subject in which certainty is difficult to achieve. They tend to be easily overawed by their peers in physics and chemistry and loath to argue against them on geological grounds. Thus, when Lord Kelvin spoke in his usual peremptory manner, geologists did not like his conclusion, but neither did they wish to violate the laws of physics. One of the few challengers was an American by the name of R. T. Chamberlin, who, in 1899, calmly announced that if so brief a history was prescribed by physics, then physics must be wrong. Some fuel other than the original heat would, he felt, be found to explain why the engine of the earth had already run much longer than 30 million years. A courageous statement indeed.

Chamberlin was fortunate, because a serendipitous event proved him almost immediately right. This event was the discovery of radioactivity in the last decade of the century. That supplied not only the missing source of energy for the earth's engine but also the means of calibrating geological history in years. When the fuss died down, Lord Kelvin was shown to have been right when he derided the age estimates of the geologists, but wrong when he thought them too long.

A chemist at Yale, Bertram Boltwood, a participant in the lively discussion that followed the discovery of radioactivity and the development of early atom models, was the first to realize the opportunity that was presented to the earth sciences. In 1907, only a few years after the Kelvin-Chamberlin argument, he used a simple technique of radioactive dating to show that the earth was more than 400 and perhaps as much as 2,000 million years old. In subsequent years, as the method was refined and the number of analyses grew, so did the estimated age of the earth. Just after World War II, I was taught that it was 2 billion years; over the next 15 years, that was raised to 3.5 billion, and today we accept 4.5 billion years. Compared with the age of the universe, currently thought to be about 15 billion years, that is still not very much.

ISOTOPIC AGE DATING

In principle, the concept of dating rocks with radioactive elements is simple. The nucleus of an atom consists of protons and neutrons. The number of protons, the atomic number, determines the chemical behavior of the element. The number of neutrons can vary, giving rise to isotopes with virtually identical chemical behavior but slightly different mass. The sum of protons and neutrons is the mass number. Oxygen has 8 protons and hence an atomic number of 8. It has three isotopes, of which two (with 8 and 10 neutrons) have considerable interest for geology. These isotopes have mass numbers 16 and 18 and are usually written as ^{16}O and ^{18}O. Carbon has three isotopes: ^{12}C, ^{13}C, and ^{14}C.

Most isotopes are stable, but not all. The isotopes of oxygen and two isotopes of carbon are stable, but ^{14}C is not; it is said to be unstable or radioactive. A radioactive isotope decays; that is, it changes to some other element or isotope. Uranium isotopes decay to lead, ^{14}C to nitrogen. The rate of decay is characteristic for each isotope and cannot be changed by any known force. It is customary to express the rate of decay as the *half-life* of the isotope, the time it takes for half the number of atoms originally present to decay to daughter atoms and to one or more kinds of radiation (Figure 2.1). Half-lives vary enormously: 26.8 minutes for an unstable isotope of lead, 1,600 years for the radium in a watch, and 4.5 billion years for the uranium isotope ^{238}U.

With a simple equation we can calculate the age of the rock using one or more of several convenient isotopes. Each isotope has its advantages and disadvantages. Some are widely available, whereas others, although less common, are attractive because their age ranges are con-

Table 2.1. *Some radioactive isotopes widely used to determine the ages of minerals and rocks*

Parent	Daughter	Half-life	Usable for	Comments
Rubidium-87	Strontium-87	47 b.y.[a]	>100 m.y.	Good in granite
Thorium-232	Lead-208	13.9 b.y.	>200 m.y.	
Uranium-238	Lead-206	4.5 b.y.	>100 m.y.	Widely used
Potassium-40	Argon-40	1.3 b.y.	>0.1 m.y.	Good in basalts
Uranium-235	Lead-207	0.7 b.y.	>100 m.y.	
Samarium-147	Neodymium-147	106 b.y.	>1 b.y.	New, promising
Carbon-14	Nitrogen-14	5,370 y.	<40,000 y.	In archeology

[a]b.y., billion years; m.y., million years; y., years.
Note: There are many other isotopes that can be used, such as steps in the uranium decay series, which, under favorable circumstances, can date sediments ranging in age from a few thousand to more than 300,000 years.

venient or because their chemical behavior in rocks and minerals is straightforward. Obviously, we are well equipped to date older rocks, especially beyond 100 million years, but for younger ones ^{14}C is one of a very small set of possibilities.

This unstable isotope of carbon is different from all others because it has not, as they have, been around from the beginning. Instead, it is created continuously in the upper atmosphere by collisions between cosmic rays and the nuclei of nitrogen. The small amount of radiocarbon so produced diffuses through the atmosphere as carbon dioxide, is dissolved in the oceans, converted by plants into organic matter, and ingested by animals. Radiocarbon begins to decay to nitrogen as soon as it is formed, but each living organism maintains an equilibrium with the atmosphere or the ocean until it dies. Then its amount of ^{14}C decreases by radioactive decay, and the age of any dead tissue, wood, or shell can be determined from whatever activity remains. At the moment, the limit of usefulness of the method lies at about 40,000 years, or a little more than six half-lives, unless new and still experimental machinery is used.

Dating a rock does require more than some chemistry and the application of a simple equation. The rock or mineral must be tight; neither parent nor daughter must have been lost or gained since the mineral was formed, except by radioactive decay. Leak the daughter away and the date will be too young; leak the parent and it will be too old. Lose or gain both and nothing useful can be said.

If a rock is metamorphosed, the radioactive clock is reset. Consider

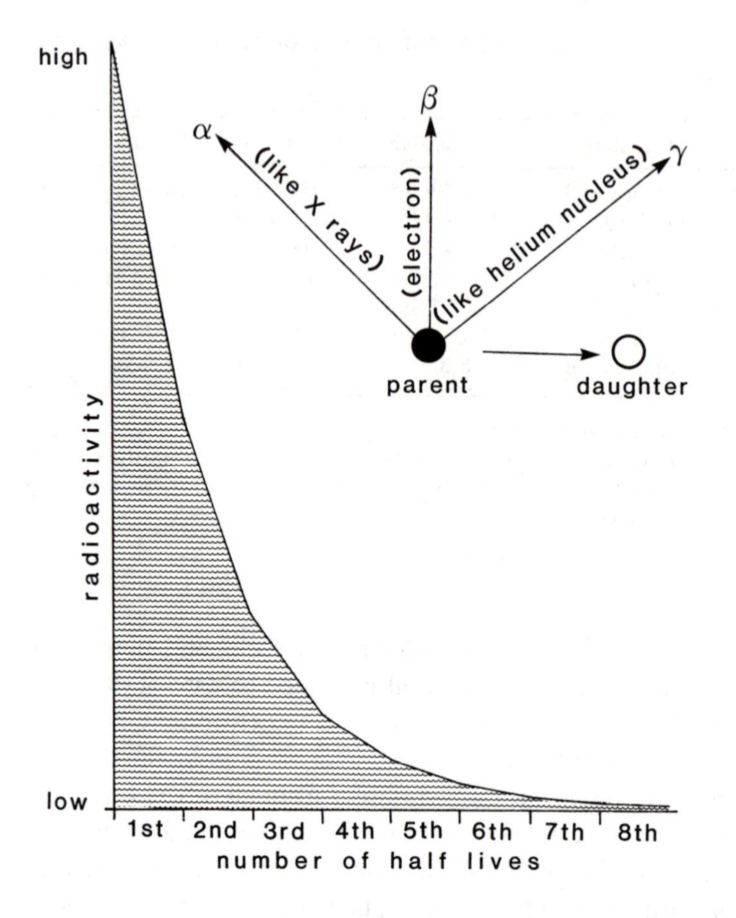

Figure 2.1. Radioactive isotopes decay to stable daughters while emitting radiation or particles or both. The rate of decay is expressed as the half-life, the time over which half of the amount present decays to the daughter isotope. This produces an exponential decay curve that at first drops very rapidly, then much more slowly.

granite: The magma cools; minerals that contain uranium, but not the daughter isotope lead, crystallize; the clock ticks. However, before we arrive to read it, the granite is metamorphosed to gneiss by heat and pressure, its minerals are converted into different ones, and the lead and uranium are redistributed. If we analyze one of the new minerals, the date we find is that of the metamorphic event. On the other hand, if the rock as a whole did not leak radioactive parents or their daughters, an analysis of it in bulk will still give us the time when the original granite was formed.

Isotopic age data are subject to errors. Many errors cannot be spotted easily, and despite the errors the calculation always yields a date.

Radioactive decay is a statistical phenomenon, and repeated measurements give slightly different answers. Thus, if we read that a date is 100 ± 6 million years, we know that if we were to repeat the measurement many times, two-thirds of the answers we would find would lie between 94 and 106 million years.

One notes here an interesting feature of the table of isotopes. In the previous chapter, using formations and fossil zones, we found it easy to place sedimentary rocks in the framework of earth history, but we encountered difficulties with igneous and metamorphic rocks, which bear no fossils. Radioactive dating, on the other hand, applies almost exclusively to igneous and metamorphic rocks. Although the ages of granites and basalts have, of course, great intrinsic interest of their own, this leaves us with the problem of how to dovetail the relative ages of sediments with the absolute ages obtained from igneous rocks. We are reduced to the inverse of the procedure illustrated in the previous chapter in Figure 1.8, inferring dates "older than" and "younger than" from igneous rocks associated by good fortune with the sediments that need to be placed in the timescale. It is a slow and painstaking job for which some geologists seem to have a special aptitude (and tolerance). To them the rest of us owe a great deal of gratitude.

THE GEOLOGICAL TIMESCALE

Today, then, we accept that the earth is 4.5 billion years old. It is not easy to come to grips with this astonishing number. Still, we should not accept it without trying to understand what it means, or everything that follows will make little sense. How shall we understand evolution, for example, if we have no feeling for the one-million-year lifespan allotted to many species? And numerous are the geologists who have constructed histories for a ten-million-year period that would have fitted comfortably into a few thousand!

Let us attempt to transform the 4.5 billion years into something we can cope with on a more human scale. Suppose we set the entire 4.5 billion years equal to a single year (Figure 2.2). One second will then represent 143 years. At midnight, when the new year starts, the earth is born. At midnight on December 31 we reach the here and now. The experience is rather sobering. The length of recorded human history, if we stretch it, comes to 30 seconds. The diverse life of higher plants and animals did not develop until the last of November. The part of earth history we know fairly well, the Phanerozoic, is less than two months long.

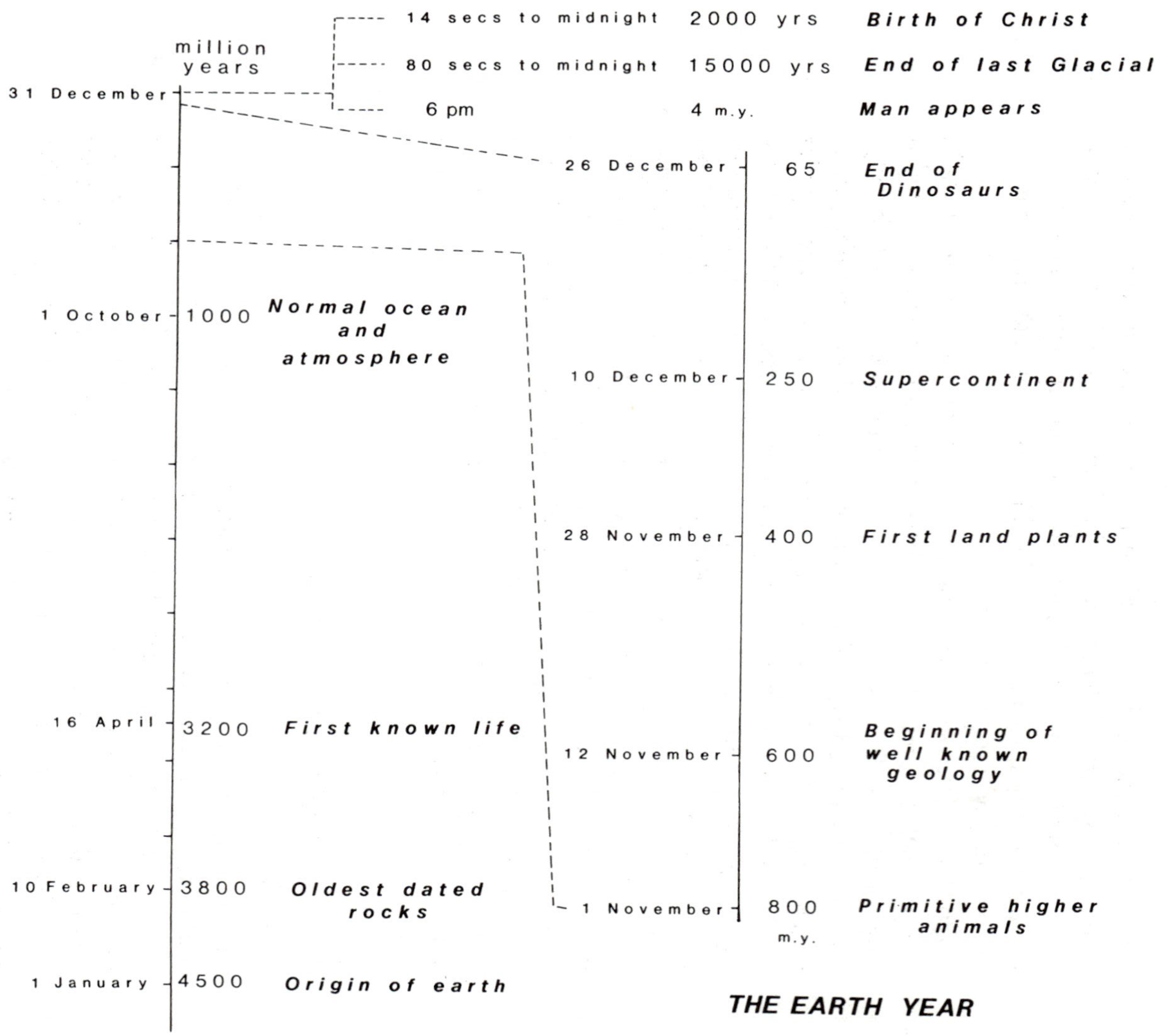

THE EARTH YEAR

million years

31 December

14 secs to midnight 2000 yrs *Birth of Christ*

80 secs to midnight 15000 yrs *End of last Glacial*

6 pm 4 m.y. *Man appears*

26 December 65 *End of Dinosaurs*

1 October 1000 *Normal ocean and atmosphere*

10 December 250 *Supercontinent*

28 November 400 *First land plants*

16 April 3200 *First known life*

12 November 600 *Beginning of well known geology*

10 February 3800 *Oldest dated rocks*

1 November 800 *Primitive higher animals*

m.y.

1 January 4500 *Origin of earth*

Figure 2.2. It is essential that one grasp the meaning of a billion years, but it is very difficult. To assist in this adjustment this diagram sets the age of the earth equal to a single calendar year. This makes each second equivalent to 143 years and each day equivalent to 12.33 million years of earth history.

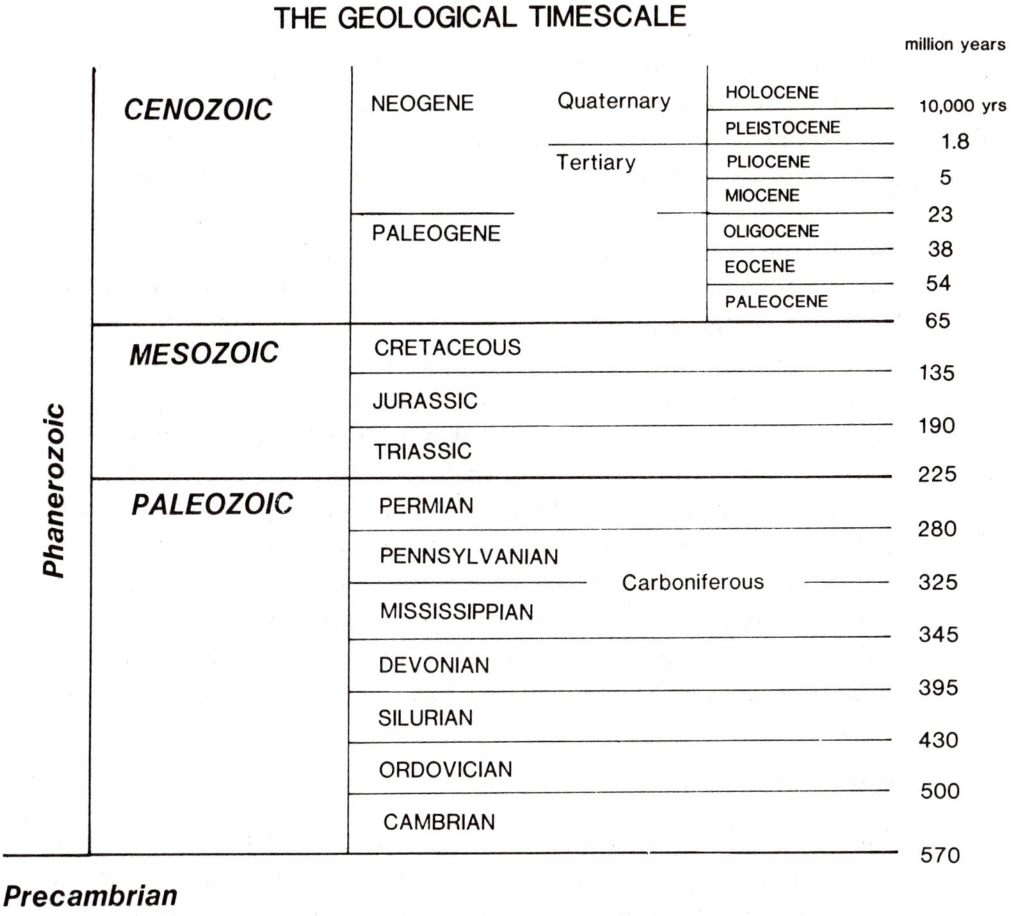

THE GEOLOGICAL TIMESCALE

	Era	Period		Epoch	million years
Phanerozoic	CENOZOIC	NEOGENE	Quaternary	HOLOCENE	10,000 yrs
				PLEISTOCENE	1.8
			Tertiary	PLIOCENE	5
				MIOCENE	23
		PALEOGENE		OLIGOCENE	38
				EOCENE	54
				PALEOCENE	65
	MESOZOIC	CRETACEOUS			135
		JURASSIC			190
		TRIASSIC			225
	PALEOZOIC	PERMIAN			280
		PENNSYLVANIAN	Carboniferous		325
		MISSISSIPPIAN			345
		DEVONIAN			395
		SILURIAN			430
		ORDOVICIAN			500
		CAMBRIAN			570
Precambrian					

Figure 2.3. The geological timescale shown here is abbreviated, but contains the names and the ages of the boundaries between units that are necessary for the discourse in this book.

For more than a century the geological timescale evolved without benefit of absolute ages, and its subdivisions are therefore of very unequal lengths (Figure 2.3). Even today, geologists usually talk about time not in years but in names, names designating intervals of greatly varying lengths, and say "during the Eocene" rather than "between 38 and 53 million years ago." Consequently, the task of familiarizing oneself with the history of the earth involves a not inconsiderable vocabulary. The system is hierarchical, beginning with a small number of units so broad that they are rarely used, and ending with levels so finely subdivided that few but specialists have heard of them. The general categories, however, become quickly familiar, with the scholarly, Greco-Latinized flavor of names like Paleozoic ("time of old life") and Proterozoic ("time of early life"), or the whiff of the Welsh countryside in Cambrian.

The units and major boundaries of the timescale are well established, and the ages of the main divisions are no longer much debated. For the finest subdivisions, however, the dating is still in flux, and the uncertainties are larger than we would like. For the late Cenozoic, we trust our boundaries to within 100,000 years or so, but as we go farther back the uncertainty increases to more than a million in the Paleocene and as much as a few million years in the middle of the Mesozoic. Inevitably, with increasing age, the statistical uncertainty of the isotopic dates also increases, limiting the precision we can achieve.

Time is, of course, continuous and has no gaps. The geological timescale, therefore, also has no gaps. The information on which it is based, however, does not come from a clock that always runs, but from the rock record, which is discontinuous. Many of the major boundaries, such as those between the Devonian and the Carboniferous, or between the Permian and the Triassic, have been placed at very widespread, major unconformities. At such points, the continuity of time and timescale is not matched by the continuity of the rocks. This distinction between time and the rock record needs to be kept firmly in mind.

TOO MUCH TIME, TOO FEW EVENTS?

The length of geological time has expanded more than 40-fold beyond what was envisaged before the means existed to calibrate it. Do we now have enough time? Does this enormously enlarged perspective accommodate all that we know to have happened in the history of the earth? It does, and it does more than that: It leaves us with time unaccounted for.

Let us try an experiment. If we know the thickness of some ancient sequence of deposits and their facies, we can estimate the time needed for its deposition, if we use the proper sedimentation rate. Furthermore, the geological timescale will tell us how much time was available. We have to take account of various possible uncertainties and errors, but in principle the exercise is simple and the answer ought to be reasonably accurate. In Wyoming, a series of early Cretaceous sandstones and shales closely resemble the coastal deposits of the present Gulf of Mexico. Going through the proposed exercise, we find that a mere 100,000 years would suffice to lay down the entire sequence; yet it occupies a stratigraphic interval 6 million years long. We can repeat the experiment elsewhere; in the majority of cases, we discover that the rock record requires no more than 1–10 percent of the time interval assigned to it. Evidently sedimentation, unlike work in Parkinson's law, does not expand to fill the time available.

What happens the rest of the time? Unconformities can sometimes be seen, or at least can be invoked to explain part of the deficit. Are they common enough to explain it all? In the deepsea, a quiet environment in which we would a priori expect a minimum of disturbance and erosion, the campaigns of the drill ship *Glomar Challenger* have shown that sometimes as much as half the record is missing. In shallow seas it might even be worse, but a shortage of more than 90 percent usually exceeds by far the sum of all visible hiatuses. Perhaps the geological record consists of rare major events separated by long intervals during which either nothing happened at all or, alternatively, each minor event left an imprint so faint that it was erased entirely by the next one. That, however, is not easy to accept, because one would expect to see some evidence for long times without deposition, a bit of soil that formed, or some burrowing by organisms. More often than not we see nothing of the kind.

Whatever the explanation, the notion that the archives of the earth mainly record brief major events, not the long quiet intervals, is not by itself disquieting. An atom also consists mainly of empty space, yet matter has much solidity, and the elimination of daily trivia from history need by no means be harmful. Even if earth history proceeded mainly by leaps and bounds, the interpretation we have achieved after two centuries of study certainly seems to be consistent, and makes a good deal of sense.

More disturbing is that this kind of record, if real, cannot be relied on to retain widespread traces of, say, a catastrophic event, such as a hurricane or the impact of an asteroid. Moreover, if we are forced to rely on a highly intermittent account to document a phenomenon that is itself pulsating, such as a stepwise evolution of life, our chance of finding the information we seek should be small indeed.

PERSPECTIVE

Rocks have their stories to tell: Faults and folds describe deformation, and environments of deposition can be deduced from the fossils their sediments contain. Fossils can also be used to arrange the rocks in the proper sequence, and simple ground rules permit us to connect a sequence of strata in one place with others elsewhere. The present is generally, though not always, the key to the past, and the younger strata rest on top of the older ones. To place the geological history so deduced in the context of time, we use dates based on radioactive isotopes. That, in turn, enables us to discuss the duration of events and the rates of processes, important clues in understanding history.

Other approaches will be discussed later: the use of stable oxygen isotopes in defining the history of the oceans, or the magnetism of rocks as a key to the movement of continents. The assortment of tools available to the geologist is large, varied, and growing.

This is enough as far as methods are concerned. Let us turn now to the first case of change: the changes in climate during the Great Ice Age. Its last cold phase is recent and vivid and still has direct consequences for our present world. Another similar event might even be in store soon. To understand the Great Ice Age, we need a grasp of what makes weather and climate – a little tedious, perhaps, but useful, and it will be needed again later when we examine how the ocean works. Having considered the Great Ice Age, we shall find that its causes lie deeper in the past, where they are connected with the drift of continents, a subject that will take us straight into the geological revolution.

FOR FURTHER READING

ON THE HISTORY OF GEOLOGY

The two newest books on this subject are also the two most interesting:

Hallam, Anthony, *Great Geological Controversies,* Oxford University Press, New York, 1982.

Albritton, Claude C., *The Abyss of Time,* Freeman, Cooper and Company, San Francisco, 1980, 251 pp.

The classic references, very readable, are as follows:

Adams, Frank D., *The Birth and Development of the Geological Sciences,* Dover, New York, 1938, 506 pp.

Geikie, Sir Archibald, *The Founders of Geology,* London, 1897; 2nd edition, MacMillan, New York, 1905.

A fascinating study of the impact of the battles over catastrophism versus uniformitarianism and the theory of evolution on society during the nineteenth century is the following:

Gillespie, Charles C., *Genesis and Geology,* Harper, New York, 1959, 306 pp.

BASIC GEOLOGY AND THE EARTH

The next two books are good basic texts on the principles of geology, the first one orderly and systematic, the second more topical but also broader, including a good deal of earth history:

Press, F., and Siever, R., *Earth,* Freeman, San Francisco, 1980, 649 pp.

Gass, I. G., Smith, Peter J., and Wilson, R. L. C., *Understanding the Earth,* M.I.T. Press, Cambridge, Mass., 1971, 354 pp.

ON EARTH HISTORY AND ITS METHODS

Three volumes in the series *Foundations of Earth Science* present a good deal more than has been discussed here; they do it interestingly and at a beginner's level without sacrificing quality:

Eicher, D. L., *Geologic Time,* Prentice-Hall, Englewood Cliffs, N.J., 1976, 150 pp.

Laporte, Leo F., *Ancient Environments,* Prentice-Hall, Englewood Cliffs, N.J., 1979, 176 pp.

McAlester, A. L., Eicher, D. L., and Rottman, M. L., *The History of the Earth's Crust,* Prentice-Hall, Englewood Cliffs, N.J., 1984, 224 pp.

The following are readable college-level texts intended for the more serious student, but they also serve as suitable reference works and have much local and stratigraphic detail:

Dott, R. H., Jr., and Batten, R. L., *Evolution of the Earth,* McGraw-Hill, New York, 1981, 572 pp.

Matthews, Robley K., *Dynamic Stratigraphy: An Introduction to Sedimentation and Stratigraphy,* Prentice-Hall, Englewood Cliffs, N.J., 1984, 512 pp.

TO ILLUSTRATE GEOLOGY

The following book stands out for its rich collection of photographs that supplement this text well:

Shelton, John S., *Geology Illustrated,* Freeman, San Francisco, 1966, 434 pp.

Climate past and present: the Ice Age

If the glacial period were uniformity, what was catastrophe?

Henry Adams, *The Education of Henry Adams*

THE SNOWS OF YESTERYEAR

An observer viewing the earth from space 20,000 years ago would have seen a different world. The outlines of the continents would have appeared subtly, and sometimes drastically, altered. The Gulf of Mexico was much smaller, and the North Sea not present at all. Indonesia, where blue seas now wash tropical islands, was a large and rather dry extension of southeast Asia. The jungles of the Amazon and the Congo were as green as they are now, and tradewinds were whipping whitecaps on the oceans, but an enormous shining mass of ice concealed much of Europe and North America, and the North Atlantic was covered with pack ice down to the latitude of Spain. Glaciers crowned the big island of Hawaii and many another mountain in the tropics, and huge lakes shimmered in the sunshine between the rainclouds over California, Nevada, and Utah.

The Ice Age is the classic example of major climate change. It took place not very long ago, and it changed the world drastically. Such places as Greenland and Antarctica have never left the Ice Age condition, helping us to visualize what the world of the Ice Age must have been like. Yet the climate change that brought this different world about was not really very large, nor was it the only example of its kind. Other climate changes, smaller but nevertheless of considerable impact, have taken place more recently.

For millions of years glaciation has followed glaciation, the last one having left us only a little more than 10,000 years ago. Could the odd weather conditions of the last two decades – the droughts in Africa, the severe winters in New England, the recent excessive rains in California – perhaps be harbingers of another one? The record of the Ice Age speaks for itself. Some 30 times already have glaciers come down from the north, and nothing indicates that we happen to live just past the end of the very last one. Another advance of the icecaps on the northern continents seems beyond doubt. The important question is not whether or not another glaciation will arrive, but when. That question cannot yet be answered.

The world has seen several ice ages, and ours has by no means been the longest. No ice age, however, should be seen as an unrelieved state of glaciation and arctic weather. Times of severe cold and extensive ice cover, called *glacials,* alternate with *interglacials,* times of milder climate like our own. Glacials and interglacials, collectively forming an ice age, have not only affected the high and middle latitudes; they have left their mark, albeit in a less dramatic and less well known manner, on tropical and subtropical lands and seas as well.

3

Climate and climate change

Our limited memory causes us to think of climate as more or less constant. It has been winter and it will be spring; the differences from year to year, although sometimes noticeable, are small. Yet it was not long ago that the climatic backdrop to human history was quite different indeed. I do not refer here to the glacial days of early prehistoric cavemen, but to the time of the Vikings, to the golden seventeenth century, and to Manifest Destiny and the Oregon Trail just a little more than 100 years ago.

Climate varies, and we shall find it easier to understand large variations if we first consider some smaller ones: what they do, what they mean, and, as far as our current knowledge permits, how they come about.

THE INCONSTANT CLIMATE

Few of us give much thought to the possibility of climate change, because our climate has been so constant for the past 50 years and, in general, so favorable. In India before 1920, for example, drought and famine could be counted on to occur about once every 8.5 years. Between 1920 and 1960 that risk was reduced by half, and the population, always controlled largely by periodic starvation, rose accordingly. Early in this century, Californians expected a really dry winter about once in 7 years, but since then this has happened only about once in 15 years. Recently, however, it seems that we may have returned to the previous, less favorable state. Because our planning for water resources has been based on the smaller risk of drought experienced between 1920 and 1960, this may well have serious consequences. Clearly, not only a change in climate can imperil a humanity straining against the limits of its resources, but also an increase in its variability, even if the average remains the same.

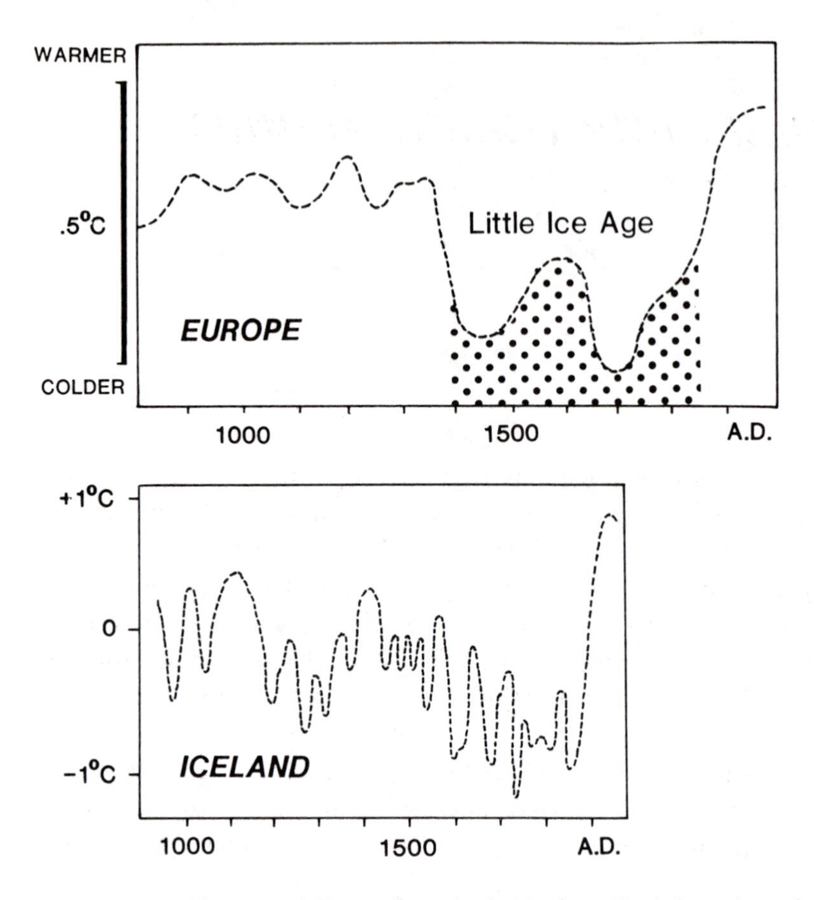

Figure 3.1. A millennium of climate change in the Northern Hemisphere. Around 1,000 A.D., conditions were very favorable, then deteriorated to those of the Little Ice Age.

During the past millennium our climate has changed repeatedly, and the changes have been rapid and at times quite noticeable. All over Europe, the fourteenth, seventeenth, and eighteenth centuries were much colder than the present one (Figure 3.1). In fact, the period from about 1650 to 1850 has been called the Little Ice Age; lost harvests and famine, with wars as aftermath, mark its record. In Iceland, where human subsistence is always marginal, the cold spell of the fourteenth century and the Little Ice Age had a particularly large impact. The earlier cold eliminated the Greenland colony, which had been settled in the warm years between 800 and 1,200 A.D., by destroying agriculture and seriously impeding travel across a sea sometimes ice-bound even in summer. Though the Little Ice Age itself involved a temperature drop of only 1.5° C, it permanently converted Iceland from a wheat-growing to a sheep-farming economy.

Not everywhere was the Little Ice Age a bad time. The prairies of central North America in the early nineteenth century were wetter than they are now, and the tales of the mountain men that lured large groups of settlers west around 1850 were founded on reality. It was not their fault that the end of the Little Ice Age arrived together with the new immigrants, who discovered land and climate to be far less suitable than had been promised. The decrease in rainfall would soon have decimated the buffalo herds had not the white man taken his toll a little earlier and somewhat more thoroughly.

Reconstruction of the climates of the recent past rests on a remarkable variety of sources: logs maintained by deck officers of the British navy and held for centuries in Admiralty archives, records of the dates and qualities of wine harvests kept (where else?) in French monasteries, and even tree rings. Most trees annually produce a layer of new wood that by its thickness reflects the weather conditions of the preceding and present growing seasons. When calibrated with weather records of the past century, tree-ring sequences allow us to extrapolate climate back in time. In the southwestern United States, the tree-ring record extends almost 8,000 years.

Was the Little Ice Age a fluke, not soon to be expected again? To answer that, we must look back further than written history permits us. A suitable record, longer even than that of the longest tree-ring sequence, comes from the advances and retreats of mountain glaciers as the climate oscillated between wet and cool and warmer and drier. The most recent climate change, for example, freed Glacier Bay in southeastern Alaska of the ice that still filled it entirely when John Muir visited there early in this century. Since the end of the last glacial period, the glaciers of Alaska, of the Alps, and of Scandinavia have advanced and withdrawn several times within their valleys, leaving a record of soil stripped away and trees felled by advances, and of glacial deposits left by retreats. This record, dated and correlated across the Northern Hemisphere, shows advances of the glaciers, each marking a colder period, at 10,500, 8,100, 5,300, 2,800, and 200–300 years ago (Figure 3.2). The last one was the Little Ice Age. Measuring their spacing with a slightly elastic yardstick, we conclude, with that whiff of periodicity* so pleasing to those who prefer nature to be orderly, that "little ice ages" are spaced about 2,500 years apart and last only 300–700 years. Although colder than the inter-

* The term "periodic" will be used for regularly reappearing and therefore predictable phenomena. We reserve the word "episodic" for irregularly returning events.

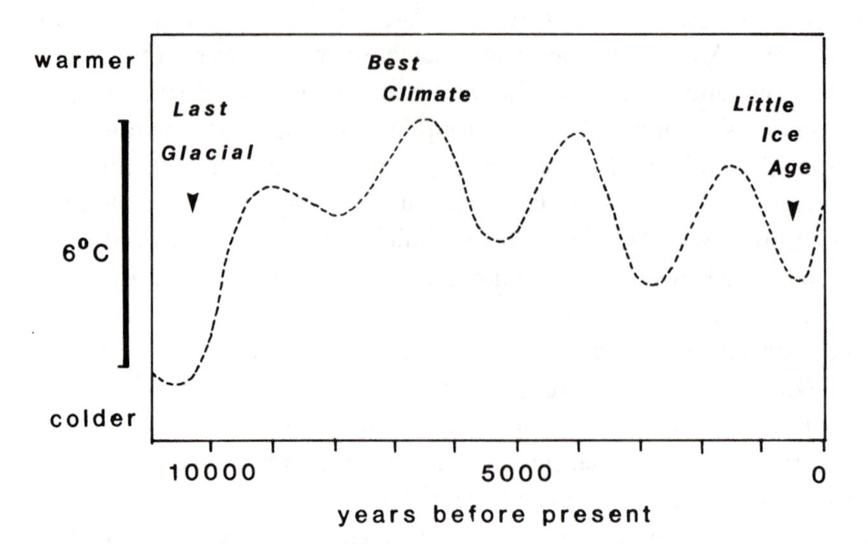

Figure 3.2. The Little Ice Age was but one of four similar cold spells in the last 10,000 years. This greatly generalized climate curve shows that we are already well past the best climate; even the mild years of the first half of the twentieth century were one or two degrees colder than the period between 6,000 and 7,000 years ago.

vening times, they are not very cold. In fact, on the whole, the post-glacial climate has been fairly level. In Sweden, for example, the mean annual temperature has never deviated by more than 3.5° C from the present value.

CAUSES OF LITTLE ICE AGES

The succession of cold and warm periods during post-glacial time suggests that we should look for some kind of episodic, perhaps even periodic, cause superimposed on the shift from glacial to interglacial conditions. What might this be? A period of 2,500 years is rather short and rules out variations in the orbit of the earth around the sun (see Chapter 5). Continents drift on a timescale of millions of years – much too long. The circulation of the oceans affects climate, but it varies in years or decades – too short. Let us reach out in desperation. Sunspots, those large disturbances on the solar surface, have long been suspected to influence climate, to cause droughts especially, although no one knows how this might work. Sunspots are periodic, with cycles of 11 and 22 years, which are too short. Recently, however, it has been discovered that there are times during which there are few sunspots or none at all. The last such occasion was during the Little Ice Age, tempting us to wonder whether sunspots might appear and disappear with the right periodicity to hold them responsible, in some unknown way, for the little ice ages. Clever, but how shall we test the idea?

Enter serendipity in the form of radiocarbon dating of tree rings. We used to assume that ^{14}C was formed at a constant rate by a never-varying flux of cosmic rays coming from the sun. This fundamental assumption, which if false would invalidate every radiocarbon date ever obtained, was finally tested some years ago by sampling the wood in separate rings of an 8,000-year-long tree-ring sequence. If ^{14}C production had been constant over time, radiocarbon years should have corresponded with calendar years obtained by counting the rings. They did not, and, what is more, the discrepancies varied with time. As a result, most radiocarbon dates needed corrections, some rather large, and upsetting to archaeologists who had just become accustomed to the new chronology ^{14}C had provided them with. Actually, the corrected dates make better sense than the old ones. The assumption that the cosmic-ray flux was constant was also discredited. Most interesting, the periods of minimal flux seem to be spaced about 2,500 years apart, the last one coinciding with the sunspot minimum of the Little Ice Age. This rather neat convergence of data from such disparate sources might cause us to regard the problem of why there are little ice ages as solved, were it not that we still do not understand how sunspots can influence climate. Moreover, the dating of the cold intervals still needs more precision and leaves room for skepticism.

BRIEF DISCOURSE ON THE WORKINGS OF CLIMATE

For more than a century geologists have concentrated mostly on the landforms and deposits left behind by glaciers and icecaps. In recent years, however, oceanographers, climatologists, geochemists, and many other specialists have combined forces to increase our understanding of the climates of our glacial past. Some of this heightened interest has resulted from the aberrant weather patterns of the past decades and from the recent awareness that a change in climate is not inconceivable, even on a human timescale. It is evident that a change in climate, or even increased annual climate variations, could have serious economic, social, and political consequences. Therefore, anything that might enhance our ability to foresee what the climate is likely to do in the next years or decades will be most useful, perhaps even essential.

Our understanding of how the atmosphere works is still incomplete, as our qualified success at forecasting the daily weather demonstrates.

Possibly there is enough randomness in the behavior of the atmosphere to preclude permanently any forecasting beyond a few months. Climate, on the other hand, is the average weather, from the human scale of decades to geology's eons. Tomorrow's rain is weather, this autumn's excess of it constitutes a wet season, but several wet seasons in a row may spell a climate change. Climate averages out much of the short-term uncertainty, and influenced strongly by processes such as ocean currents that have greater constancy than airflow, it has what might be called a longer memory. We can therefore hope to understand climate from a long record of observations even while the behavior of the atmosphere remains imperfectly known. If such a long record should be found to contain persistent or periodic elements, we might be able to forecast climate sooner than we can next season's weather. It is for reasons like this, as well as from normal curiosity, that the study of climate has become a lively and somewhat crowded field.

The earth's climate is an engine whose parts are land, ocean, and atmosphere. The engine runs on heat from the sun. Some of the heat is reflected back to space, but much reaches the surface and warms the soil, the air, and the water. Over the oceans, the heat causes evaporation; clouds form, travel, and condense, and rains fall elsewhere.

At high latitudes the sun's rays strike the earth at a low angle and heat the surface much less than do rays near the equator. The hot equatorial air, being light, rises and flows north and south (Figure 3.3), distributing warmth across the planet. Rising air expands and cools and is less able to hold moisture, and so we have an equatorial rainbelt. Aloft, the warm air gradually cools, and between 25° and 35° latitude, part of it begins to sink, returning along the surface to the equator. The sinking air is compressed and produces the subtropical high-pressure zone, which is dry, because compressed air can hold more moisture. The rest of the upper air, at middle and high latitudes, continues toward the poles, cooling all the while. It ultimately sinks in the polar regions and returns as a cold surface wind.

The seasons complicate this simple picture because of the uneven heating through the year in temperate and polar regions. It is further modified by a force resulting from the rotation of the earth, the Coriolis force, which affects the direction of any current on earth, whether air or water. It is proportional to the mass times the velocity of the current; hence, it has a large effect on air and sea currents, but is imperceptible to a walker. Looking downstream, the Coriolis force turns a current to the right on the Northern Hemisphere and to the left

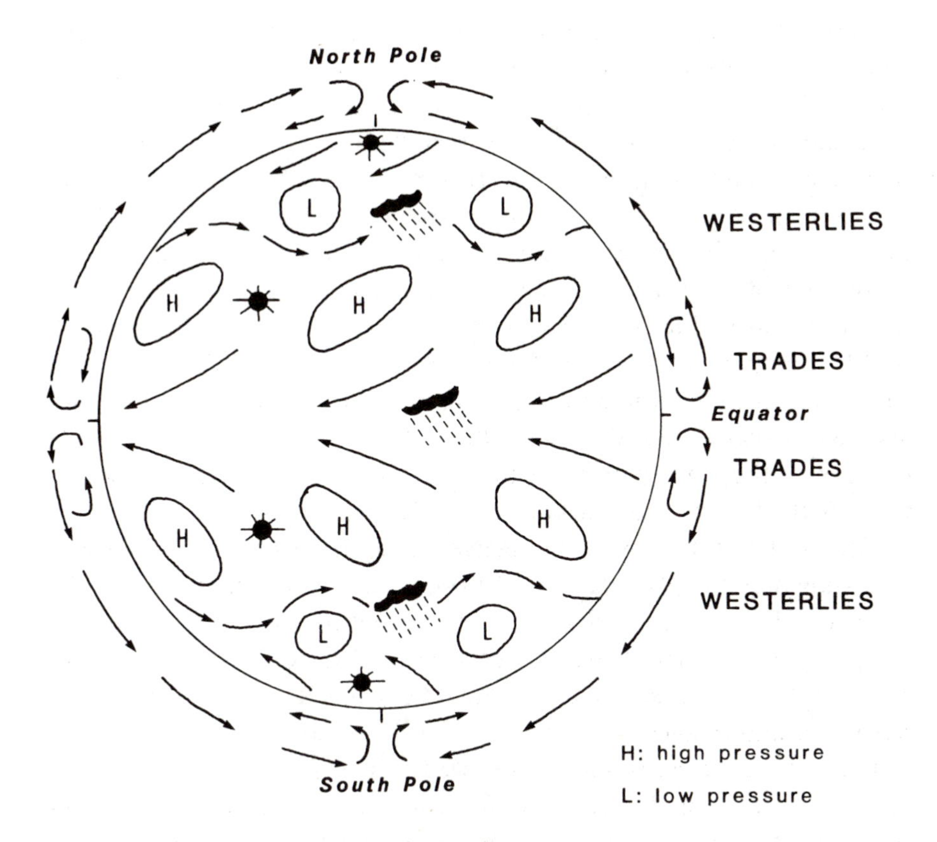

Figure 3.3. The planetary circulation of the atmosphere begins with warm air rising at the equator and spreading north and south. As it does so, its capacity to hold moisture diminishes, and tropical rains fall. In the subtropics, some of the now cooler air returns to the equator. The Coriolis force turns the flow into easterly trade winds, and the sinking air is dry and makes for a sunny climate. At mid-latitudes, a zone of westerly winds carries low pressure systems and their rains eastward; its strongest expression is the jetstream at high altitude.

on the Southern Hemisphere. It can be visualized (although not fully understood) by imagining oneself standing at the North Pole: Throw a ball south with great force. While the ball flies, the earth turns eastward under it, and when the ball finally hits the ground, it will do so well west, or to the right of the intended target.

The Coriolis force converts the surface airflow from the subtropics toward the equator into two sets of easterly winds. These *tradewinds* blow all year between 10° and 25–30° latitude. Similarly, the equatorward flow at high latitude is converted into polar easterlies. In between, the air that continues northward at greater heights is deflected to form winds from the west that are especially noticeable between 40° and 60° north and south. These *westerlies* transport the seemingly endless trains of weather systems that bring the rain to western Europe and to the Pacific coast of North America. Part of their driving force is a high-altitude wind of great velocity, the *jetstream,* well known to west-bound air travelers who find their flights delayed when it blows with unexpected strength. The jetstream does not flow straight, but in a pattern of large horizontal curves, about six on its path around the earth. Where the jetstream swings north it brings warm and often moist air from low latitude; where it turns south it is cold and often dry. The geographic positions of the curves in its path are fairly fixed, but their amplitude is not, and neither is the intensity of the flow. Both vary from year to year and over longer periods, and when they do, anomalous seasons are the result (Figure 3.4). Possibly, very long term changes in the wave pattern of the jetstream may have had something to do with the onset and departure of glacial periods.

This is, in simple form, the general circulation of the atmosphere, also known as the planetary circulation, because it depends only on the presence of a sun, an atmosphere, and a rotating earth. Consequently, the planetary circulation has been a feature of the earth from its beginning, and we can be sure that tradewinds, temperate westerlies, a jetstream, and tropics and subtropics have existed at all times.

If we add continents and ocean basins, then the airflow is modified. Anyone living near the sea is familiar with the sea breeze, which comes up in midmorning, vanishes in the late afternoon, and is replaced at night with a wind from the land. During the day the land heats more rapidly than the sea; the warm air rises and is replaced by air blowing in from the sea. At night, the sea retains its warmth longer than the land, and the airflow is reversed. The same process, but with seasonal rather than daily reversals, and on a much larger scale, produces the *monsoons.* Over Asia in summer, for example, the air heats, and, in, rising, induces an influx of moist air from the Indian Ocean to the south. Coming from the ocean, the winds, their direction altered by the Coriolis force, bring the much-needed rain of the southeast monsoon. In winter, Asia is cold, the ocean warm, and the monsoon, now blowing from continent to sea, is dry and cool. Monsoons, being products of the configurations of land and sea, must have appeared early in earth history, as soon as there were continents, but their patterns have varied with time as continents have drifted and oceans have opened and closed.

Not all solar energy goes toward heating the air and the surface of the earth. Some of it is reflected back immediately by clouds, or farther down by warm earth surfaces. Clouds are good reflectors, which return

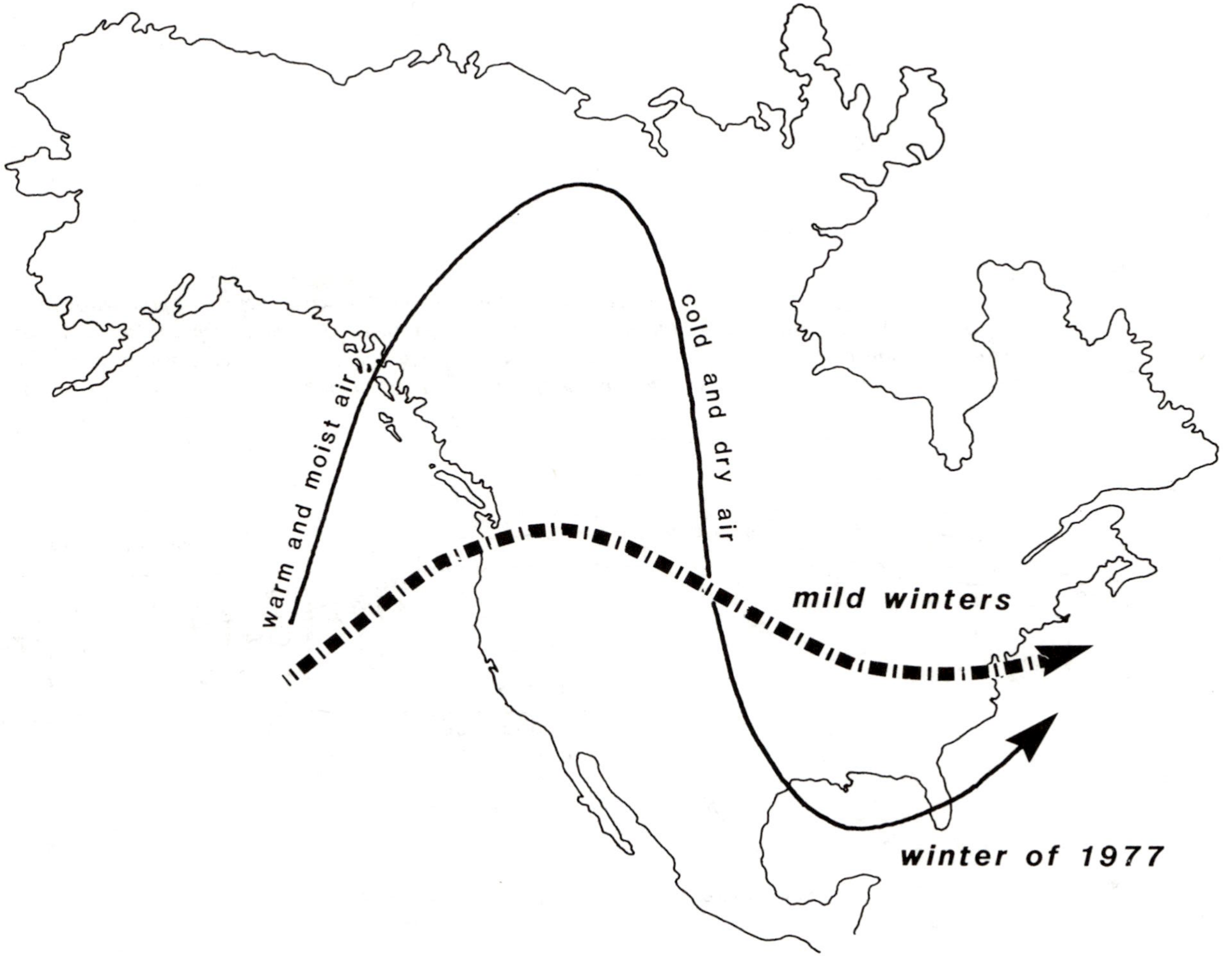

Figure 3.4. The jetstream flows around the earth in a pattern of some six large waves. The positions of these waves are fairly well fixed, but their amplitude varies greatly from year to year, and so does the intensity of the flow. These changes in the path of the jetstream have considerable influence on the weather of the temperate zones, especially in the winter.

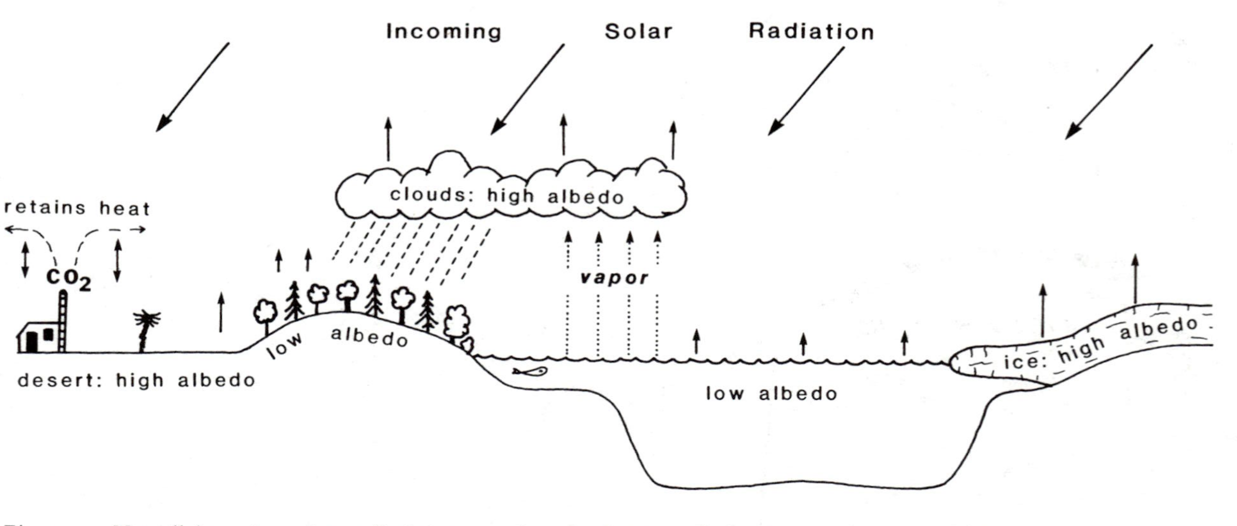

Figure 3.5. Not all incoming solar radiation goes to heat land, sea, and air. A great deal is reflected by clouds, by snow and ice, and by desert, which have a high albedo, whereas forests and water, with their low albedo, absorb most of the heat. Carbon dioxide, at present generated copiously by our burning of fossil fuels, refuses passage to the reflected heat waves and so exerts a warming influence, a greenhouse effect.

radiation, and snow and ice 50–85 percent. An earth with large deserts at low latitude, or with vast icecaps, will be cooler than one that is mostly green or covered with water. This reflective capacity is called *albedo*. Ice has a high albedo, the ocean a low one. The albedo of the earth and its variations with time have had a large influence on climate.

A warm ocean evaporates more than a cold one and is therefore liable to have a thicker, more widespread cloud cover. This reduces the amount of radiation that reaches the water; the ocean cools, and the cloud blanket diminishes. This mechanism, called *negative feedback*, works toward stabilizing the climate.

Not only clouds reflect radiation back into space; dust in the atmosphere does the same. A volcano sending a plume of ash up into the stratosphere can lower the temperature measurably, until all the ash has fallen back to the surface. Some volcanic gases have the same effect. The "year without summer" of 1816 in the eastern United States followed an enormous eruption of the Indonesian volcano Tambora. Burning forests, the clearing of land, and many other human activities besides shuffling feet also stir up lots of dust, but it does not go high enough; the next rain washes it out in days or weeks. Only volcanoes, perhaps a huge meteorite, and clusters of nuclear bombs are candidates for a climatic role by virtue of the dust they raise.

The opposite effect, a warming of the earth, is produced by carbon dioxide in the atmosphere. Carbon dioxide allows sunlight to pass through, but the heat reflected back from the surface has a different wavelength and is absorbed. Even a little carbon dioxide forms quite an effective blanket, retaining heat that would otherwise be lost to space. This phenomenon is known as the *greenhouse effect*. The carbon dioxide content of our atmosphere is now only about 0.03 percent, but it is increasing as we burn coal, gas, and oil, strip forests away, and cultivate old soils rich in organic matter. Since the industrial revolution began early in the last century, the carbon dioxide in the atmosphere has increased by 15 percent. That is not yet much, and a good part of the addition has dissolved in the ocean, but we can confidently predict that as we continue prodigally to burn fossil fuels and cut our jungles, the figure will double in something like 50 years.

The effects of such a CO_2 increase will not be trivial. The mean annual temperature might rise several degrees, and more in polar regions. Much of the ice in Greenland and the Antarctic would melt, and sealevel would rise enough to flood many coastal plains and cities. The climate would change in ways we still argue about, but none of them are likely to be regarded as improvements.

4

Portrait of a glaciation

The realization that there was once an Ice Age is relatively new and did not come easily. Louis Agassiz, a young Swiss geologist, later at Harvard, accepted in 1830 the evidence he had seen in the Swiss Alps, but it took him many years to convince his colleagues that it indeed meant that glaciers were once much more extensive than they are now. Today that seems rather surprising, because the tracks left by the icecaps of Europe and North America are anything but inconspicuous. The walls of debris left by melting icefronts, for example, eloquently testify to depositional and erosional processes that no longer occur in the temperate parts of the Northern Hemisphere. Once the concept of so large a climate change had been accepted, however, the subject proved fertile and became a meeting place for many specialists, from oceanographers to archaeologists.

ICECAPS ON THE WORLD

The appearance of the world at the height of the last glacial vividly demonstrates what a climate change is capable of accomplishing. From perhaps 25,000 to 16,000 years ago, the cover of ice was extensive over North America and Europe, but strangely patchy in Siberia (Figure 4.1). A dome of ice perhaps 3 km thick rested on Canada, extending south to the Ohio River valley and New York, and bordering in the west on an icesheet over the northern Rocky Mountains and Cascades. Another icecap covered Scandinavia and spread southward across Britain, Germany, and European Russia. On both continents, treeless plains traversed by innumerable meltwater streams lay south of the ice margin, and beyond that came the boreal forest of birch, pine, and spruce. The temperate broadleaf forest had retreated far to the south, to North Africa and the Near East from Europe, and in America

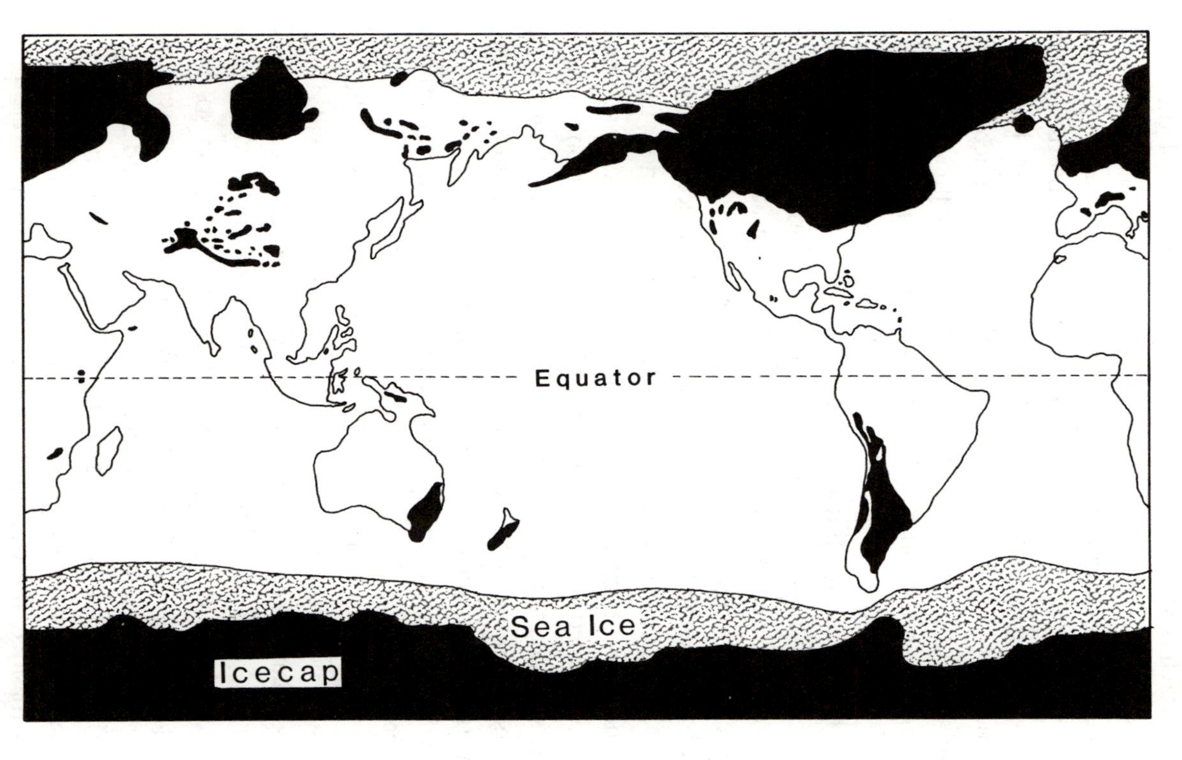

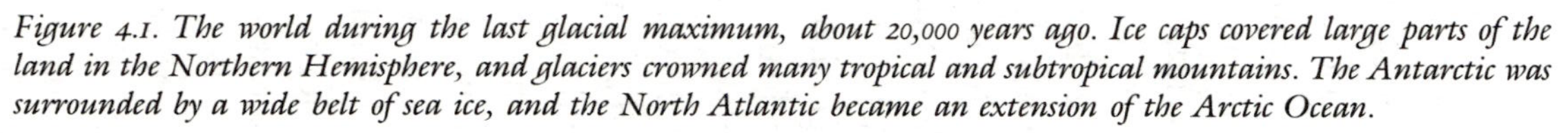

Figure 4.1. The world during the last glacial maximum, about 20,000 years ago. Ice caps covered large parts of the land in the Northern Hemisphere, and glaciers crowned many tropical and subtropical mountains. The Antarctic was surrounded by a wide belt of sea ice, and the North Atlantic became an extension of the Arctic Ocean.

to such favored climes as Florida and the southern edge of the Gulf states.

In the Southern Hemisphere, the Antarctic icecap was somewhat larger than it is now, and there were mountain glaciers on the higher ranges of Australia, Africa, and South America. In the main, however, the vast icesheets were a feature of the Northern Hemisphere. In all, some 40 million km^3 of ice were piled on land, and one imagines polar conditions, familiar today only to Eskimos and Alaskan oil drillers, for the central United States and middle and western Europe.

In reality, however, the climate at the edge of the ice was not so severe. Forests grew not far beyond, especially in North America; the summer days remained long and bright, as they must at that latitude; and the winters, although worse than now, were not truly arctic. The edge of the ice extended so far south not so much because a polar climate prevailed there but because a mound of ice flows under its own weight like molasses (Figure 4.2). As long as the ice moved faster than it melted, it continued on its way south, coming to a halt only where ice flow and summer thaw were in balance. Under such conditions, a slightly warmer summer at the ice margin would cause a quick retreat, whereas an increase in snowfall thousands of kilometers to the north would thicken the ice there and push its edge farther south, sometimes into the boreal forest.

A floating icecap, on the other hand, will sink deeper when it thickens; it will not flow outward to the same degree as will a cap based firmly on land. That is the reason why the frozen Arctic Ocean did not form the nucleus of the northern icecaps. The Antarctic ice does rest on land in the center, but its edges enter an ocean so deep that a floating ice shelf is formed. In contact with the warmer sea and attacked by waves, this ice shelf melts and fractures, and icebergs break off and drift away. The largest, thickest icecap will be stopped by even a narrow sea as long as it is more than a few kilometers deep. In this manner, the circum-Antarctic ocean has effectively protected the southern continents from widespread glaciation. Some areas, seemingly well suited to accommodate an icecap, such as Siberia, did not. Northern Alaska also remained free of ice. To form glaciers and icecaps, ample precipitation in the form of snow is as necessary as is a low temperature. Icecaps form where the snowfall in winter exceeds the thaw of summer. Copious Atlantic storms fed the icecaps of eastern and central Canada and northern Europe. Winds from the Pacific, then as now, brought snow to southern Alaska

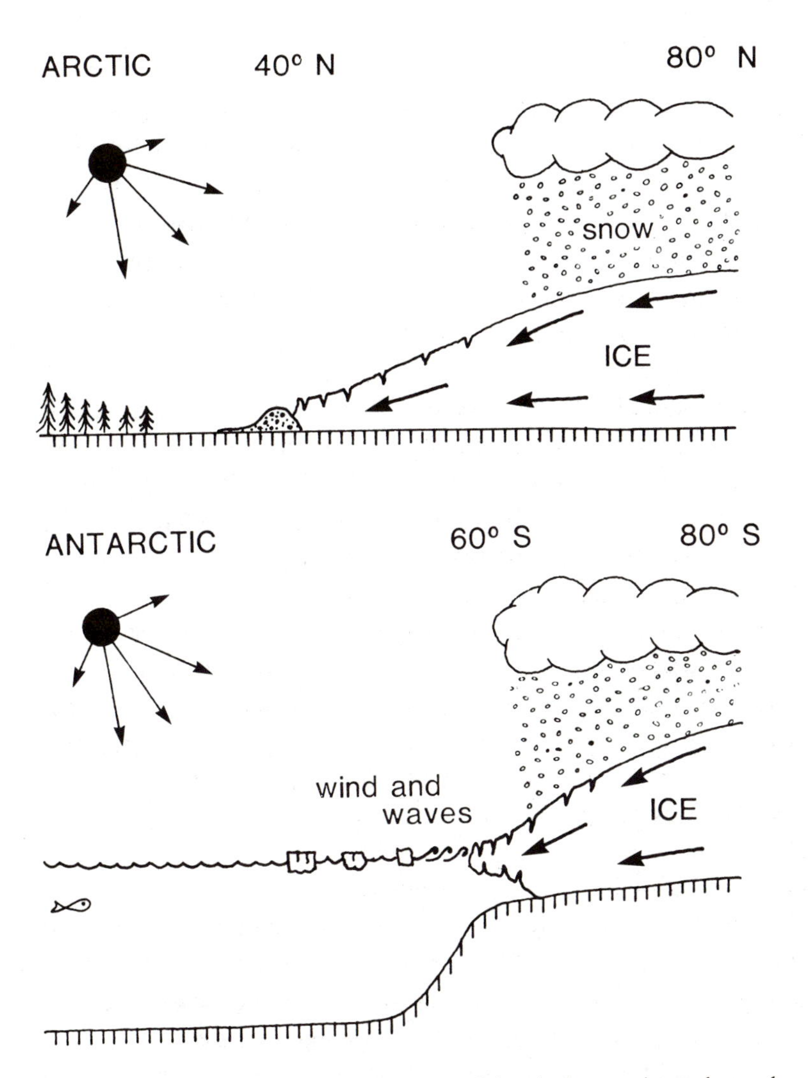

Figure 4.2. Because ice flows under its own weight, the ice margin on the northern continents was able to advance far south into regions that even then did not have an arctic climate. In the Southern Hemisphere, such an advance was not possible. There, the deep ocean surrounding Antarctica presents an impassable barrier, because waves and currents erode the floating edge of the ice.

and British Columbia, but they lost their moisture in the mountains and were dry when they reached the northern slopes. Storms reach Siberia only after they have passed across the width of Asia and have dropped their rain and snow in the mountains of the south. The Arctic Ocean, permanently frozen, did not evaporate and so furnished almost no precipitation.

GLACIAL OCEANS

The question of the nature and climatic role of the Ice Age oceans was, somewhat surprisingly, not seriously raised until recently, when oceanographers, climatologists, and Quaternary geologists banded together to find out how the two-thirds of the world that are under water had fared during the Ice Age. They clarified many issues and obtained many answers, although, as is too often the case, some of the questions that seemed simple at the start have remained elusive or become more complex.

Deep-sea sediments far from shore consist mainly of the microscopic calcareous or siliceous shells of planktonic organisms. These fossils record, and on examination reveal, the condition of the surface sea in which they lived and died. This information can be extracted if we assume, reasonably so for a past no more remote than a few hundred thousand years, that most species have not appreciably changed their environmental requirements. If we have at our disposal many widely distributed and dated cores of ocean sediments, we can obtain estimates of salinity and temperature, and even nutrient levels, for the surface waters of Pleistocene oceans. If we use species that flourish only in one season, we can even construct maps of the temperature and salinity distributions at the sea surface for summers and winters now long gone. Dating the core samples is, of course, critical and remains tricky, though various methods are at our disposal.

Consider the sea surface temperature for the summer of the year 18,000 before present (B.P.). The North Atlantic shows the difference most dramatically (Figure 4.3). It was frozen in the north between Greenland and Great Britain, and full of drifting ice as far south as a line between Cape Hatteras and Spain. The Gulf Stream, which today warms the shores of Ireland, England, and Norway, crossed directly from Florida to the Azores and warmed only Africa. A day at a Carolina beach in water of 6–8° could hardly have been called a summer holiday.

Compared with this huge southward shift of arctic conditions, the changes in the North Pacific seem rather trivial (Figure 4.4). More surprising, the equatorial ocean was much cooler, and many tropical seas were not much warmer than the present North Sea inshore during a hot spell in August. On the other hand, at mid-latitude there was almost no change in temperature; California, Madeira, and the Azores were much as they are now.

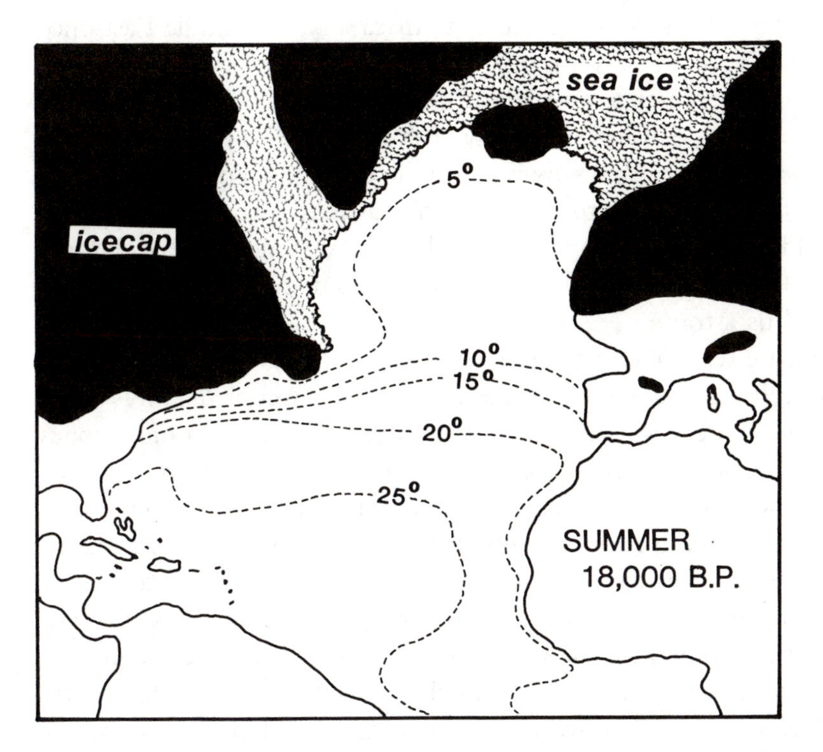

Figure 4.3. During the last glacial maximum the North Atlantic was an arctic sea. The edge of the permanent sea ice ran from Britain to Nova Scotia, with drifting pack-ice to the south. Even in summer the temperature between Spain and the Carolinas did not exceed a frigid 10°C, and the Gulf Stream, today responsible for the mild climate of northwestern Europe, crossed straight from Florida to north Africa.

THE WORLD BEYOND THE ICE

What was the rest of the world like when ice covered so much of the Northern Hemisphere? What were the conditions in Ice Age Greece, California, Brazil, or central Africa? The question is not easily answered, because as we compare evidence from different parts of the world, we must be certain that we look at quite precisely the same instant in time. To correlate an ice advance in Minnesota with a deposit in Colombia that might be several thousand years younger or older simply will not do, because the fluctuations in climate were numerous and often brief.

Obviously, the large climatic swings in the high and middle latitudes could hardly have been without repercussions in the subtropics and tropics. Indeed, we have long known that closer to the equator our present climate was preceded in many places by a distinctly different, much wetter one. Lake levels in East Africa and the Middle East once were well above those of today. Rock paintings in the Sahara bear witness to the presence of a flourishing population in what was an almost lush land, rich in game. Logic has suggested to many geologists that such wet periods, or *pluvials,* ought to have coincided with the glacials of the higher latitudes, but logic has told others that the pluvials should correspond with warm and hence wet interglacials. Unfortunately, hypotheses resting only on logic, no matter how compelling, often lead us astray.

What little reliable data we possess informs us that the world outside the glaciated regions was by no means uniformly wetter than it is today (Figure 4.4). Because the temperate, subtropical, and tropical zones were compressed toward the equator by the expanding cold belts, the Sahara benefited from the rain that today falls in the Mediterranean. California also received more rain during glacial times, because a high-pressure zone over the icecap in the north deflected Pacific winter storms southward. With a climate in that state much like the climate of Oregon today, pinyon woodlands and lakes fringed by pine and juniper were found in interior valleys, and the coastal zone was green with dense forests. The lakes are now largely gone, but the groundwater stored during those wet millennia still supplies the needs of Los Angeles and makes possible the semi-desert agriculture that enriches California, both uses ill-adapted to the water resources of the region.

Most of the world, however, was drier than it is today because the lower ocean temperatures reduced evaporation. The moisture supply was accordingly much less, and so was the average global precipitation. On land it was everywhere cooler, but the effect was felt most strongly at higher latitudes. After all, a drop of 2–3°C below a yearly average of 27°C is felt less sharply than a decline of 5–10°C below a present mean of only 15°C, as was the case in central North America.

ONSET AND DECLINE OF A GLACIAL PERIOD

There have been many glacials, but because glaciers are so destructive of their own and older deposits, we know much only about the last

Figure 4.4. Not everywhere is a glacial period a time of extreme cold. Temperatures over most of the world were only a few degrees lower, except near the edge of the icecaps (top). The surface waters of the oceans were colder not only in the extreme north and south but also, surprisingly, in the equatorial region. The pattern of changes in precipitation (bottom) is complex and not fully understood. Thus, we are not yet in a position to forecast changes in rainfall should another glacial period arrive.

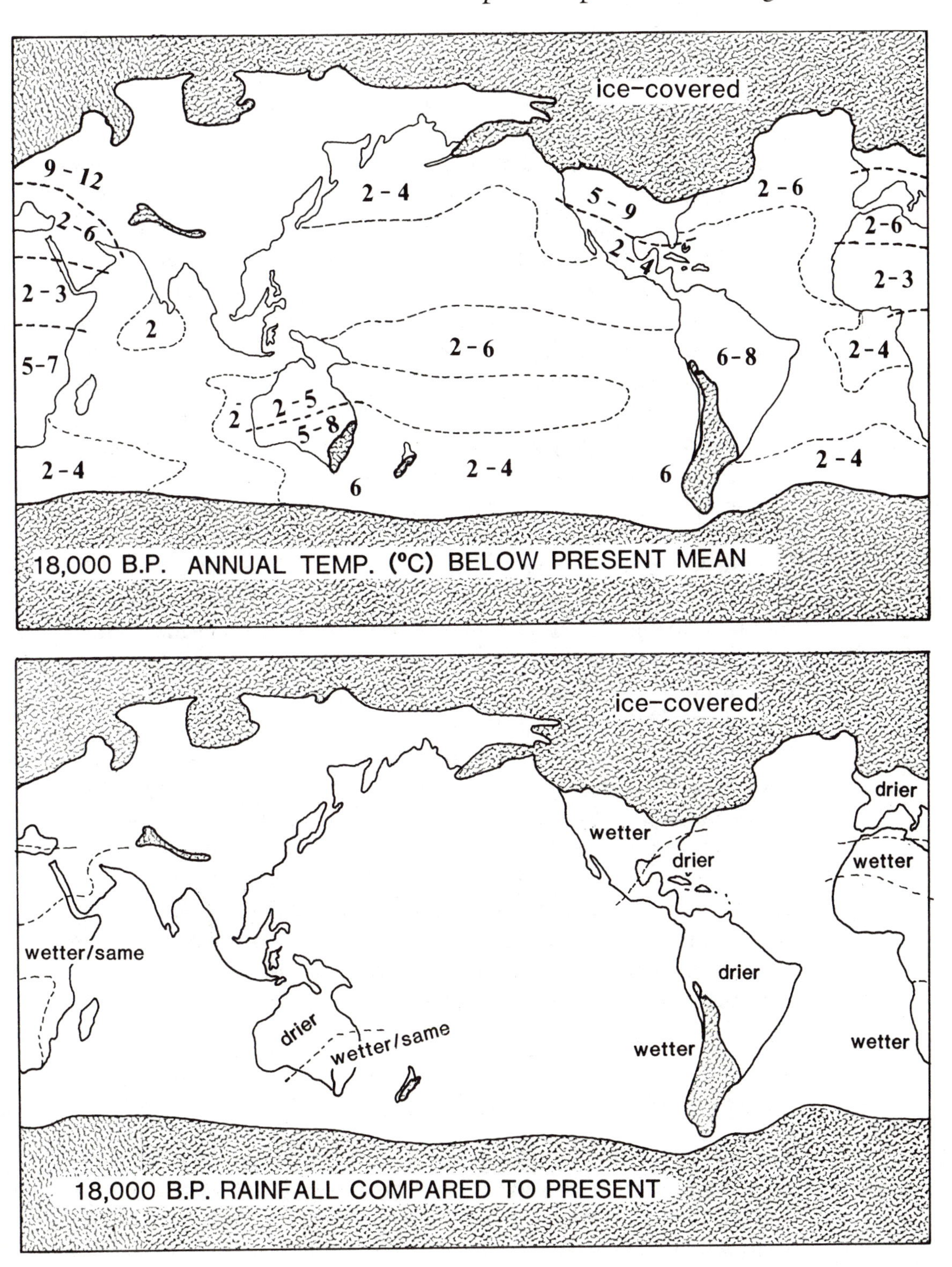
ice-covered
9 - 12
2 - 6
2 - 3
5-7
2
2 - 4
2 - 4
2 - 6
2 - 5
5 - 8
6
2 - 4
5 - 9
2 - 4
2 - 6
6 - 8
6
2 - 6
2 - 3
2 - 4
2 - 4
18,000 B.P. ANNUAL TEMP. (°C) BELOW PRESENT MEAN
ice-covered
drier
wetter
drier
wetter
wetter/same
drier
wetter/same
drier
wetter
wetter
18,000 B.P. RAINFALL COMPARED TO PRESENT

one or perhaps two. The most recent glacial period announced its arrival about 115,000 years ago with two brief, sharp cold spells, but it got going in earnest some 40,000 years later (Figure 4.5). It had several cold *stadials* separated by times of slightly improved conditions called *interstadials*. The last and most severe of those stadia began around 28,000 years ago and reached its peak after some 10,000 years. Just a few millennia later the ice began to melt quite suddenly, withdrawing 2,000 km in less than 6,000 years, from Ohio to Hudson Bay. In the west, the Cordilleran ice sheet left Puget Sound 14,000 years ago and was completely gone 4,000 years later. During the same period the ice in Europe cleared Britain, the Netherlands, and northern Germany and retreated into the mountains of Scandinavia. It had disappeared entirely 7,000 years ago. Compared with the 60,000-year duration of the entire glacial, its demise in much less than 10,000 years can certainly be called sudden. The onset of each cold phase was no less abrupt; it has been estimated that the ice thickened at a rate of 20–60 cm, and advanced about 1 km per year.

How sharp such climate changes can really be is illustrated by a brief event that took place about 11,000 years ago. At that time the ice had been melting in northwestern Europe for several millennia. Birch and pine woods had replaced the icecap and the mossy, shrubby tundra that fringed it. Around 10,500 B.P., however, a sharp cold spell brought the tundra back from Sweden to Britain and the Low Countries, wiping out the forests and restoring the herds of reindeer and their Paleolithic hunters to their old domain, all within one century. Dramatic as this event must have been, it lasted only 700 years and ended as abruptly as it had begun. Already 9,000 years ago (Figure 4.5) the summers in Europe were almost as warm as they are today, and a forest of oak and hazel covered England and Germany where only tundra had existed 1,000 years before. This optimal climate endured until about 5,000 years ago, when one of the little ice ages heralded a more variable and gradually cooler and wetter climate in the countries of the eastern North Atlantic.

THE LEVELS OF LAND AND SEA

One after the other, glaciations came and went during the Pleistocene, each drawing enough water from the ocean to lower sealevel greatly and to alter the geography of the world's coasts. Each time the ice melted, the sea rose and inundated the former coastal plains. Such changes in sealevel, which are simultaneous all over the world, are

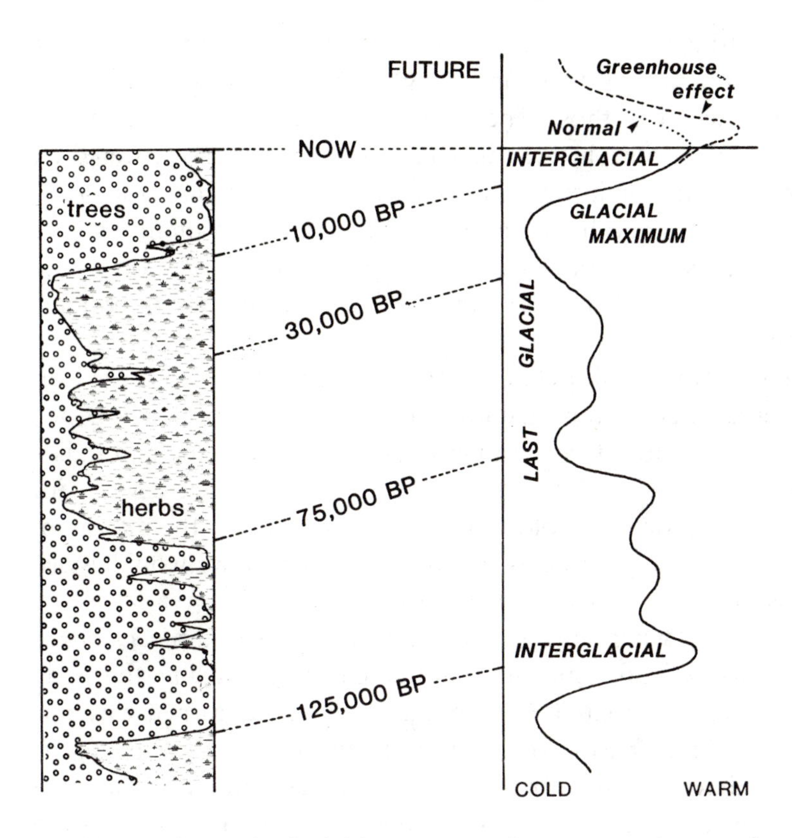

Figure 4.5. Pollen grains buried in peat or wet clay are very resistant to decay and tell us much about the response of the vegetation to climate changes. Even something as simple as the ratio of tree to herbaceous pollen clearly defines cold and warm spells. Combined with other information, pollen records (left) can be converted into a climatic history for the last 100,000 years (right). A speculative extrapolation to the future by John Imbrie of Brown University assumes that the inevitable return to another glacial period will be delayed by the greenhouse effect of our burning of coal and oil.

called *eustatic*. The next to the last glaciation, a large one, accumulated about 75 million km^3 of ice and so produced a eustatic drop in sealevel of between 150 and 200 m. If all of the ice remaining in Greenland and Antarctica were to melt, a eustatic rise in sealevel of about 70 m would ensue. During the last glacial maximum, 18,000 years ago, our best estimate is that the level of the sea fell between 100 and 150 m. Why is this estimate so crude? Could we, should we, be more precise?

A glacial eustatic drop or rise in sealevel can be estimated in several ways. We might, for instance, calculate the total volume of ice from the area and thickness of the icecaps. The area can be determined with some confidence from the end moraines, the walls of debris left at the outer fringes of the ice, but the thickness presents more of a problem. Mountaintops that pierce a glacier are frost-splintered and look jagged, whereas those once covered entirely by ice are worn smooth. Unfortunately, in Scandinavia and Canada the ice covered all but a very few mountaintops, and our thickness estimates lack precision. In Antarctica, both the extent and the thickness of the ice are uncertain, because the edges of the icecap, resting as they do in the sea, have left only diffuse traces on the seafloor, and few mountains rose above the ice.

One might look for shorelines of the right age, now submerged on the continental shelves, the shallow platforms that surround the continents. Shorelines often have distinctive relief and deposits, and radiocarbon dates on enclosed wood or shell can tell us their ages. With dates and depths for many old, now submerged shores, the maximum lowering of the sea might be ascertained. In addition, we would establish the history of the subsequent rise of the sea.

Unfortunately, when this simple idea was pursued, the evidence turned out to be confusing and contradictory, producing the wide range of estimates just cited. Only gradually have we learned why this problem is so difficult to solve. As it turns out, most of the difficulty is due to the behavior of the earth's crust when a large load is placed on it or taken away. Because we do not possess a cosmic tide gauge to measure sealevel, we must define it relative to the adjacent land, and both the land and the seafloor refuse to maintain fixed levels.

Vertical movements of the earth's crust due to loading or unloading are called *isostatic,* in contrast to vertical *tectonic* movements, which accompany earthquakes and mountain building. Icecaps are large weights, and we have long known that they depress the surface of the earth, which rises again when the ice melts. Yet, curiously, few realized until just recently that adding or subtracting ocean water has an analogous large effect on the seafloor.

Continents and ocean floors, the components of the earth's crust, can be regarded as rafts floating on a denser fluid, the underlying mantle (Figure 4.6). These rafts obey the buoyancy law of Archimedes, and if weighted down with an icecap, they will sink until they have displaced a volume of mantle material equivalent to the added weight. Conversely, if the ice melts, the raft will rise. The seafloor responds in the same way to removal or addition of water. As a rule of thumb, every 100-m layer of ice will depress a continent by about 27 m, and every 100 m of water removed from the sea will cause the seafloor to rise 30 m. Such isostatic compensation also applies to other kinds

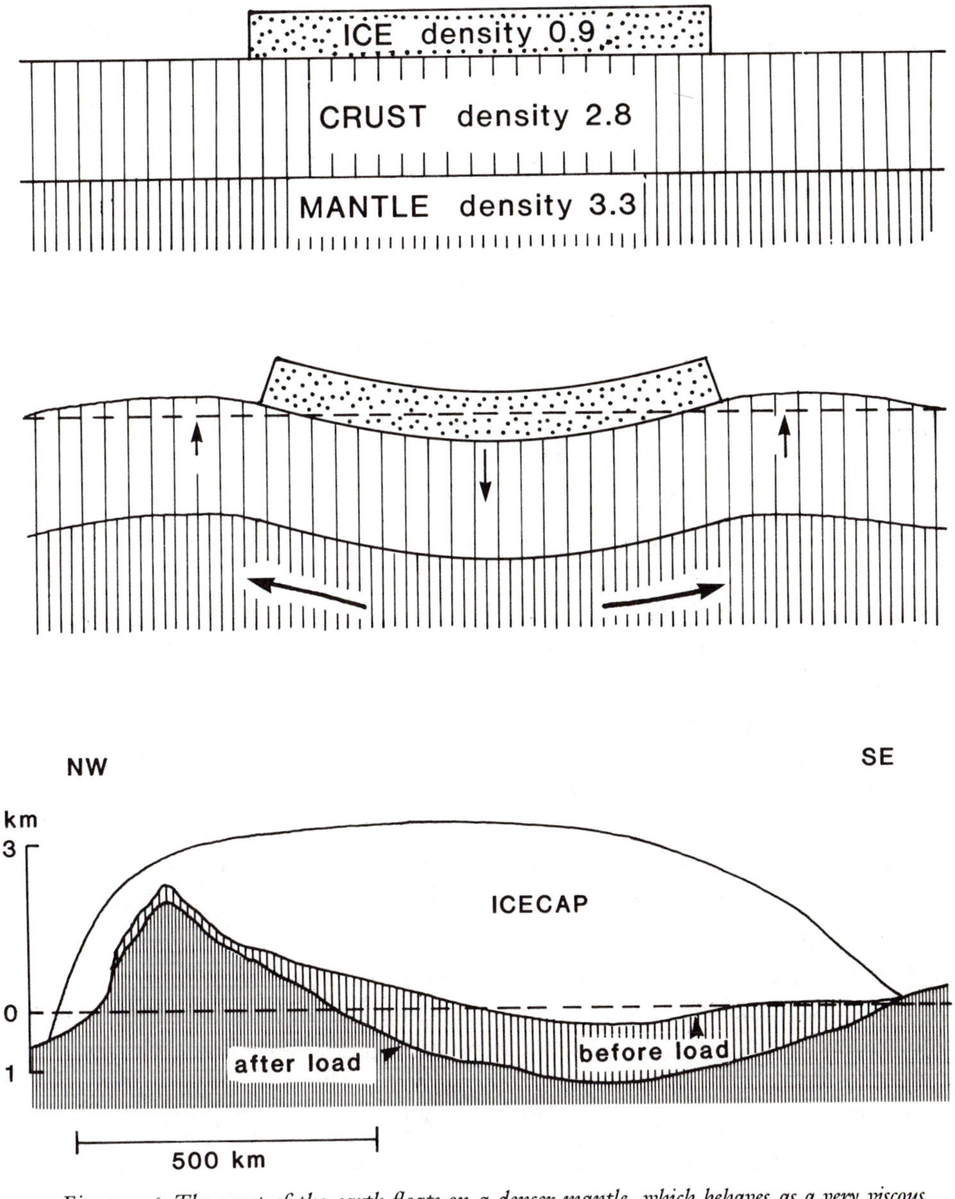

Figure 4.6. The crust of the earth floats on a denser mantle, which behaves as a very viscous liquid. If we place an icecap on the crust, it will bend and displace an amount of mantle material equivalent to the mass of the ice. The displaced mantle flows outward and raises a low but broad welt around the icecap. Because the mantle is so viscous and flows so slowly, equilibrium in either direction will be delayed compared with the growth or melting of the ice.

of loads: a new volcano, a large delta, or the removal of a mountain range by erosion.

There are, as always, complications. Because the crust has strength, it bends like a beam and subsides for some distance beyond the edges of the load. When the crust sinks, it displaces mantle material, which flows away to either side and raises a low welt around the depression. The mantle flows back when the load is removed. Because isostatic compensation depends on the flow rate of the very viscous mantle material, it is a slow process, and subsidence or uplift wil persist for quite some time after a load has been emplaced or removed. Even though the northern icecaps disappeared 8,000 years ago, Canada and Scandinavia have thus far risen only a little more than half of their eventual 500–800 m. Before the formerly ice-covered region is back in equilibrium several thousand years from now, Hudson Bay will be dry land.

Obviously, if we attempt to measure the rise in sealevel near formerly glaciated regions, we should be prepared for messy corrections. Oxygen isotope ratios for marine fossils (see Chapter 5) show that the amount of water removed from the sea during the last glacial maximum was equivalent to a layer 165 m thick, considerably more than even the deepest submerged shorelines would suggest. Of course, 165 m of water represents the removal of a considerable weight, and we must make a correction for the isostatic effect. If we do that, we find a compensating rise in the seafloor of about 50 m, so that the net drop in sealevel was only 115 m, a value comfortably within the range of the other estimates.

Let us assume, then, that the late glacial sea was about 100–120 m lower than the present one. This produced a very different coastal geography, except where the continental shelf is narrow and the sea deepens rapidly, as is the case along the coast of northern California or the Mediterranean. In the Gulf of Mexico, on the other hand, more than 200 km of coastal plain were added (Figure 4.7). The islands of Indonesia were welded into an extension of southeast Asia large enough to influence the equatorial climate and to enable prehistoric people to make their way into Australia across a sea much reduced in width though still redoubtable. The Bering Sea was mostly dry, serving as a landbridge for the first Indians to migrate into North America. The late glacial coastal geography and the manner in which it changed when the ice began to melt and sealevel rose have a large role to play in our understanding of climate and of the migrations and behavior patterns of early man.

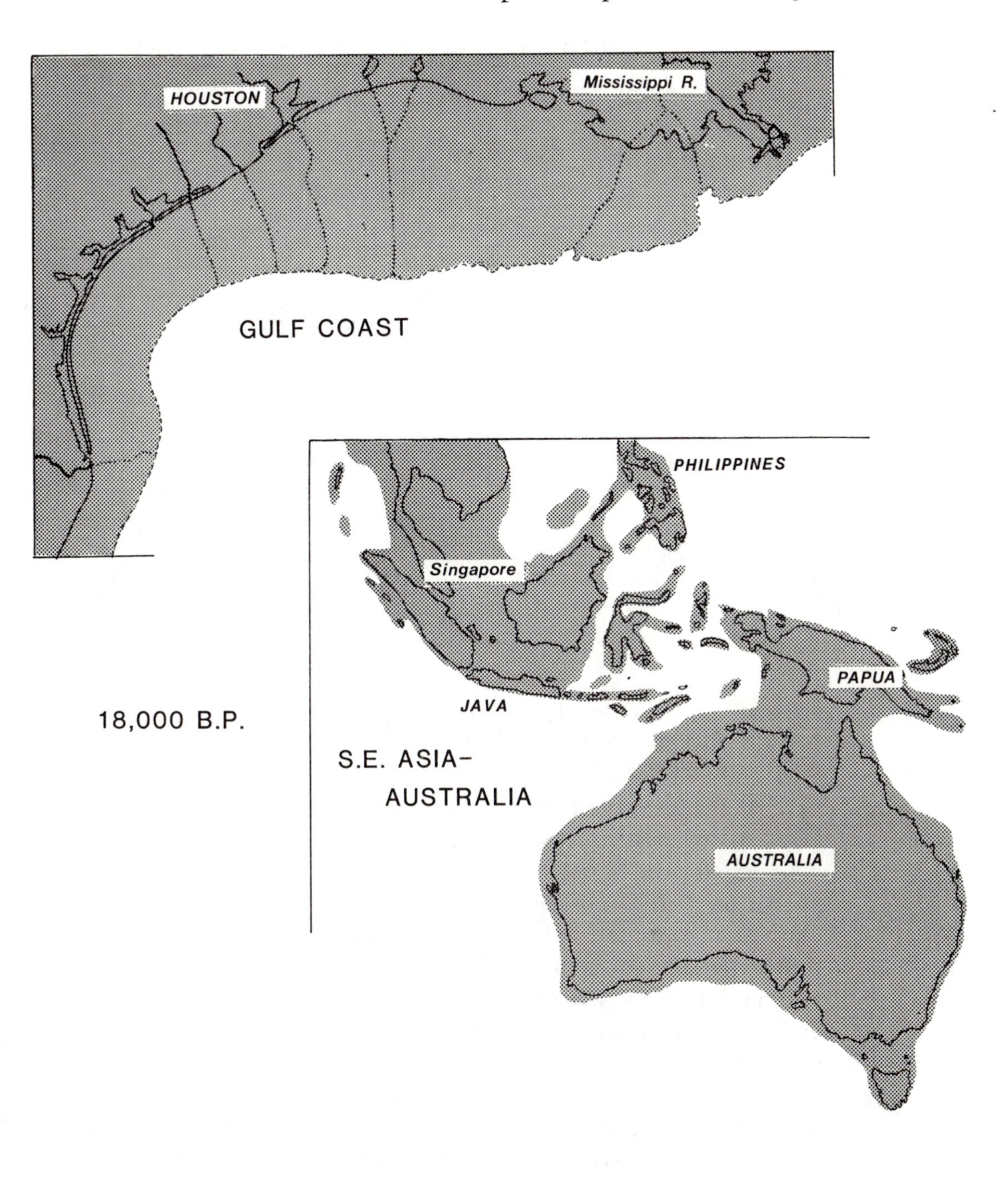

Figure 4.7. The formation of icecaps removed a great deal of water from the ocean. During the last glacial maximum this caused sealevel to fall by about 120 m and exposed most of the continental shelves, greatly altering the geography of the world in some places.

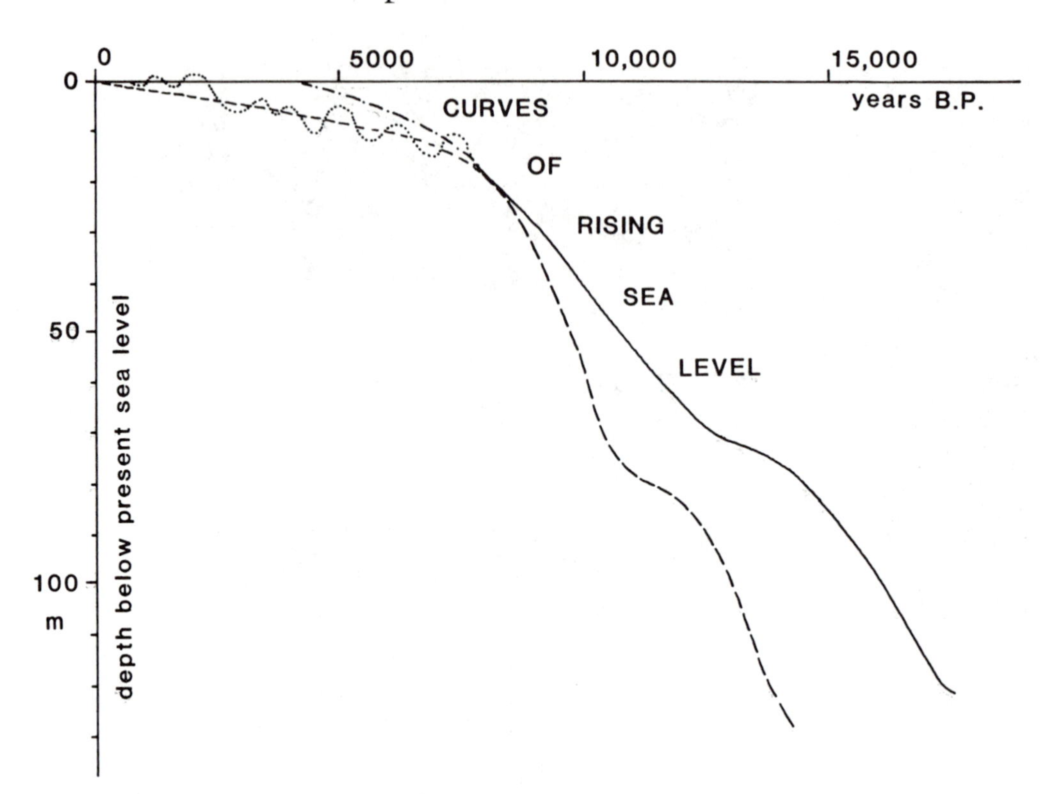

Figure 4.8. At the end of the last glacial period the ice melted and sealevel rose again. Our knowledge of this rise still leaves, as this diagram shows, room for controversy. Each of the curves has its enthusiastic proponents and firm antagonists.

When the last glacial period ended, the rise of the sea was initially rapid, as much as 2 m per century (Figure 4.8). Then, about 8,000 years ago, the rise began to slow, with the sea still about 25 m below its present level. It is not clear where the water for the remaining rise came from, because the icecaps on the northern continents were by then completely gone. Our best guess is that the ice of West Antarctica continued to melt. If that is so, and if the West Antarctic is really unstable, it might still be melting, and another 7–10 m rise will be in store for us.

To assess that distressing possibility, we need more precise knowledge of the most recent part of the post-glacial sealevel rise. Unfortunately, the data are conflicting, and opinions on the subject are sharply divided. Some believe that the rise stopped several thousand years ago, others that the sea is still rising, whereas a third group thinks that it has for some time been oscillating by 1 or 2 m around the present mean. The problem is that when we deal with a few meters' difference and a few centuries, we are pressing the limits of the accuracy of the data. Obviously, more work is in order, work that might have very real value in devising ways to protect low-lying cities and lands.

The rate at which the shoreline moved during the early post-glacial rise was far from trivial. The shore on the wide shelf of the Gulf of Mexico, for example, proceeded landward at about 3 km per century. In fact, almost everywhere the coast moved so rapidly that there was no time to form coastal deposits of any consequence. Even the Mississippi River, large as it is and was then, was unable to build a delta; it simply retreated ahead of the rising sea within its own valley. Only when, around 8,000 B.C., the rise slowed to much less than 0.05 cm per century did the horizontal march of the shore decrease and coastal deposits begin to build. From that moment on, in many places, beaches, barrier islands, and lagoons grew upward as fast as the sea rose. Quite often the position of the present shore is close to where it was 8,000 years ago, even though sealevel was then some 25 m lower. As time passes, all of those lagoons and estuaries trapped by the rise of the sea will fill with sediment and vanish. In a few thousand years, all the beautiful coastal wetlands that have escaped conversion to condominium villages and yacht harbors will be converted into coastal plains fronted by straight beaches.

5

The Cenozoic and other ice ages

We used to think that the Great Ice Age consisted of four glacial periods, the oldest of which began about 2 million years ago, so defining the start of the Pleistocene. That belief has turned out to be wildly wrong, but the fourfold subdivision of glacials and interglacials has stubbornly retained its foothold in books and minds, even in the minds of geologists who should know better. Today we can distinguish some 30 severe cold spells separated by brief warmer times. In the Northern Hemisphere the oldest ones go back about 3 rather than 2 million years, thus starting well before the Quaternary, and Antarctica has been in a full ice age for at least 15 million years. There is evidence that even earlier the world's climate fluctuated between colder and warmer periods, though the differences were not as great as those between glacial and interglacial times. Moreover, as will be seen in Chapter 11, the Ice Age came upon us rather gradually. It feebly announced itself about 40 million years ago and advanced in steps. Finally, ours is not the only ice age the world has ever seen, and if we wish to understand the phenomenon, we must take into account that such cold excursions occurred with some frequency throughout most, if not all, of the earth's history.

GLACIATIONS AND OXYGEN ISOTOPES

The comings and goings of glacials and interglacials have been most completely recorded in the sediments of the deep ocean, thanks to the behavior of two stable isotopes of oxygen. Of the two, the lighter one, ^{16}O, is by far the more abundant. When a marine organism takes up carbon, either by itself or in the form of carbon dioxide or carbonate, it acquires ^{16}O and ^{18}O in a ratio reflecting the relative abundances of the two isotopes in the surrounding seawater. When the organism dies and its calcareous shell sinks to the seafloor, its $^{16}O/^{18}O$ ratio is preserved and can be measured with great precision.

The light isotope evaporates from the sea more readily than the heavy one, so that atmospheric moisture and rain are enriched in ^{16}O, whereas the sea retains a small excess of ^{18}O. Normally, if rain falls on land, rivers return the water to the sea, and wind and currents mix it so well that, except in isolated basins such as the Mediterranean, the oxygen isotope ratio is everywhere the same. During a glacial period, however, large amounts of water evaporate from the sea and are locked up in ice with their excess ^{16}O, thereby leaving the ocean with a $^{16}O/^{18}O$ ratio that is lower than that during an interglacial. When the icecaps melt, the water is returned to the sea, and the original ratio is restored. Therefore, fossils in deep-sea sediments contain a record of glaciations and deglaciations in the form of oscillations in the oxygen isotope ratio. All we need to do is to find a long ocean sediment core that encompasses sufficient time, analyze it, and date it, and we shall have the climatic history of the Ice Age.

Nothing is, of course, ever quite that simple. First, the core must not contain any hiatuses. Also, some organisms are selective about oxygen isotopes and do not accurately reflect the isotope ratio in their environment. Most important, the oxygen isotope ratios of fossil carbonates also depend on the temperature at which the organism prepared its shell. Organisms living in cold water take up more ^{18}O than those in a warm sea. Thus, the $^{16}O/^{18}O$ ratio for microfossils from a glacial ocean is lower than that from an interglacial one, not only because much ^{16}O was stored in ice, but also because the waters were colder. About one-third of the difference between glacial and interglacial oxygen isotope ratios is due to temperature. We are able to correct for this temperature effect and so obtain both ice volume and temperature-change curves.

Suppose we have analyzed a core that yields the variation of oxygen isotope ratios with time shown in Figure 5.1. How shall we date the events it displays for tens to hundreds of thousands of years in the past? We can use radiocarbon dating on the calcareous fossils, but only to about 40,000 years ago. There is also the complex decay chain of uranium, which, under favorable circumstances, can take us back to 300,000 years. Beyond that, no good isotopic dating technique exists for this type of sediment. We can, however, make use of the fact that the earth's magnetic field frequently reverses direction, changing magnetic north into magnetic south and vice versa (see Chapter 6). Most sediments contain small magnetic grains, which, as they settle gently to the deepsea floor, orient

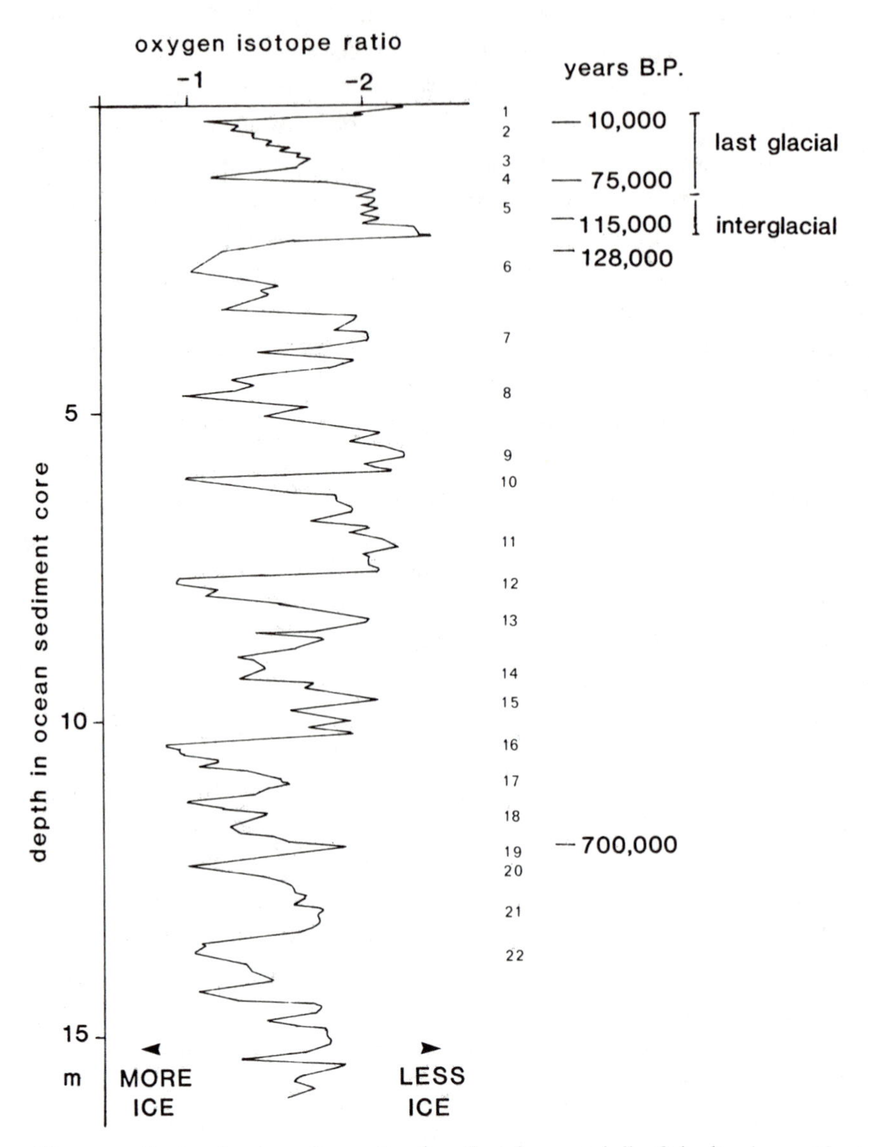

Figure 5.1. Oxygen isotope ratios measured on the calcareous shells of planktonic organisms are an index for the temperature of the sea and for the amount of water stored as ice on land. Conventionally, the oxygen isotope ratio is shown as a deviation from an arbitrary standard. The curve shows the fine details of Quaternary climatic change that an oxygen isotope curve is able to reveal, including 22 major fluctuations in 700,000 years.

themselves in the direction of the magnetic field. The faint magnetization thus imprinted on the sediment can be measured and its reversals dated with reference to the known reversal history. Between reversals and isotopic dates we must interpolate on the assumption that the rate of deposition did not vary. That is, of course, a somewhat dubious assumption, even in the deepsea, and the chronology of the oxygen isotope history is still its weakest point.

The glacial record of the ocean shows some remarkable features (Figure 5.1). Each of the many glacial stages lasted about 100,000 years and was separated from its neighbors by much shorter warm periods of some 10,000 years. Each glacial began very suddenly and, punctuated by several maxima and minima, grew steadily more severe until it ended as abruptly as it had begun.

BEFORE THE PLEISTOCENE

When, by fiat of stratigraphers, the Pleistocene began about 2 million years ago, glacial conditions had been around for quite some time. The normal state of the world, however, seems to be one of ice-free poles, a rather smaller temperature gradient from equator to pole than we experience today, and smaller seasonal contrasts. Only once in a while does an ice age interrupt this rather warm and equable climate.

Unmistakably glacial sediments turn up quite early in the geological record, between 3 and 2 billion years ago, but their preservation is poor, and we can say little of the geographic distribution of the "icecaps," not even whether they belonged to a true ice age or merely were laid down by the polar icecaps of a climate more like our present one.

There is, however, little doubt about the reality of the next several ice ages, those of the latest Precambrian, the Ordovician-Silurian, and the Carboniferous and Permian (Figure 5.2). Scratched rocks, boulder clays typical of the deposits of the sole of a glacier, and extensive ice-rafted sediments in the sea permit no other conclusion. The Paleozoic ice ages left their marks mainly on the southern continents, and they were related to the position of the South Pole at that time. The late Precambrian ice age, on the other hand, seems to have been exceptionally widespread and to have descended to a very low latitude, perhaps even to the equator. It also lasted much longer than any of the others (see Chapter 14).

There is in the dates the suggestion of a crude 200–250-million-year spacing of the ice ages, with an average duration for each of 50 million years or so. That, of course, may well be an illusion, but the great lengths of past ice ages clearly imply that we must still be far from the end of our own. What about the future? Will ice once again cover New York and London? Will huge lakes inundate the southwestern United States? Scholars have ventured various guesses

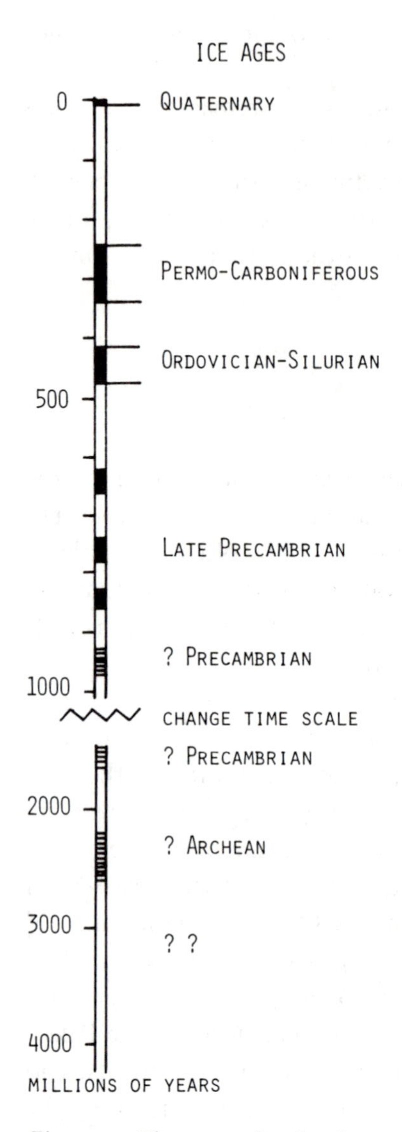

Figure 5.2. The present Ice Age is not the only one the world has known. At least four and perhaps five others have occurred during the past 1 billion years. Particularly severe was the late Precambrian Ice Age, which appears to have had three peaks and lasted abnormally long (see Chapter 14).

regarding this interesting question, but guesses is all they are. The most reassuring is John Imbrie of Brown University, who believes that the next round will be delayed by a man-made carbon dioxide blanket and is still 23,000 years away. Others, however, using essentially the same data, have recommended that we brace ourselves for a cold year 3,000.

CAUSES OF GLACIAL AND INTERGLACIAL CLIMATES

Why should the climate of the earth have oscillated for several million years between two extremes, spending very little time in transition between glacials and interglacials? And what brings on an ice age in the first place, if the normal climate of the earth is rather evenly warm? Most current attempts to address this last question rely on the concept of continental drift, and we must defer that discussion to Chapter 11. For the phenomenon of alternating glacials and interglacials, on the other hand, the answers are to be sought elsewhere. No wholly satisfactory theories exist at this moment, but the various approaches to the problem are illuminating.

Before we consider some of the proposed explanations, it is well to take careful note of just what must be explained. Any proposed mechanism must provide for a climate that oscillates between severe and mild conditions. It must deal with the possibility that these oscillations are periodic. Moreover, because climatic variations of shorter duration are superimposed on the glacials and interglacials, we must seek causes that operate on many timescales, from hundreds of thousands to hundreds of years. Finally, because the evidence for the abrupt onset and decline of glacials is becoming quite strong, any proposed mechanism should permit sharp transitions from one state to another.

The theories, many of which deal simultaneously with the causes of the ice age as a whole and of its internal oscillations, are legion, but they fall into just a few categories. Some seek the cause in variations in the output of the sun. Others look toward heat filters interposed between donor sun and receptor earth. Volcanic ash, cosmic dust clouds, and a carbon dioxide blanket are examples of such filters. Yet another class of hypotheses rests on variations in the orbit of the earth around the sun.

Many of these propositions are difficult to test, dealing as they do with past conditions not likely to leave an unambiguous geological record. It is not easy to see, for example, how variations in solar output or passage through a cosmic dust cloud could be detected other than through their climatic consequences, thus leading us on a merry path of circular reasoning. Such hypotheses I judge not very beneficial.

FIRE AND ICE

Other propositions rest on firmer ground. Volcanic eruptions were proposed long ago as a possible cause for ice ages and have intermittently returned to the center of the stage. The effect on temperature of a large plume of ash and gases at high altitude is real enough, as has been demonstrated for many eruptions. Such eruptions, however, like that of Krakatoa in 1883, have also shown that for any cooling to occur, the ash cloud must be very large and raised very high, and that the effect lasts at most a few years.

The geological record is inconclusive. Data from boreholes in the ocean floor have convinced several geologists that the cold periods of the Middle Miocene, the Pliocene, and the Pleistocene were accompanied by intensified volcanism, and less dramatic post-glacial climatic events have also been attributed to volcanism. Others argue, with some justification, that if one wants it enough, a correlation can always be achieved between two series of events that are both frequent and poorly fixed in time. If, as is inevitable in studying the late Cenozoic geological record, the timescale is in millions of years, it is good to remember that during 1 million years, 5 to 10 glaciations may have occurred, as well as numerous volcanic eruptions; yet none need have coincided.

The volcanic-dust hypothesis suffers from weaknesses other than lack of chronological persuasiveness. Few eruptions have a global impact on temperature; to have the desired effect, a rapid succession of violent events, continuing for millennia, would be needed. There is no evidence that more intense volcanic activity and exceptionally violent eruptions were more common in the past than they are now. Moreover, no one has dared to use volcanic activity to explain the apparent episodicity of glacials and interglacials. Given this particular stumbling block, it is not surprising that some blithe spirits have suggested that the shoe is on the other foot, that the loading and unloading of ice and water on the earth's crust as glacial follows interglacial are the causes of volcanic eruptions. For the time being it is probably best to regard volcanism as a secondary and at best minor factor, and to reserve judgment on the postulated correspondence between fire and ice.

We are no better off with another filter: the heat-preserving carbon dioxide blanket. If it were to vary much with time, as recent work suggests, it certainly might have a major impact on climate, but at present we lack the data to judge the plausibility of this mechanism for creating glacials and interglacials.

AROUND AND AROUND THE SUN

A fundamentally different approach links climatic variations to periodic changes in the way the earth circles the sun. This astronomical

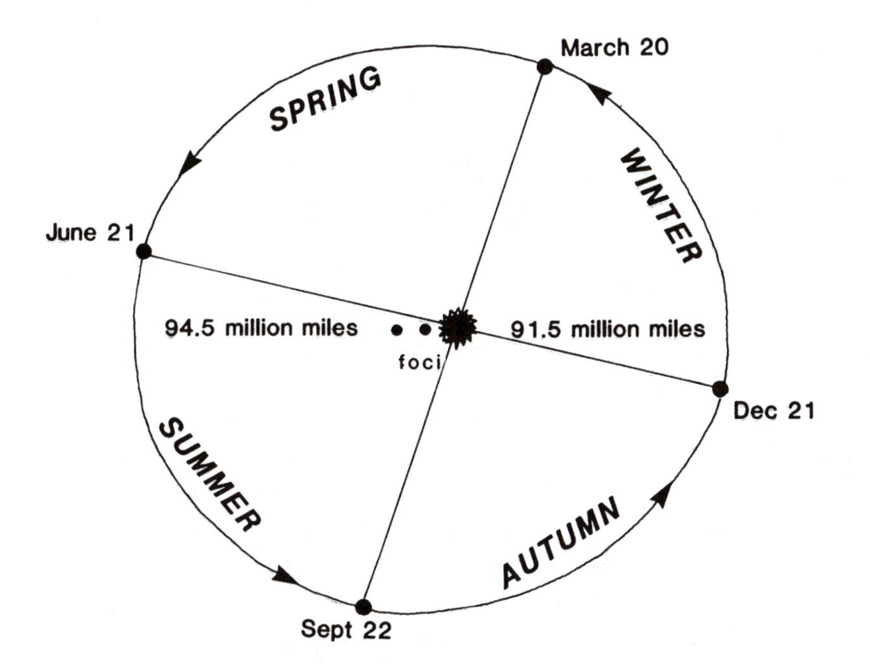

Figure 5.3. Over time, the long axis of the orbit of the earth changes its length, the tilt of the earth's axis varies, and so do the points on the orbit where winter and summer solstices and the equinoxes occur. At the present time the earth is closest to the sun during the northern winter, which is therefore somewhat warmer than it would be if the winter solstice were placed at the opposite end of the orbit.

theory first surfaced just a few years after the existence of the ice ages was recognized. It was given its final form by the Yugoslavian mathematician M. Milankovich, and says the following:

The orbit of the earth is an ellipse with a small eccentricity that varies by about 2 percent over a period of 100,000 years (Figure 5.3). The distance between sun and earth varies accordingly. Furthermore, the axis of the earth is tilted with respect to the orbital plane. Because of this tilt, the Southern Hemisphere receives more heat for half of the year, the Northern Hemisphere for the other half. The tilt, the obliquity, varies by 3° over a period of 41,000 years, and with it vary the seasonal length of the day and the amount of heat received at higher latitudes. Finally, the earth's axis wobbles around a mean position with a period of 21,000 years. This wobble, the precession, shifts around the orbit the points where midwinter and midsummer solstices and the equinoxes occur. At present, winter begins just before the earth is closest to the sun, but 10,000 years ago it was winter when the distance was at its maximum. The periods of the eccentricity, obliquity, and precession add up to a complicated curve. This curve depicts the

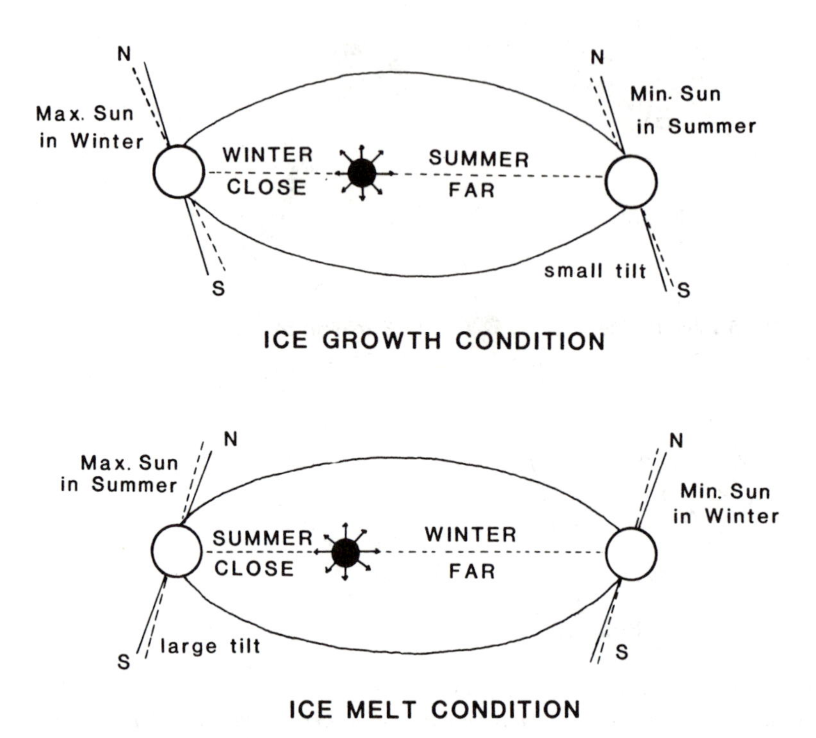

Figure 5.4. The impact on climate of the change with time in the orbital behavior of the earth is illustrated here for two extreme conditions. The first is a state conducive to the growth of icecaps, the second to their melting. What takes place in the Southern Hemisphere is the opposite of what is shown here.

variations with time of the heat received by the earth, the times of year when maximum and minimum heating occur, and the parts of the earth that are heated most. If, for example, the tilt were to be minimal while the sun was farthest away in the summer and the eccentricity was as large as possible, the Northern Hemisphere would be considerably cooler (Figure 5.4).

This much is indisputable, but there is uncertainty about the actual climatic consequences of these changes in solar heat supply, which on the whole are not very large. It is necessary to test the climatic impact by other means, for example by showing that the periodicities of the Milankovich curve coincide with those of the geological record. This was, until recently, the point at which the whole hypothesis collapsed, because the curve could by no stretch of the imagination be matched to the traditional scheme of four glacials separated by interglacials. Today, however, oxygen isotope ratios show that the sequence of warm and cold intervals was far more complicated. A rigorous mathematical analysis by John Imbrie and collaborators has recently shown that the oxygen isotope variation with time is itself the sum of three periodic components with periodicities of 100,000, 41,000, and 23,000–19,000 years. The match with the Milankovich curve is close and difficult to dismiss as mere coincidence.

To me, the evidence seems persuasive, and I am inclined to think that we are close to the explanation for the glacial-interglacial rhythm, or most of it. At the same time, the theory fails to account for the Ice Age itself, because the orbital variation continues far back into the past, whereas ice ages appear only occasionally.

A FUNDAMENTAL INSTABILITY AND OTHER THOUGHTS

Twenty-five years ago, Maurice Ewing and W. F. Donn proposed that glaciations and deglaciations result from instability in the Arctic climate. If one begins with an ice-free Arctic Ocean, seawater evaporates, snow falls on adjacent lands, and an icecap forms. The increased albedo causes the temperature to drop, the Arctic freezes, and the supply of moisture is cut off. Deprived of snow, the icecap shrinks, the temperature rises, and the Arctic Ocean once more becomes free of ice. The cycle might have started when the Arctic basin drifted into a polar position, and the climate became colder. Today, this hypothesis seems less plausible, because although the Arctic Ocean appears to have become frozen no earlier than the Pliocene, it has been centered on the pole for very much longer. Moreover, the arctic ice cover has probably been permanent for at least 3 million years.

The idea of a fundamental climatic instability, however, remains intriguing. Perhaps climates do not respond to stimuli in proportion to their strength. Some may react so slowly to change that conditions change again before the adjustment is complete. Others, more stubborn, may not respond at all until the push for change exceeds a certain threshold; then they flip fast and far. The sudden onset and decline so typical of glacial periods would make sense in this light.

Most explanations we have considered do not require, and some do not even permit, a rapid change from one state to its opposite, and yet that is what we seem to see. Perhaps we should look for a self-reinforcing process, *positive feedback.*

Such a process might be as follows: A modest temperature drop occurs, and less snow melts in summer, thus extending the area of white ground and raising the albedo. More heat is reflected, next summer even less snow melts, and the albedo continues to increase. During the 1960s a cooling trend reduced the mean annual temperature of the Northern Hemisphere by a mere 0.5°C, but the area of permanent snow cover at high latitudes grew disproportionately. After the especially miserable summer of 1971, the area covered with snow all year increased by 20 percent, and the effect lasted for several years. A similar positive feedback would result from an increased snowfall, but it seems that several cool, overcast summers have a larger effect than the same number of very snowy winters. The albedo-based feedback hypothesis suffers, of course, from the usual complications, but it has considerable merit.

We should also consider the role of the ocean. Water holds much more heat for a given volume than air, and releases it far more slowly. The ocean is thus a gigantic reservoir and transporter of heat from low to high latitudes, albeit a slow one – on a scale of years rather than the days and weeks of the atmosphere. Fossils in ocean sediment cores show that when the last glacial began, the North Atlantic retained its pre-glacial warmth for a long time. As a warm ocean in an already cold world, it furnished abundant moisture to storms, which took the moisture north and dropped it as snow. When the ocean finally cooled, its role as a snow machine diminished considerably. The North Pacific functioned in a similar way. Thus, the time lag in the adjustment of the oceans to a global cooling contributed to the rapid growth of the icecaps. On the other hand, once the adjustment was complete, the low level of evaporation from cold oceans may have reduced the supply of moisture to the point that a decline in the ice cover began.

There are other theories, and one might consider other feedback mechanisms, but the evidence will remain inadequate for some time. In science, one is always torn between the wish to modify or validate a fine hypothesis proposed by someone else and the urge to generate clever new ideas that are entirely one's own. Pushed by such contrary desires, science tends to slow down when a concept is very alluring or very strongly supported, and to speed up when there is clearly room for competing ideas. The continental-drift theory, as we shall see next, seems to be in the first stage, but, at least for the moment, the study of ancient climates is a lively example of the second.

PERSPECTIVE

Climate, it seems, is a most variable condition, and even small changes – a few degrees cooler, a slight rise in rainfall – can alter the balance of erosion and deposition, modify the setting of life, and leave a distinct imprint on the rock record.

The ice ages, of course, are extreme among the vast number of climate changes the world has seen. Because they are so dramatic, they illustrate most clearly the remarkable degree to which we can interpret the sedimentary archives. Inevitably, as we go further into the past, the detail must diminish, and speculation will assume a greater role, but not even the Precambrian entirely lacks a decipherable record.

Glacials and interglacials appear to have been triggered by changes in the orbit of the earth, but for whole ice ages we must invoke causes better matched to their long durations and to the rarity of their appearances. The timescale is long enough to have continents drift, mountains build, and sealevel change, long enough to see the geography of the earth profoundly altered. The forces that drive such changes are internal to our planet, and before we can proceed toward a deeper understanding of major climate changes, we must turn to the geological revolution and to global tectonics.

FOR FURTHER READING

Somewhat surprisingly, there is not at this time a large choice of books on climate, climate change, and the ice ages. This is true not only for the general reader but also for the specialist, who must rely mainly on the primary journal literature, which is not very accessible.

ON WEATHER AND CLIMATE

The following book, small and inexpensive, provides a well-written brief overview of how climate works, followed by a very readable discussion of the impact of climate and climate change on human affairs:

Roberts, W. O., and Lansford, H., *The Climate Mandate,* Freeman, San Francisco, 1979, 197 pp.

More thorough, yet at a good level for the interested layman, is this introduction into weather and climate:

Battan, L. J., *Weather,* Prentice-Hall, Englewood Cliffs, N.J., 1974, 136 pp.

Real college-level texts in climatology are more demanding. The following are particularly good and useful examples, and the last is quite rigorous:

Critchfield, H. J., *General Climatology,* Prentice-Hall, Englewood Cliffs, N.J., 1983, 453 pp.
Neiburger, M., Edinger, J. G., and Bonner, W. D., *Understanding Our Atmospheric Environment,* Freeman, San Francisco, 1973, 293 pp.

ON CLIMATE HISTORY AND ICE AGES

Next is a charming account of the search for the causes of the ice ages. It includes a nice personal account of the research that led to the present popularity of the orbital theory and is most worthwhile:

Imbrie, J., and Palmer, K., *Ice Ages, Solving the Mystery,* Enslow Publishers, New York, 1979, 224 pp.

Useful at times is a large reference work, even though meant mainly for the professional:

Nilsson, T., *The Pleistocene – Geology and Life in the Quaternary,* Reidel, Holland, 1982, 651 pp.

Finally, this is the current richest work about climate history and its impact on the post-glacial world and human history:

Lamb, H. H., *Climate History and the Modern World,* Methuen, London, 1983, 387 pp.

Drifting continents, rising mountains

Our views have constantly to be revised and adjusted, but occasionally new aspects of far-reaching consequence shed such unexpectedly different light on existing problems that the effect might be compared to that of a revolution. All that had hitherto been sacrosanct crumbles to the ground; hardly anything is left untouched.

H. H. F. Umbgrove, *The Pulse of the Earth*

A GEOLOGICAL REVOLUTION

Like children, the sciences alternate brief periods of rapid growth with long intervals of stability. Examples of such bursts of renewal are the introduction of the Bohr-Rutherford atom model in physics and the deciphering of the genetic code in biology. Consolidation follows, and the startling ideas of yesterday become the conventional wisdom of today, until once more the intuitive and sometimes seemingly irrational processes of a revolution replace the traditionally orderly approach of scholarship. Each time, the field emerges refreshed and matured from the turmoil.

The triumph of uniformity over catastrophism early in the nineteenth century and the Darwinian theory of evolution half a century later were geology's first revolutions. Long stability followed; the education of a geologist in the 1950s was not much different from that of the 1920s. Today, another revolution lies just behind us; tempers have cooled, and the new basic concepts are well enshrined. In many areas, however, notably in earth history, the consequences of the revolution in large part still remain to be explored. We shall be concerned with that subject throughout the remainder of this book.

The revolution had its precursor, and his fate is informative. Beginning in 1912, and culminating in a book in 1915, *Die Entstehung der Kontinente und Ozeane,* the German meteorologist Alfred Wegener noted, as many had done before him, the similarity of the opposing coasts of the Atlantic, and he undertook to test with geological evidence the notion that a single great continent had shattered and drifted apart. He especially desired to do away with the need to postulate sunken lands and vanished landbridges, commonly invoked then to explain intercontinental fauna connections, but incompatible with isostasy. This incompatibility was known at the time to geophysicists but was widely ignored by geologists and paleontologists. Wegener, who amassed an impressive set of geological matches across oceans, received a surprised but sympathetic hearing in Europe, but on the other side of the Atlantic the climate was cold. In 1926, at a meeting in Atlantic City, American geologists, almost to a man, read him out of the company of acceptable thinkers.

Evidently, continental drift was an idea whose time had not yet come, and it languished for decades. The reluctance among geologists to consider drift seriously was instinctive and emotional rather than factual. Wegener's failure to provide a suitable mechanism for drift, for example, had been severely criticized, but when the great British geologist Sir Arthur Holmes provided one shortly afterward, no one listened. Geologists on the southern continents continued to turn up supporting evidence, but their remoteness denied them proper attention. It did not help that Wegener was not a certified geologist, but more important was that, to most geologists, the flaws and cracks in the existing scientific structure still seemed small. Having just learned to live with the abyss of time, they were not yet prepared to deal with the uncertainty of place that is inherent in a world of drifting continents.

6

Continental drift and plate tectonics

In the middle of the present century, interest in continental drift was revived by some surprising results from study of the magetic properties of rocks. A little later, our growing knowledge of the geology of the ocean floor culminated, via a remarkable series of hypotheses, in the plate-tectonics revolution and brought a belated victory for Wegener's ideas.

WANDERING POLES OR WANDERING CONTINENTS?

Certain igneous rocks become magnetized according to the magnetic field of the earth that exists when and where they cool and crystallize. Basalts especially preserve the field of the time of their formation well, but many other rocks, including some sediments, do so too. We measure the intensity and direction of the ancient magnetic field on samples whose orientation we record before they are pried loose from the rock in the outcrop. This measurement gives us the azimuth to the magnetic pole; with suitably spaced samples, we determine by triangulation its past position.

The rock also retains the magnetic record of the paleolatitude at which it was formed. If we march poleward carrying a magnet suspended on a string, the bar remains parallel to the lines of magnetic force and will thus point more steeply downward the closer we approach the magnetic pole (Figure 6.1). The angle, the inclination, depends only on the latitude. Alas, the azimuth to the pole and the paleolatitude are all we get; we know the direction of the meridian and the sample's position on it, but not the paleolongitude. With enough samples we can orient two ancient continents with respect to the pole and restore them to their initial latitudes, but we shall not know

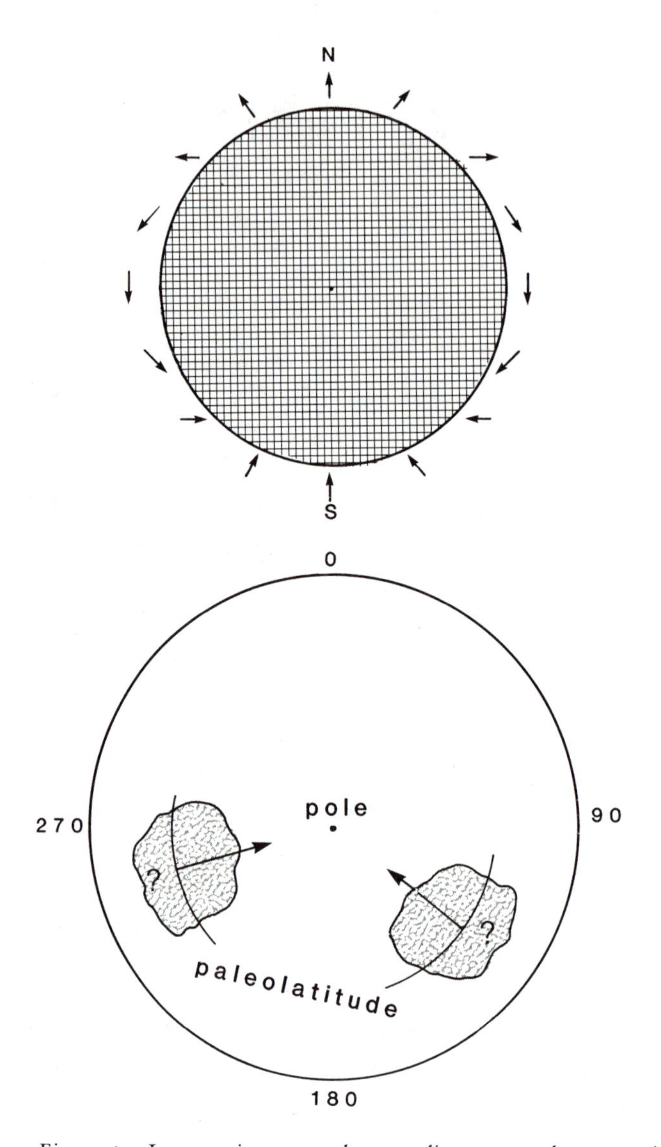

Figure 6.1. In many igneous and some sedimentary rocks, magnetic minerals become aligned with the earth's magnetic field at the time the rock forms. This enables us to determine the direction, the azimuth, of the magnetic pole at that time, and because a magnet becomes progressively more steeply inclined as we move it from the equator to the pole (top), we can also obtain the paleolatitude. The paleolongitude, however, cannot be determined, and we are unable to say whether two continents, correctly oriented and placed at the proper latitude, existed side by side or were separated by an ocean (bottom).

whether they were adjacent to one another or were separated by a vast ocean.

By the late 1950s, some curious facts had begun to emerge from measurements of this kind. If one obtained on one continent, but for different moments in time, a set of pole positions, they did not cluster around a single point, but traced a line on the globe, a polar path. This forced the conclusion either that the pole remained fixed while the continent drifted or that a wandering pole was observed from a stationary continent. The data did not permit a choice between these two options, but to most geologists a wandering pole seemed easiest to accept.

Once polar paths had been obtained for several continents, however, it was clear that this could not be so. Consider, for instance, the polar paths for Europe and North America (Figure 6.2). Beginning far apart in the early Phanerozoic, they converge with time on the present pole. It would strain our credulity to assume that in the past there were two sets of magnetic poles, one for each continent, that miraculously have joined at the present time. Similar separate but converging polar paths have been determined for all continents, and if we fit the continents back together in approximately the arrangement proposed by Wegener, all polar paths are found to coincide. Clearly, this supports the concept of continental drift rather than the wandering of multiple poles.

Except for a few geophysicists, however, the magnetic evidence persuaded few that Wegener deserved a new hearing. The more conservative geologists took refuge behind those paleomagnetic measurements that would not fit this simple picture. Later, these discrepancies were largely eliminated by a better understanding of paleomagnetism and its measurement, but by then oceanographers had reopened the question of continental drift from a very different angle.

CONTINENTS AND OCEAN BASINS

Continents and ocean basins, because of their sheer size, exude an aura of permanence. They also consist of very different types of crust, and even among the oldest rocks the distinction between those of continental origin and those formed on the ocean floor is usually clear. As far as we know, no ocean basin has ever been converted into a continent, nor is it physically possible that an entire continent might founder to become an ocean.

Together with their submerged margins the continents occupy slightly more than 40 percent of the earth's surface, and their mean

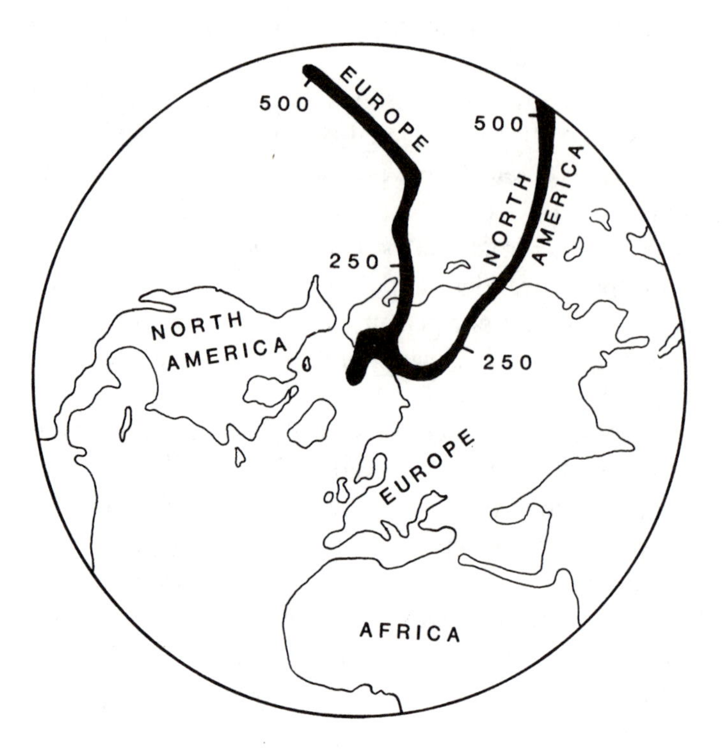

Figure 6.2. If we determine the positions of the magnetic pole for successive times in the past for a single continent, we find that they fall on a line, a polar path, thus indicating either that the pole moved and the continent was fixed or that we observe from a stationary pole the drift of the continent. If we make the same measurements for other continents, each will have its own polar path, and all paths will converge on the present pole for the present time (the numbers are millions of years). The various paths can be made to coincide by reuniting the continents in the manner proposed by Wegener.

elevation is only a little above sealevel (Figure 6.3). The ocean floor lies about 4,000 m deeper. Isostasy explains this remarkable two-level earth: The continents, consisting in large part of rocks resembling granite in composition, are light and thick. The ocean floor, made of basalt and other dense rocks, is heavy but thin. Both types of crust float on the underlying mantle, which is denser than either. The boundary between the mantle and the overlying oceanic or continental crust, called the Mohorovicic discontinuity or Moho, after its Yugoslavian discoverer, is sharp and easily detected with seismic waves. With continents and oceans in isostatic equilibrium, the difference in buoyancy (density times thickness) accounts for the observed difference in mean elevations.

It is therefore with good reason that continents and oceans seem immutable to us, but one remains uneasy. There is, for example, the

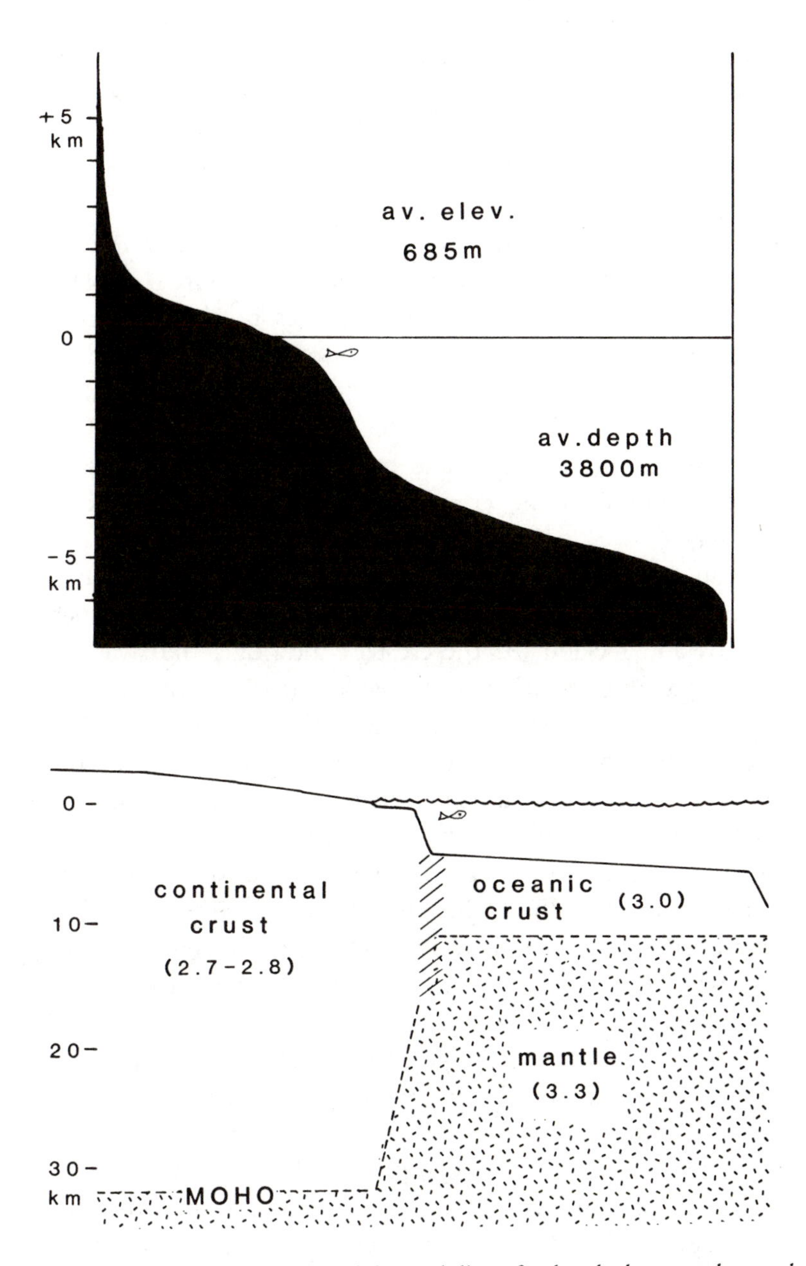

Figure 6.3. Most of the surface of the earth lies a few hundred meters above sealevel or a few thousand below (top). This two-level earth can be understood by realizing that a thick, light continental crust and a much thinner but slightly denser oceanic crust float like rafts on a mantle denser than either of them (bottom: in parentheses are the densities in grams per/cubic centimeter). Isostatic compensation produces the two dominant levels.

curious fact that the rock record on the continents implies the presence of adjacent oceans for more than 3 billion years, whereas the sediment blanket of the ocean floors is much too thin to represent so long a service as the world's ultimate sanitary landfill. Besides, no one has yet found any part of the ocean floor as old even as 200 million years. A most intriguing paradox.

The most striking feature of the ocean floor is the *mid-ocean ridge,* which girdles the earth like the seam on a tennis ball (Figure 6.4). This ridge has some unusual properties. It always rises to about 2,500 m below sealevel. Except in parts of the Pacific, its crest is always cleft by a narrow, longitudinal, fault-bound rift valley that is the site of shallow earthquakes of the sort that indicate tension. At frequent intervals, the ridge crest is offset by linear ridges and troughs, called *fracture zones,* that are at right angles to the crest and lend the ridge a striking orthogonal pattern.

The mid-ocean ridge is built of basalt flows and intrusions and is dotted with submarine volcanoes and scattered volcanic islands, such as the Galapagos and Iceland. At the crest of the ridge, the lavas are freshest and seem youngest, whereas the volcanic islands increase in age as one draws away from the ridge axis. Within the rift valley, sediments are thin and patchy, but they thicken down the flanks. Because in the open ocean the sedimentation rate varies little from one place to another, this suggests that the age of the crust increases with distance from the axis, as do the ages of the volcanic islands. Finally, the flow of heat, being very high through the floor of the rift valley, rapidly decreases downflank and away from the axis. We shall see shortly that these features played a major role in the geological revolution, but they are not the only ones.

The shallow and hot mid-ocean ridge contrasts sharply with another set of features of the deepsea floor: *trenches,* the narrow, elongated troughs often 5–7 km deep. Most of these trenches are found in the Pacific, which they encircle almost completely. Trenches have a very low heatflow and lack active volcanoes. Trenches bordering on continents may contain thick sediments, but because a prolific source is nearby, this does not imply great age. Like the mid-ocean ridge, trenches are associated with numerous earthquakes, but in this case of a kind that indicates compression of the crust rather than tension. These earthquakes occur along a plane, called the Benioff zone, which descends steeply (at 45–60°) to a depth of 300–700 km.

Two chains of islands accompany most trenches. The first chain, or arc, lies close by and consists of deformed and sometimes metamor-

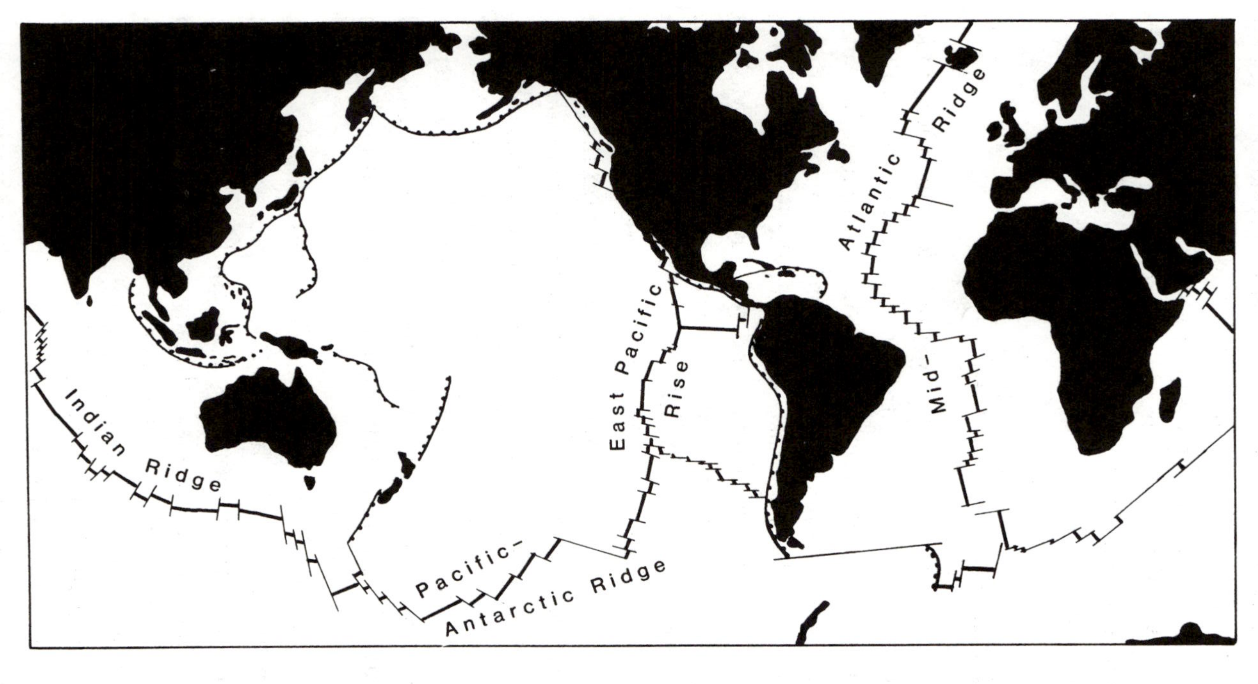

Figure 6.4. The most prominent features of the ocean floor are broad rises, called mid-ocean ridges, and deep, narrow troughs, the trenches. The mid-ocean ridges, offset in many places at right angles by fracture zones, form an almost globe-girdling system, but trenches are mainly restricted to the Pacific Ocean.

phosed sediments. Active volcanoes make up the second one. Rather than the *basalt* typical of the mid-ocean ridge, those volcanoes pour out *andesite* lavas intermediate in composition between oceanic and continental rocks. Where a trench borders on a continent, the sedimentary arc is seldom distinct, and the volcanic arc lies well inland, as it does, for example, in the Andean range of western South America.

A DAISY CHAIN OF HYPOTHESES

It is obvious that the ocean floor is a curious domain where much demands an explanation. The first attempt at a universal hypothesis made use of an idea suggested in 1929 by Sir Arthur Holmes, but since then buried in the depths of libraries: the idea that the mantle is stirred by convection currents. Suppose that, far down, the mantle is heated somehow. It expands, becomes lighter, and rises. Below the Moho it spreads sideways, and when sufficiently cooled, it sinks once more. Such convective flow can readily be observed, as in a pan of boiling soup. Above the rising column (Figure 6.5) the mid-ocean ridge forms, because the crust heats and, expanding, forms a rise on the seafloor. Stretched by the diverging flow underneath, it might crack now and then, to the accompaniment of earthquakes and volcanic eruptions. Away from the rising column the crust cools and contracts, and the depth of the seafloor increases. Where the cold limbs of the convection current sink, they drag the crust down, thus forming trenches and causing earthquakes. The depth of the Benioff zone seems excessive in this context, but that is a detail.

This model explains many things, but it fails to deal with the change in age away from the ridge crest. It is also not helpful with the details of the trench and arc complex, nor does it say anything about fracture zones. That is bothersome, especially because these features are so large, so common, and clearly so important. Fracture zones, in addition, sometimes behave strangely. In the eastern Pacific, for example, they offset the crest of the East Pacific Rise by hundreds of kilometers but terminate against the coast of California without distorting it. Patently, these are not ordinary faults with simple horizontal slippage.

In 1960, H. H. Hess of Princeton University improved considerably on the simple convection model, changing it only a little, but enough to make the publication a true classic. Hess allowed the crust itself to ride passively on the flowing mantle like a raft on a river. At the ridge crest, this produces a gap that must continually fill with lava, solidify, and then crack again. At the opposite ends there will be excess crust; it

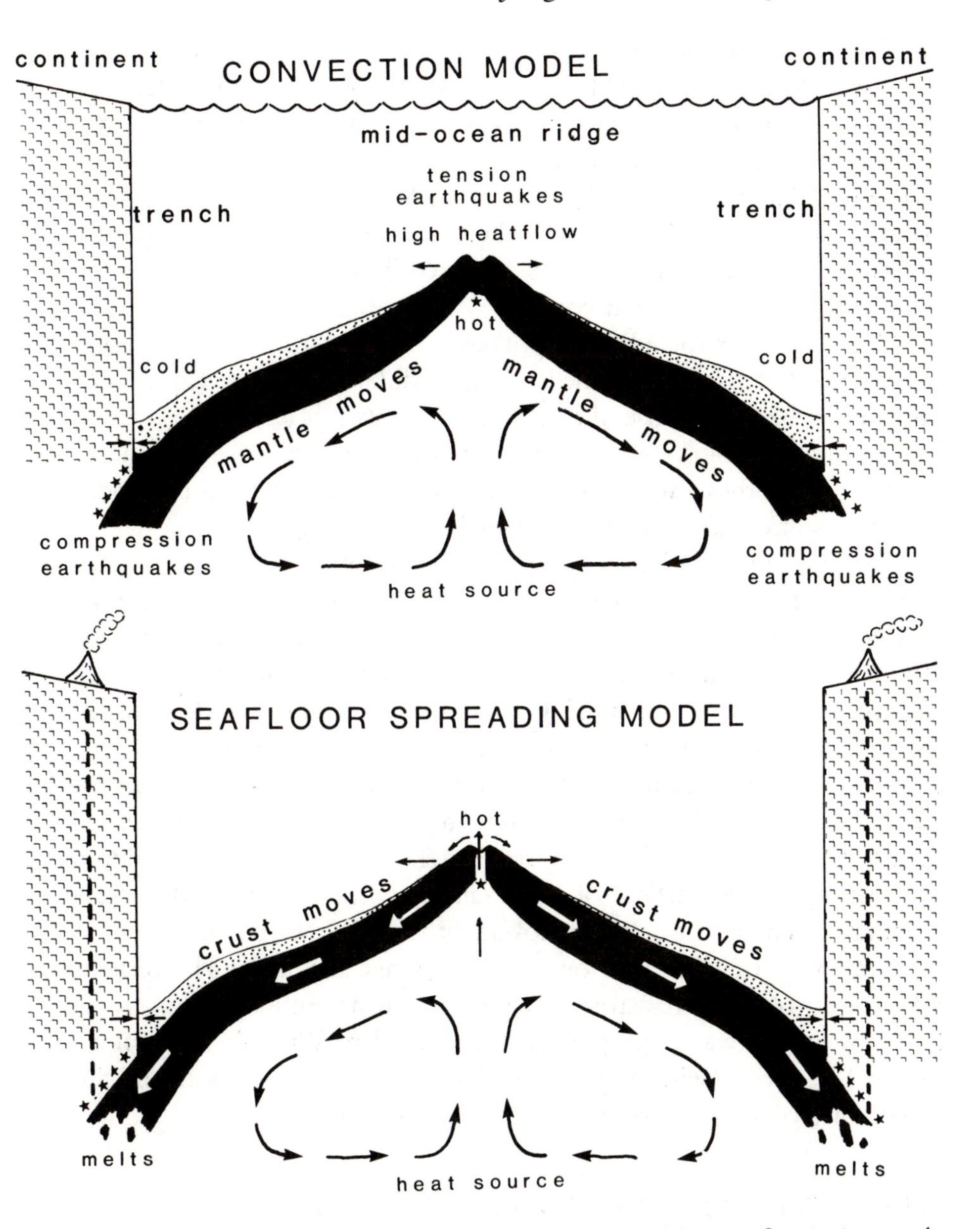

Figure 6.5. The earliest explanations for the curious properties of the ocean floor (top) assumed convective flow in the upper mantle underneath a stationary crust, but failed to explain, among other things, why the seafloor increases in age away from the ridge crests. The seafloor-spreading model (bottom) solved this problem by allowing the ocean floor to ride with the mantle. This requires the formation of new crust on the ridges and disposal of the old crust in trenches. This model fails to account for the absence of trenches in the Atlantic, which has, nonetheless, a distinct mid-ocean ridge.

is dragged into the mantle by the descending limbs, melted, and recycled. Now we do need the entire Benioff zone; the sedimentary arc becomes the crumpled edge of the opposing slab, and the volcanic arc is seen as the result of melting of the sinking slab by the heat of friction. Andesite can be obtained by melting basalt adulterated with oceanic sediments and seawater. The model also explains the apparent age increase away from the crest indicated by volcanic islands and the thickening of the sediment cover.

Hess's seafloor-spreading hypothesis was well received by marine geologists because it explained so many different things at once. On the other hand, because the whole cycle takes place in the oceans, continental geologists did not feel particularly involved. Soon, however, the hypothesis was to receive impressive support from two independent sources and could no longer be ignored by anyone.

The first confirmation came from an explanation for fracture zones proposed by the Canadian J. Tuzo Wilson. Traditionally, fracture zones had been assumed to be faults with horizontal slip that offset an initially continuous ridge crest and gradually increased the distance between the crest segments (Figure 6.6), as *strike-slip faults* do on land. Wilson suggested that the ridge never was continuous but always consisted of straight segments offset by what he chose to call *transform faults*. As the seafloor spreads away from each segment, the excess crust at the ends sinks into the trenches. Earthquake activity should be restricted to the portion of the fault that lies between ridge crest segments. There, the relative movement of the rocks on both sides is the opposite of what it is on a normal strike-slip fault. The relative directions of motion across a fault can be determined from earthquakes, and it was a simple matter to validate Wilson's suggestion. Lynn R. Sykes, a seismologist at Columbia University, did so, and the credibility of seafloor spreading rose considerably.

MAGNETIC ANOMALIES AND POLARITY REVERSALS

Almost simultaneously, strong support for the seafloor-spreading model arrived from an entirely independent source: the data accumulated during years of patiently towing magnetometers behind research ships to study the magnetic field at sea. The magnetic field of the earth, fairly simple in principle, is locally distorted by the magnetism of certain rock formations. On land, such distortion tends to be irregular and complex, but at sea there are bold patterns of alternating ano-

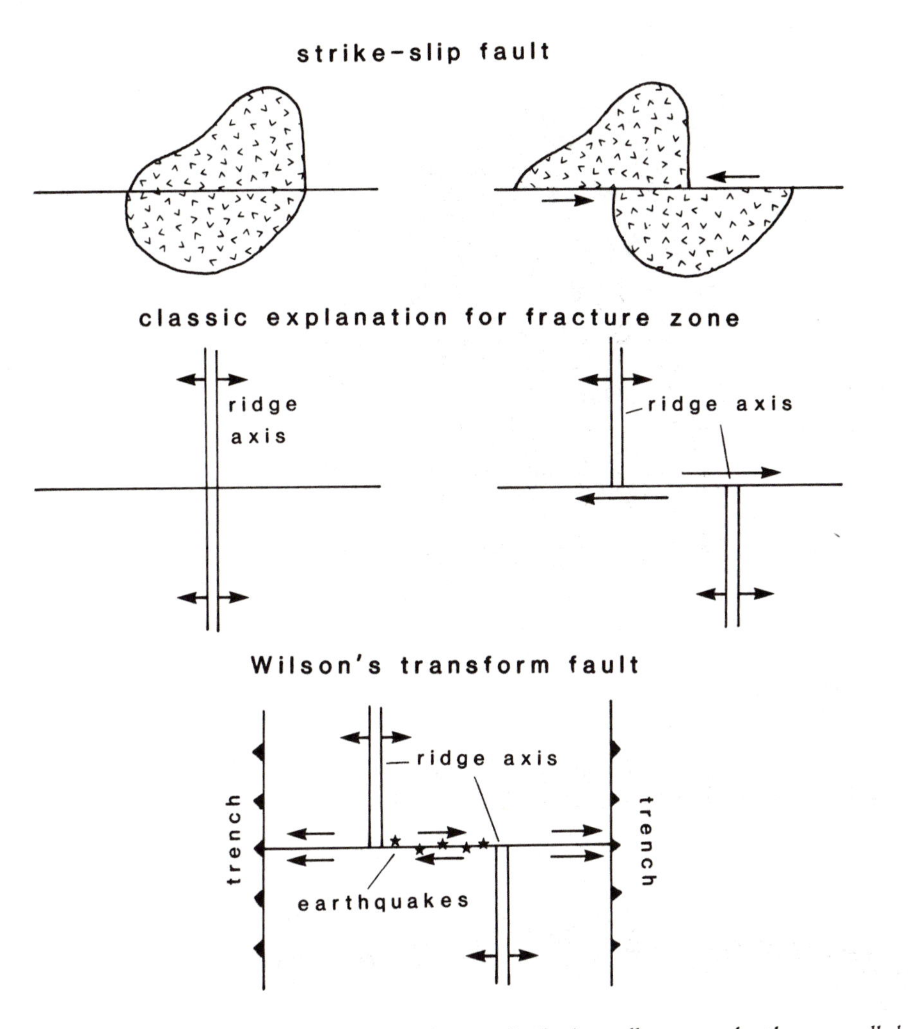

Figure 6.6. Faults along which pieces of crust slip horizontally past each other are called strike-slip faults (top). Initially, the fracture zones that offset mid-ocean ridge crests were regarded as strike-slip faults, under the assumption that the ridge crest began as a continuous feature. Accepting seafloor spreading, a pattern of motions along fracture zones can be inferred that is fundamentally different from that produced under the previous assumption (center). This pattern is confirmed by the occurrence of earthquakes only between ridge crests (bottom) and so confirms the seafloor-spreading model.

malously high and anomalously low values. If we color the values above the theoretical field, the positive ones, black and leave the low, negative values white, a marine magnetic-anomaly map resembles a zebra skin, with stripes parallel to the mid-ocean ridge crests and distorted by fracture zones (Figure 6.7). There always appears to be a high positive anomaly at the ridge axis, and the anomalies to either side are often strikingly symmetrical.

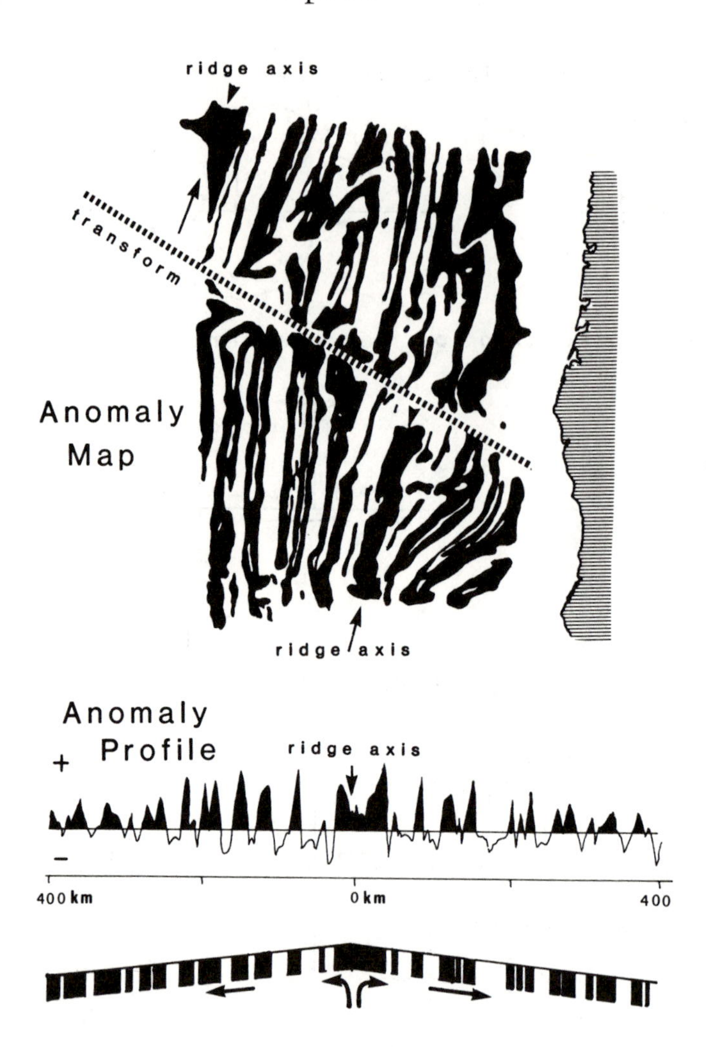

Figure 6.7. Observations at sea have shown that the oceanic crust has a pattern of alternating high (black) and low (white) magnetic values that resembles the skin of a zebra (top). Note the offset of bands along a transform fault/fracture zone. Magnetic traverses across mid-ocean ridges (center) show this pattern of anomalies above (positive) and below (negative) the earth's field. Often the anomalies are symmetric to either side of a much broader central anomaly. They can be explained as the result of seafloor spreading. Lava intruded during a normal polarity is magnetized in the same direction as the present field and adds its strength to it, whereas an older crust intruded during a reverse polarity phase subtracts from the present field (bottom).

This anomaly pattern, first published in 1961 for the northeastern Pacific, presented quite an enigma, and some extravagant explanations were initially contemplated. The correct answer was found by two Cambridge geophysicists: Drummond H. Matthews and his research student, Fred J. Vine. First put forward rather quietly in 1963, it rests on a by-product of the search for paleomagnetic pole positions. Since the beginning of this century it had been known that if one measures magnetic orientations in thick stacks of more or less continuous lava flows, the observed magnetic fields preserved in the rocks periodically reverse their polarity, that is, magnetic north becomes magnetic south and vice versa, without a corresponding change in azimuth or inclination. This can only mean that the magnetic field of the earth itself reverses its direction from time to time. Careful work by Allan Cox of Stanford University, among others, has established that reversals are common, occurring irregularly but on the average about every 700,000 years. Each reversal itself takes only a few thousand years. Numerous measurements of lava sequences on land dated by the potassium-argon method (Table 2.1) have established a polarity-reversal time-scale now extending back more than 5 million years (Figure 6.8). Beyond that, reasonably continuous lava sequences are scarce, and the uncertainties of isotopic age dating become as large as or larger than the spacing of the reversals.

Vine and Matthews, familiar with polarity reversals, reasoned as follows (Figure 6.7, bottom): Suppose the seafloor spreads and a fissure opens. It fills with basalt lava, which cools and adopts the present normal polarity. If we tow a magnetometer across this new crust, the magnetization of the rock and that of the earth's field point in the same direction, the values add together, and we observe a positive anomaly. If, some time ago, an earlier crack had filled with lava while the earth's field was opposite to what it is today, the lava would have acquired a reverse polarity. Measuring across that older block today, the strength of the rock's field, being in the opposite direction to that of the earth, would subtract from it, and a negative anomaly would be seen. Thus, like a magnetic tape recorder, the spreading ocean crust tracks the reversals of the earth's field through time in the form of positive and negative departures from the present field.

When the crust forming at the ridge axis splits in two, the pieces, bearing symmetric halves of the original magnetic anomaly, will move apart, and a new block will be inserted between them. Thus, the central block always has the present magnetic polarity, and the anomalies on the flanks are mirror images of each other. If the rate at which the seafloor spreads remains constant, the widths of successive anomalies depend only on the lengths of the intervals between reversals.

As soon as a reasonably dated polarity-reversal timescale became available, Vine, in an important article in 1966, tested his model by plotting the widths of marine anomalies against the periods of time

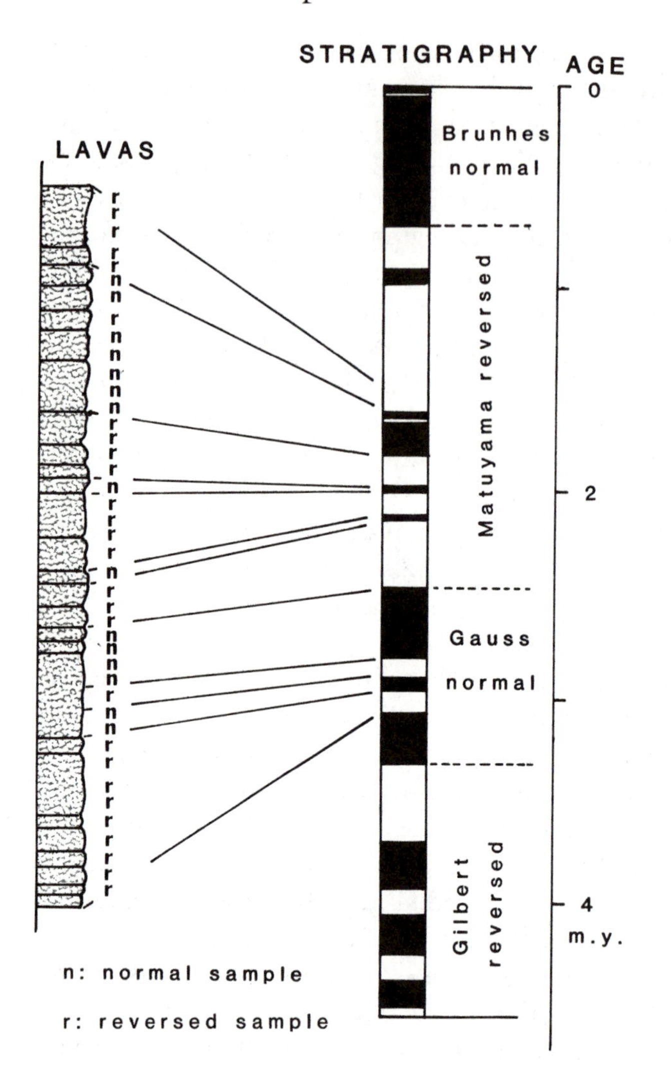

Figure 6.8. Paleomagnetic measurements on long sequences of lava flows (left) have shown that the direction of the earth's magnetic field, its polarity, episodically reverses, magnetic north becoming magnetic south, and vice versa. Isotopically dated, such measurements provided a polarity-reversal timescale, a portion of which is shown at right.

between reversals on land. The fit was excellent for several mid-ocean ridge profiles, and Vine's hypothesis and the seafloor-spreading model were simultaneously confirmed. As a bonus, he obtained the rates of seafloor spreading over the last several million years, which range from an opening rate of 2 cm/year in the North Atlantic to more than 30 cm/year in the central South Pacific. Of course, one needs luck in any discovery, and Vine had it; there is no a priori reason why seafloor spreading should be constant over any length of time, and it is not always so, but it happened to be true for the data he used.

As an aside, I note here that a Canadian, L. W. Morley, independently arrived at the same conclusion as Vine, but publication of his article was delayed, and it finally appeared in a less well known journal. When he emerged from obscurity, credit had already firmly accrued to others.

Vine's article made believers out of many, and the next logical step was soon taken. If we assume that, on some ridges at least, the spreading rate has always been constant, reversals older than those found on land can be dated by extrapolation from the magnetic anomalies observed at sea. We can check the assumption of a constant spreading rate by comparing data from many ridge profiles, and with crust ages obtained by drilling in the ocean. In this manner, the polarity-reversal timescale has been extended back to about 150 million years ago, at which point we run out of anomalies; all older crust has long since been subducted. Once the times of reversal had been fixed in this manner, and with lots of magnetic profiles on hand, a map of the ages of the oceanic crust was constructed (Figure 6.9). This map demonstrates how young the ocean floor is compared with the continents and how little remains even of crust formed as recently as 100 million years ago. Seafloor spreading has clearly resolved the paradox of old continents and young oceans by showing that the ocean floor is continuously renewed.

PLATE TECTONICS

For many geologists, Vine's article validated seafloor spreading beyond any further doubt. Simultaneously, mid-ocean ridges, fracture zones, trenches, ocean crustal ages, heatflow distributions, magnetic-anomaly patterns, and Benioff zones had been explained. Note, however, that moving continents, continental drift, was not part of the hypothesis.

All this contentment notwithstanding, some serious problems remained. For example, seafloor spreading requires a separate convection cell between each pair of fracture zones, each rising column staggered with respect to its neighbors as the ridge axes themselves are staggered. As the number of known fracture zones grew and their spacing shrank, the number and the odd flat shape of the presumed convection cells became improbable. Puzzling also was that whereas the hypothesis demanded that each mid-ocean ridge be accompanied by a pair of

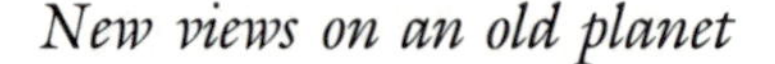

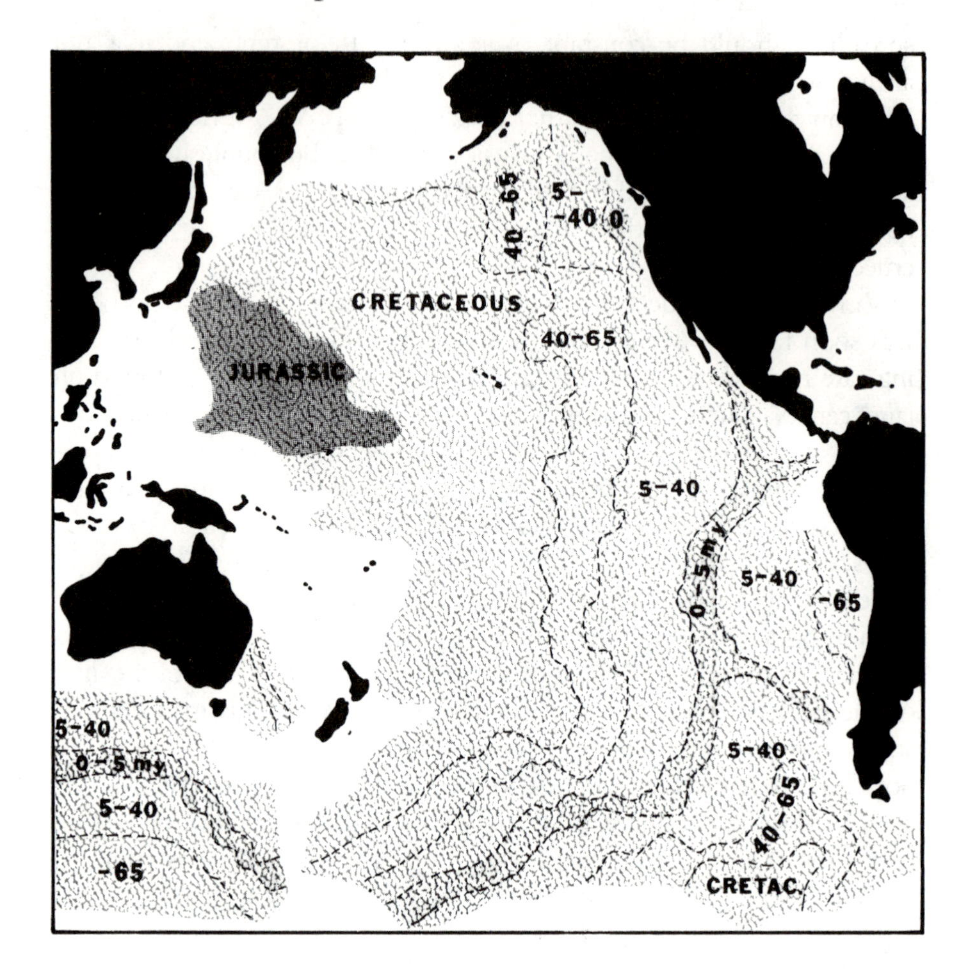

Figure 6.9. The marine magnetic anomaly sequence has been correlated with the polarity reversal timescale, extrapolated back in time, and calibrated with dated samples of the oceanic crust, permitting us to construct a map of the ages of the ocean floor. It is evident that older crust is progressively more scarce, a result of subduction. The numbers indicate the ages of the Cenozoic crust in millions of years.

trenches to dispose of the excess crust, the Mid-Atlantic Ridge, best known of all, had no trenches. Neither do large parts of the ridge in the Indian Ocean.

In retrospect, the solution to these and other problems is seen to have required only the removal of a single mental block: the belief that the Moho is the major physical boundary underneath continents and oceans. As long as one accepts this, continents and ocean floors are the major structural units, and mantle flow below and passive drift above the Moho are natural consequences. Removing such blocks can be very difficult, but the effort is often rewarded with new insight.

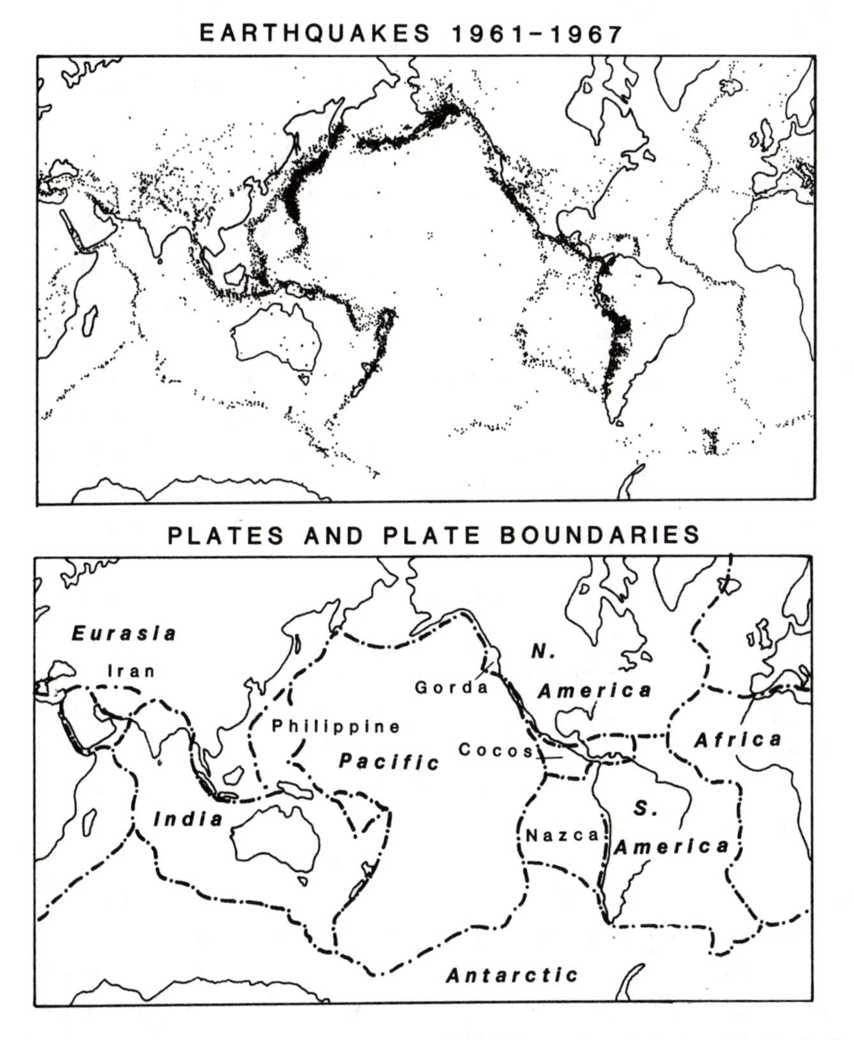

Figure 6.10. If we consider where on the surface of the earth geological activity in the form of earthquakes (top) or volcanic eruptions is located, we find a small number of narrow but distinct zones. These zones often are not located at the boundaries of continents and oceans. Setting the distinction between those two temporarily aside, we see that the earth's crust is divided into a small number of large plates bounded mainly by very active mid-ocean ridges and trenches (bottom).

Simultaneously, several young scholars, less burdened by conventional wisdom than their elders, discarded all preconceived crustal units, went on to see where the seismic and volcanic activity in the world was located, and discovered that these active zones are narrow, sharp, and continuous (Figure 6.10). They define a small number of large segments of the earth's surface now known as *plates.* Plates are circumscribed by mid-ocean ridges and trenches linked by means of transform faults such as the San Andreas fault in California. Some plates contain only ocean crust; others hold both oceans and continents. The concept was first presented by two young geophysicists: Dan P. MacKenzie at Cambridge and W. Jason Morgan at Princeton. It was soon confirmed with seismic data, and plate tectonics was born.

In a world so subdivided, ocean basins and continents are clearly secondary elements, and this means that the plates must include more than just the crust. The plates are segments of what is called the *lithosphere,* an outer shell of the earth about 100 km thick, 3–10 times thicker than the crust. The lithosphere consists of a thick slice of mantle with a thinner top layer of crust (Figure 6.11). The plates are rigid and float on a soft layer in the mantle called the *asthenosphere.* Lithosphere and asthenosphere are not new inventions; evidence for their existence had been available for decades but was long judged to be of little importance.

Plates meet at three kinds of boundaries: They may drift apart or diverge, collide or converge, or slip harmlessly past each other on transform faults where their boundaries are parallel to the direction of their relative motion (Figure 6.11). At a divergent plate boundary, new lithosphere forms as lava wells up from below and cools. Because this lava originates in the mantle, it is basaltic in composition, has a high density, and forms oceanic crust, never continental. Even if a divergent boundary forms within a continent, it will soon manufacture its own ocean basin.

When plates collide, one is forced under the other or *subducted.* When two oceanic plates converge, either one may be subducted under the other, but if one plate edge is oceanic and the other continental, the continental edge will remain on top, because it is too buoyant to sink. When two continents collide, neither can be subducted. Their edges crumple, but ultimately all movement must cease. Because the surface of the earth is covered entirely with tight-fitting plates, the movement of the entire set may be modified when that happens.

We now understand why the Mid-Atlantic Ridge needs no trenches of its own. As long as the total amount of new lithosphere created at mid-ocean ridges equals the amount destroyed in subduction zones, it is not important where the recycling actually takes place. The new crust manufactured in the Atlantic causes the Americas to override the various Pacific plates that are subducted along the west coast of the Americas. The American plates grow larger, the Pacific plates shrink, but a worldwide equilibrium is maintained.

An important rule in plate tectonics states that all plate activity

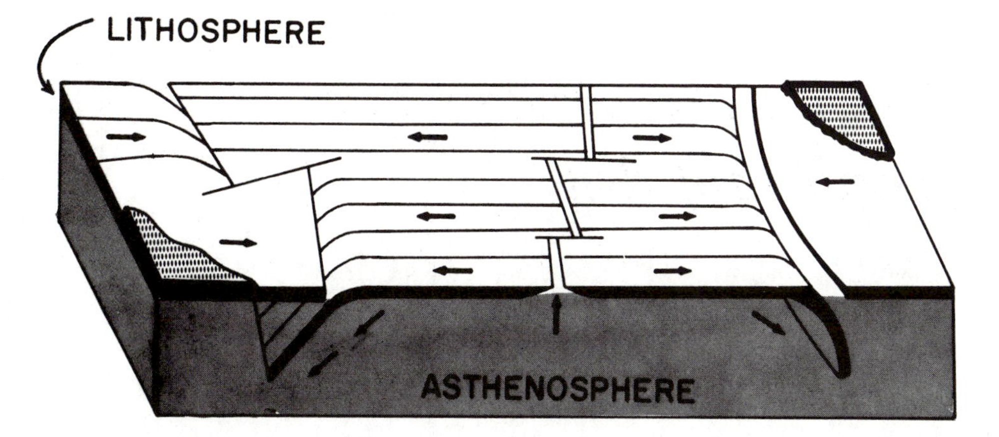

Figure 6.11. Recognition of the plates of Figure 6.10 led to the plate-tectonics model, which assumes that the earth is covered with 100-km-thick slabs of lithosphere floating on a weak asthenosphere. Lithosphere plates diverge at mid-ocean ridges, converge in trenches now known as subduction zones, and slide harmlessly past one another along transform faults. Passively on the backs of the plates ride continents and ocean basins.

occurs at their boundaries and that their interiors are not deformed. This rule is, of course, an oversimplification, because earthquakes do shake the interiors of plates, and volcanoes do occur there. The activity to which they testify, however, is not enough to alter the size and shape of the plate significantly. At least, so we trust.

Plate motions have yet to be measured directly, but we can determine their speed from the ages of the crust furnished by magnetic anomalies. By definition, their direction of motion relative to one another is parallel to transform faults. Knowing both and assuming size and shape to be constant, with due allowance for plate growth at divergent edges, we can deal with the movements of plates as an exercise in spherical geometry. In principle, at least, we should be able to reconstruct rigorously the past configurations of oceans and continents. This makes it clear why the axiom that the interior of a plate is not deformed is so important; it is a device to treat the reconstruction of plate motions as a problem in geometry, and if it were found to be seriously wrong, we would be unable to derive their past configurations. The smaller the number of plates and the simpler their outlines, the more precise this exercise in paleogeography can be. Going back in time, we eventually run out of ocean floor, and in the middle Mesozoic the game is over. We must then turn to paleomagnetic data, which tell us the ancient latitudes and orientations of the continents but remain silent about oceans.

7

Continental breakup and continental drift

The theory of plate tectonics states that the surface of the earth is modified mainly at the divergent and convergent plate boundaries. At a divergent plate boundary, seafloor spreading and continental drift create new oceanic crust and sometimes form new oceans and more but smaller continents. At a convergence, oceanic crust is lost, island arcs form, and mountains rise, or collision welds continents together. New oceans are evidence for divergence, and the scars of ancient collisions, called *sutures,* testify to the existence of former oceans.

PANGAEA AND PANTHALASSA

Wegener, then, was entirely right. Some 250 million years ago there was a single supercontinent romantically named Pangaea, the "all-land," surrounded by a superocean, Panthalassa, the "all-sea." Swarms of jigsaw-puzzle enthusiasts have tinkered with Wegener's reassembly of the continents (Figure 7.1) but have not really changed it very much, except that we now use the continental slope at 2–4 km below sealevel as a better continental boundary than his shelf edge. The present shore, of course, is far too temporary to have any meaning in this regard. The perfect fit remains elusive, because when continents divide, the crust at the margins is severely mangled, making it impossible to determine the exact position of the scar. For the global picture, fortunately, such uncertainties are not of much consequence.

Pangaea itself was assembled during the Paleozoic from smaller pieces, of which the largest, Gondwanaland, had hung together for more than a billion years. Around its bulk, numerous small pieces, each with its own history, like glass shards in a kaleidoscope, drifted in ever-changing arrangements that have by no means been fully

Figure. 7.1. Three possible, and equally satisfactory, ways to join Africa (Af), South America (SA), India (In), Antarctica (Ant), and Australia (Au) to form Gondwanaland. The white area between them consists of continental crust under the continental shelves.

sorted out. North America appears always to have looked about the same, but Europe and Asia were fragmented into many small drifting and rotating parts that swarmed around the edges of great Gondwanaland. Some were small indeed; England and Wales, for example, managed an independent existence for much of the Paleozoic.

Especially fascinating is the accordion-like history of the North Atlantic Ocean. At least two Precambrian Atlantic oceans are suggested by suture zones in eastern North America, remnants of continental collisions. Then, late in the Cambrian, North America was once more separated from the rest of the world by a third Atlantic ocean, which gradually closed as Pangaea was put together (see Chapter 8).

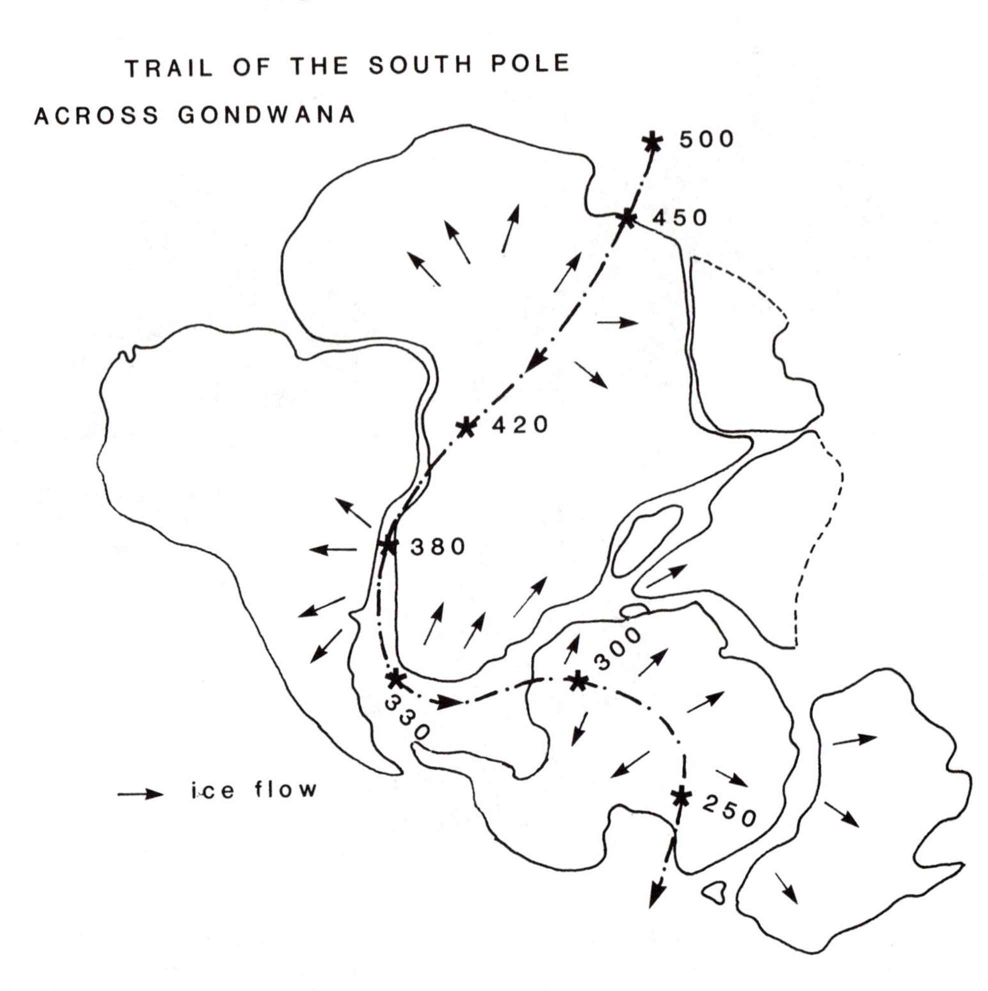

Figure 7.2. Between the late Precambrian and the Permian, the South Pole traversed Gondwanaland. The numbers along the track mark its progress in millions of years. Three times during this voyage, icecaps (arrows indicate their flow directions) formed around the poles. At other times the pole was apparently free of ice.

THE FACE OF PANGAEA

Maps of continents drifting around the planet provide an interesting starting point for speculations about the Paleozoic world, its ocean circulation, and its climate. Most of that is for future geologists to know, but we may permit ourselves at least a glimpse of Pangaea, the largest land ever.

The story begins with Gondwana's march across the South Pole (Figure 7.2), beautifully documented with paleomagnetic data. While the South Pole crossed the Sahara, glacial deposits were laid down and

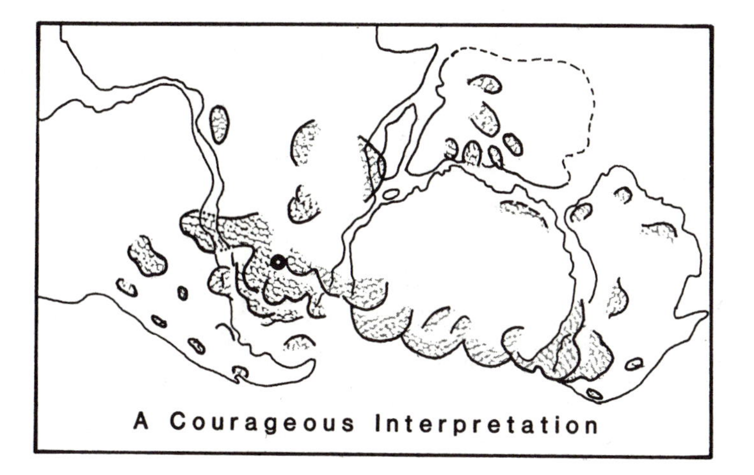

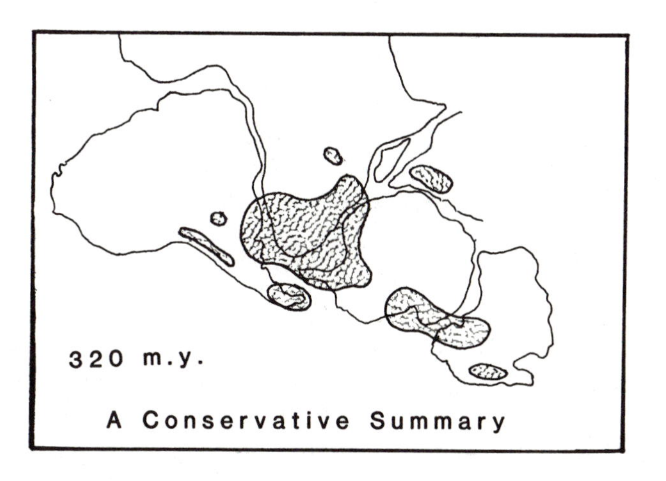

Figure 7.3. Two interpretations of the south polar icecap during the Carboniferous (Mississippian) Ice Age 320 million years ago. The upper one has much more detail and therefore seems more realistic than the lower one, which, however, in its conservatism is more commensurate with the evidence.

the rocks were grooved by moving glaciers, but for the next 100 million years such evidence of glaciation is totally lacking. Then an extensive and rather well known set of glacial deposits formed during the Pennsylvanian Ice Age, as the pole crossed southern Africa and Australia. Icecaps first appeared about 320 million years ago and soon spread to 60° south latitude (Figure 7.3). They vanished again around 60 million years later, when the edge of Antarctica cleared the South Pole. Their history would tell us much about ice ages in general and

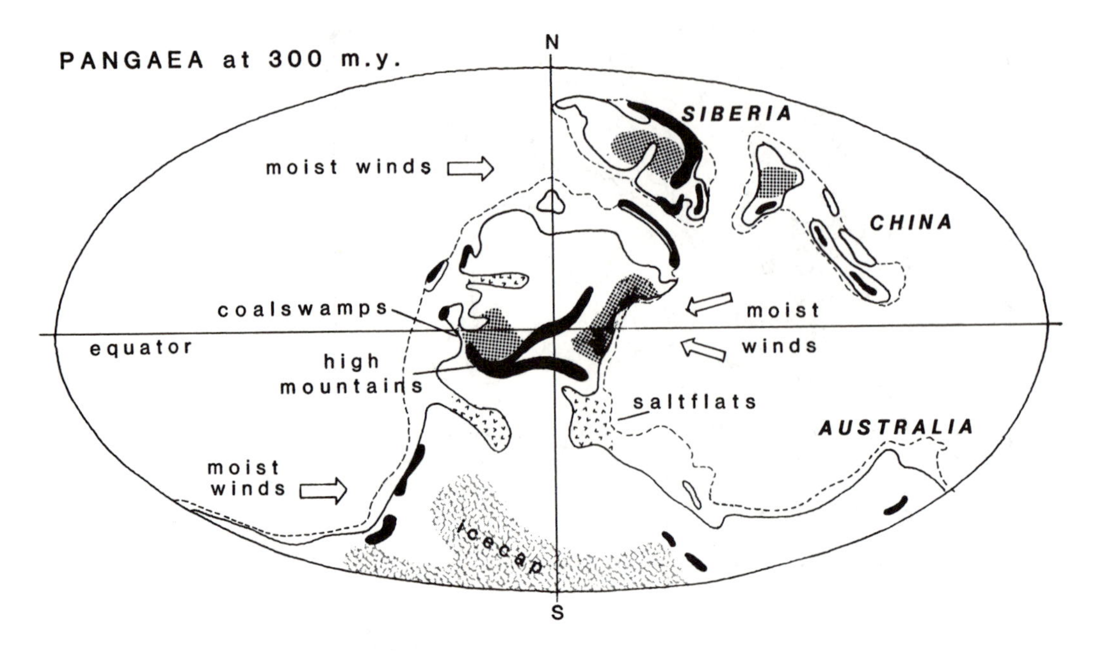

Figure 7.4. During the later Carboniferous (Pennsylvanian), Pangaea, though incomplete, with Siberia and China plus southeast Asia still waiting in the wings, was already a supercontinent. The south polar icecap, the lush equatorial swamplands in the east, the subtropical deserts, and the evaporites fit a simple and plausible climatic pattern (see Chapter 3). The odd shape of the south polar region is a result of the map projection.

conditions in Pangaea in particular, but the Ice Age lasted rather long, and the record, although often very fine, is difficult to decipher in detail.

Now consider once more Pangaea, almost, but not quite, complete 300 million years ago; Siberia plus Kazakhstan and China still had to join (Figure 7.4). Appropriately, the icecaps were located at high latitudes, and the sunny and dry subtropics had an abundance of evaporites, deposited on desert saltflats and in coastal saltpans. The recently completed continent extended almost from pole to pole, forming a barrier in the way of ocean currents and planetary winds. Based on our understanding of today's climate, we surmise wet west coasts in the temperate zones, and much rain on the equatorial east coast. The extensive coalbeds formed in swamp forests do bear out this conjecture, although a few occur at surprisingly high latitudes. Perhaps they were the Paleozoic equivalent of Scottish peat bogs and Canadian muskeg. For the rest, this huge land, bordered or traversed in many places by the still lofty mountains remaining from the collisions of its birth, should have been very dry in its interior, and dominated by monsoons to a degree not seen even in present India. The extensive redbeds, salt deposits, and dune sands of the Permian confirm that during its brief existence, much of Pangaea was a harsh, arid world not unlike present central Asia.

PANGAEA DISMEMBERED

The supercontinent, having existed for a mere 50 million years or so, began to show signs of stress as early as the Triassic. In New England and other borderlands of the future North Atlantic, faulting produced deep rift valleys into which copious basalt lavas flowed. Later in the Triassic, similar rifts appeared elsewhere. They announced that all was not well with Pangaea, but it would be hasty to conclude that disintegration and drift were about to begin.

The Triassic rifts were quite similar to those that now extend across East Africa from Mozambique through Tanzania and Kenya into the Red Sea, forming fault-bound valleys on a gentle rise a few kilometers high and up to 1,000 km wide. This African rift system, famous as the birthplace of the oldest known human beings, is more than 20 million years old; so far, it has produced a beautiful country and some spectacular volcanoes, but it has failed to sunder the continent. Only in the Red Sea and Gulf of Aden has separation begun, and that just recently.

Obviously, we must distinguish between rifting and drifting. Although cracks along the traces of future ocean shores appeared in Pangaea as early as 200 million years ago, the North Atlantic did not begin to open until 35 million years later, while the South Atlantic waited another 40 million years (Figure 7.5). For a while, the great eastern embayment of Pangaea, the Tethys, growing westward, divided the supercontinent into two, but soon Laurasia in the north itself split in two, and Gondwana, shattered like a pane of glass, sent pieces on their way to form separate continents or to attach themselves to Asia and Europe. It all took time, however, and even now new oceans are forming in the Red Sea and the Gulf of California.

What, precisely, happens when a continent disintegrates? The African rift valleys and the two embryonic oceans tell us much about the sequence of events. It all began (Figure 7.6) with a doming of the crust accompanied by high heatflow, probably from a source located below the lithosphere. As the brittle crust of the dome bulged up, it stretched and fractured. Keystone blocks sank to relieve the stress and to accommodate the increase in surface area. Three rift valleys meeting at the top of the dome are an efficient and apparently common way to

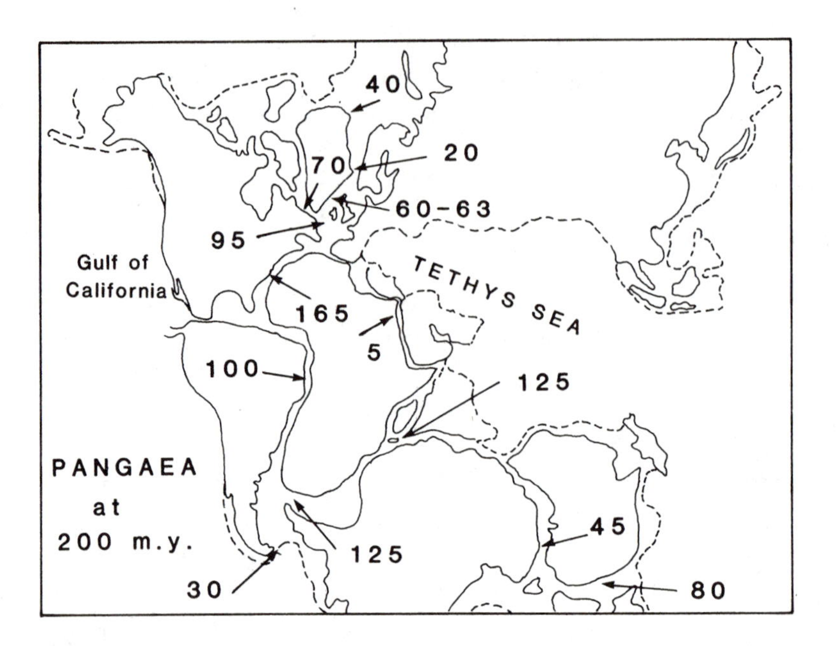

Figure 7.5. When Pangaea broke apart during the Mesozoic, drift began at different times in different places. The earliest of the new oceans was the southern North Atlantic; the South Atlantic followed considerably later. The process is not complete; the Red Sea began to open 5 million years ago, and the Gulf of California is even younger.

accomplish this. Faulting and melting thinned the lithosphere, and lava rose through fissures to build volcanoes. At this stage the rift valley was high above sealevel; the land sloped away from it, and the rivers drained outward. The first sediments in what would eventually be an Atlantic or Indian Ocean were patchy lake, floodplain, desert, and volcanic deposits. There was no divergence of the two sides of the rift valley.

Continents may remain in this stage for millions of years, and sometimes go no further. The Pangaean rifts, however, eventually began to diverge, and so much dense mantle magma was injected that the crust at the bottom of the rift valley became heavy, transformed itself into oceanic crust, and sank below sealevel. A true ocean was formed, shallow and narrow to begin with, and with its edges turned up like those of the Red Sea or the western Gulf of California. However, as

Figure 7.6. The breakup of a continent probably begins with local heating beneath the lithosphere. This produces by expansion a surface bulge, which cracks to form a rift valley, accompanied by much volcanism. Eventually, though not always, the stationary phase of rifting is followed by drift, oceanic crust is intruded, and the sea invades the rift valley to form an embryonic ocean basin.

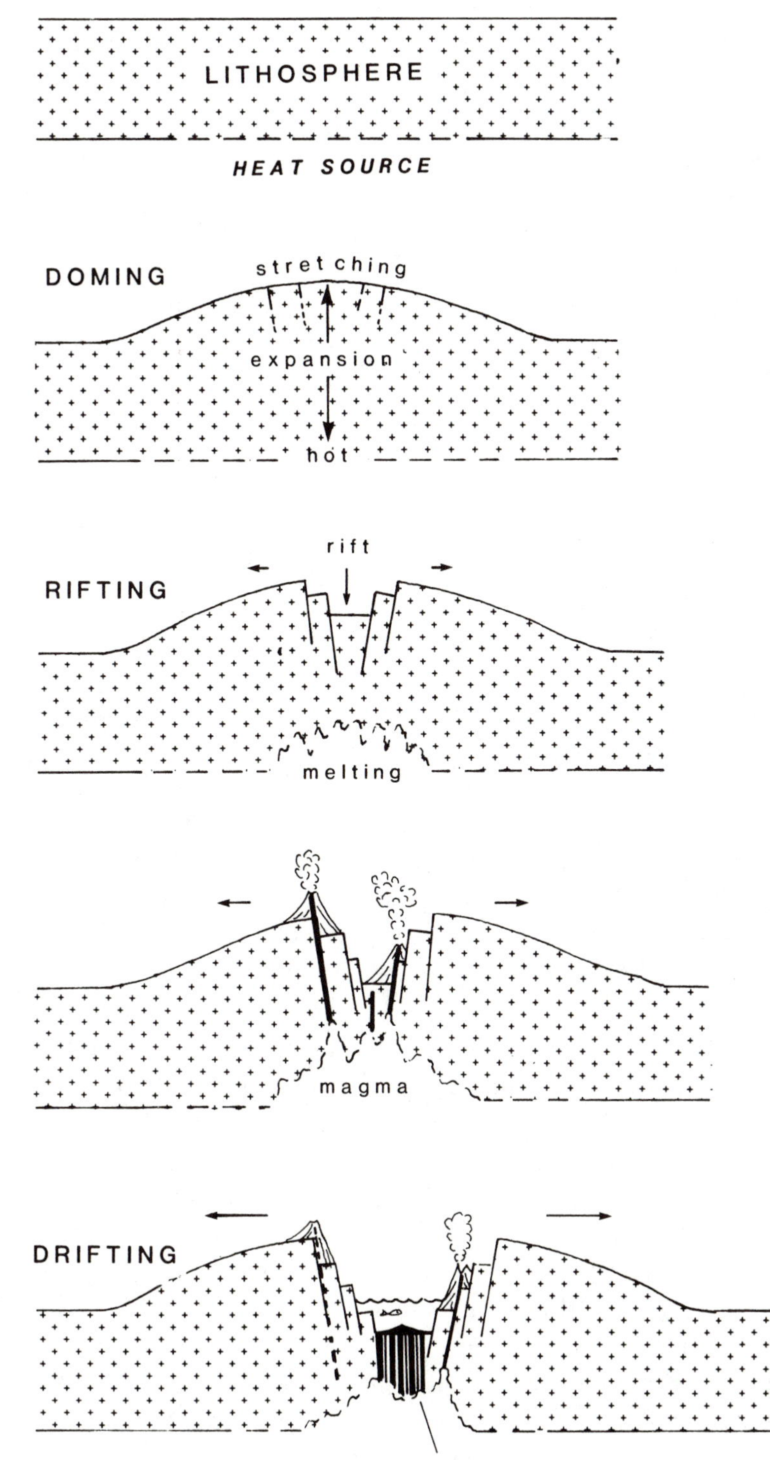
LITHOSPHERE
HEAT SOURCE
DOMING
stretching
expansion
hot
RIFTING
rift
melting
magma
DRIFTING
OCEANIC CRUST

the distance between the edges of the continents and the hot center increased, cooling caused the margins to become more dense and to sink. The slope was thus reversed, and large rivers began to carry much sediment into the new sea.

When Pangaea came under stress, many domes with triple rifts formed so close together (Figure 7.7) that their branches connected to become continuous rifts, forerunners of new oceans. For each triplet the third branch was left as a failed rift, ending blind and pointing away from the new ocean. Although of no further significance to continental drift, these failed rifts had, as we shall see, a splendid future in oil.

DOMES AND HOTSPOTS

Continents join because they happen to collide, but why they disintegrate is less obvious. There is good reason to believe, however, that hot domes are a key element and that the process that causes doming can be seen at work today. Most of the world's active volcanoes are associated either with mid-ocean ridges or with subduction zones, but some, remote from plate boundaries, are not so easily explained. In the heart of the Pacific plate lie the Hawaiian Islands, a fine string of active and recently extinct volcanoes sitting atop a broad rise in the ocean floor (Figure 7.8). The easternmost volcanoes, those of the island of Hawaii, are fully active and but a few hundred thousand years old; those of the adjacent island of Maui are barely extinct. The farther west we go, the longer the volcanoes have been dead. Beyond the last island there are others even older, now truncated by the waves, reduced to small pinnacles or entirely submerged beneath the sea. Where the chain, now deeply sunken, reaches an age of 45 million years, it turns sharply north, but the age progression continues. Other linear volcanic chains in the Pacific, although not all of them, have the same trend and are also active only at the eastern end.

To account for such an orderly and obviously meaningful pattern, let us assume a hotspot deep in the mantle, below the lithosphere, where, for reasons not yet completely clear, the temperature is anomalously high. There, part of the mantle melts, and the magma rises to the surface, heating and doming the lithosphere and fracturing it. A volcano forms, grows, then dies as the plate passes over the magma plume. The dead volcano sinks and erodes away, and a new one rises behind it, until a whole chain, young at one end and old at the other, is formed.

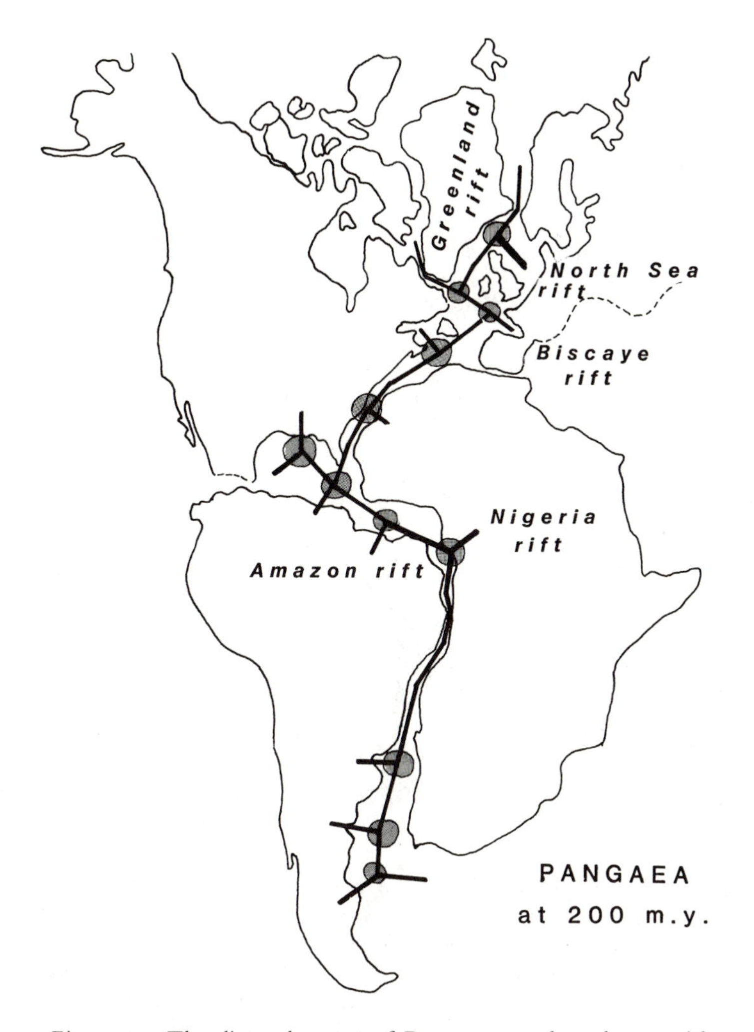

Figure 7.7. The dismemberment of Pangaea may have begun with rows of domes above sublithospheric heat sources. Such domes often crack in the pattern of a three-pointed star. Two rifts of each dome ultimately connect with their neighbors to form the new ocean, in this case the Atlantic, whereas the third, an abortive rift, dies, but often not before it becomes the site of formation of copious oil.

If the hotspot is stationary, the line of volcanoes must parallel the direction of motion of the plate; the age progression measures its velocity. Should the plate change direction, the volcanic chain also turns. Thus, the hotspot concept can be tested with plate directions and plate speeds for the recent past calculated from magnetic-anomaly patterns. The reorientation of the Pacific plate implied by the 45-million-year-old bend in the Hawaiian chain is indeed confirmed by

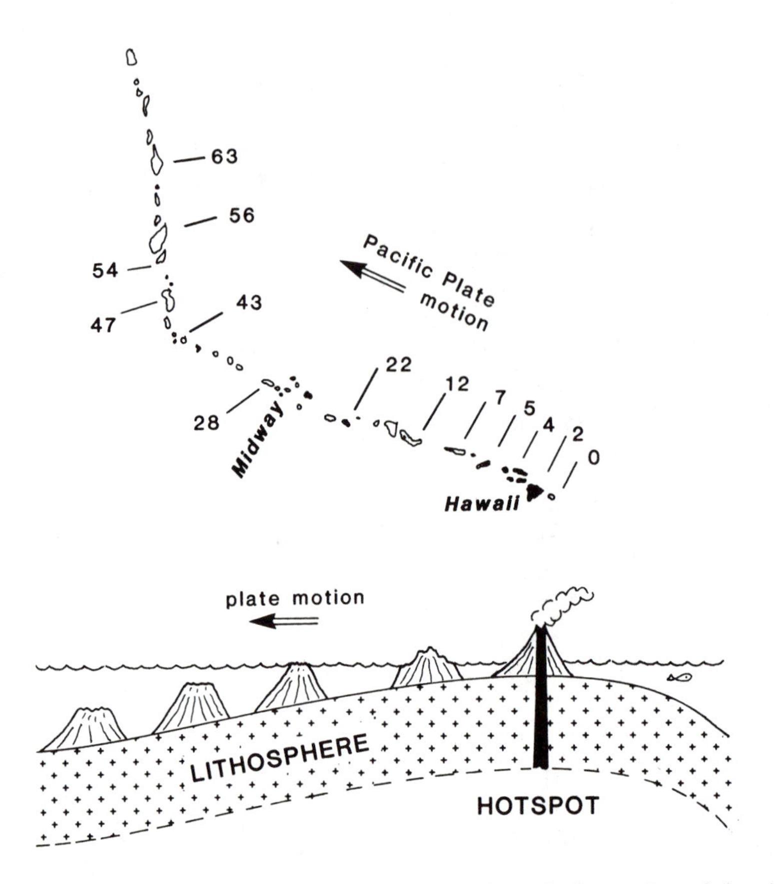

Figure 7.8. The Hawaiian Islands are the product of a hotspot located deep in the mantle below the Pacific plate. As the plate drifts across the hotspot, the lithosphere expands and fissures, and volcanoes build, only to die again when they have drifted past. A new volcano then forms upstream. The result is a volcanic chain parallel to the direction in which the plate moves. The age progression of the volcanoes, shown in millions of years (top), measures the speed of movement. Prior to about 45 million years ago, the Pacific plate evidently moved in a more northerly direction.

independent evidence. Although a few linear island chains may be located above hotspots that are not quite stationary, many have survived this test with ease and are a welcome gift from nature. Transform faults and spreading rates tell us how plates move relative to one another, but the trail of volcanoes behind a stationary hotspot relates the motion of a plate to a fixed reference and thereby to the conventional grid of the earth of poles of rotation and equator.

Hotspots are not limited to ocean basins. Yellowstone in North America, the Eiffel in Germany, and Kilimanjaro in East Africa are continental ones; Iceland, Easter Island, and the Galapagos are additional examples in the oceans. Some geologists see far more hotspots than others, but not every volcano distant from a plate edge need be one. We do not yet know why there should be hotspots at all, nor from precisely what depth they rise.

It is only a small step to postulate that rows of hotspots set Pangaea on the road to ruin. Many modern continental hotspots resemble the ancient domes we have discussed, and the model seems to fit. On the other hand, why should Pangaea have fallen victim to such a virulent case of hotspots, whereas the present ones, common as they are, do not appear to threaten the integrity of our continents? To answer this, it has been suggested that the thick continental crust acts as a thermal blanket. If the insulating patch is small, excess heat can easily escape around the edges, but if it is large, the heatflow might be impeded to the extent that many active hotspots arise to restore the equilibrium flow. Is this true? We do not know.

EDGES OF RIFTS AND MARGINS OF CONTINENTS

About half the continental margins of the world began as edges of Pangaea's rifts. Called passive margins for their lack of earthquake and volcanic activity, they have an interesting history and considerable economic importance.

When the embryonic rifts began to sink below sealevel, their flanks were like stairs of fault blocks descending to the center (Figure 7.9). The first invasions of the sea were hesitant; where suitable closed basins existed and the climate was sunny and dry, as in the early South Atlantic, thick deposits of salt were laid down on top of the earlier continental sediments. With time, however, the shallow sea became permanent, and in its clear waters reefs flourished on the uptilted edges of the fault blocks, and quiet lagoons and beaches lay behind. Such narrow seas, like the Gulf of California, are often fertile, and as the water deepened, organic matter in abundance settled to the bottom and was buried in the mud.

Eventually, however, the reefs died when the sea became too deep, and large rivers, following the reversal of the slope (Figure 7.6), began to bring silt and clay into the sea in ever larger quantities. Reefs and organic mud were buried under thick layers of clayey silts and clays. If the sediment supply was large, as on the east coast of North America, deposition easily kept up with subsidence, and a prism of shelf sediments formed that now measures from 6 to 10 km thick. On other continental margins, those of western Australia or

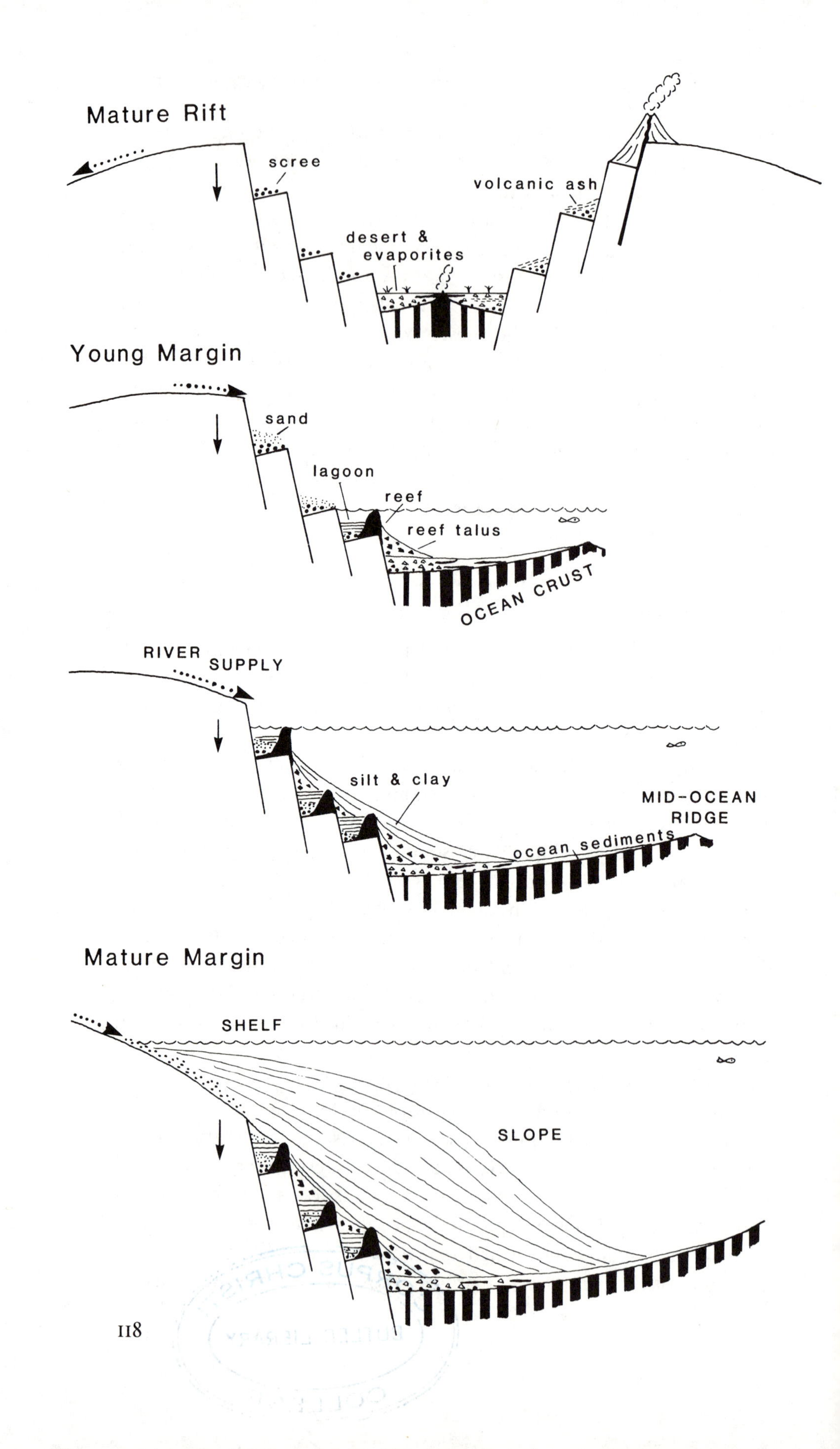
Mature Rift
scree
volcanic ash
desert &
evaporites
Young Margin
sand
lagoon
reef
reef talus
OCEAN CRUST
RIVER
SUPPLY
silt & clay
MID-OCEAN
RIDGE
ocean sediments
Mature Margin
SHELF
SLOPE

West Africa, for example, the sediments and structures of the early rift are much less deeply buried and can therefore be more easily examined.

Taken together, the passive margins of the world are the largest storehouse of sediment on earth. They also contain a great deal of oil. About three-fourths of all giant oil fields, those yielding more than half a billion barrels of oil, lie buried in modern or ancient passive continental margins, holding among themselves an estimated two-thirds of the world's oil reserves. Most of that good oil is found not on the margins of rifts that became real oceans but in the sediments of rifts that failed.

To extract oil efficiently and profitably, several conditions must be met (Figure 7.10). There must be abundant organic matter in a source bed of some kind, often decayed marine plankton preserved in dark shales. This precursor of oil is not usable, however; it must simmer at moderate temperature and depth for a long time before it will convert into oil and gas. If the temperature is too high or the time too long, natural gas will form rather than oil, or the organic matter may even turn into useless carbon dioxide. If the time, depth, or temperature is too limited, the organic matter will remain stuck in the shale, and this shale oil, although it might be mined some future day, is not now a particularly valuable commodity.

Next the oil must migrate out of the shale, where it is too finely dispersed to be extracted, and travel to a porous and permeable reservoir that we can tap with the drill and pump out. Sandstones and reef limestones make good reservoirs. In the reservoir, the oil, being lighter, will separate from the water that is always there and float on top. If there is gas, it will gather above the oil. The reservoir must be sealed with a tight cap (a clay will do), or the oil and gas will continue upward to the surface and seep away.

Rifted margins satisfy all these conditions. They have the organic matter, the reefs or beach sands for reservoirs, and the sealing by subsequent clays. In their early history they are warm but not too deeply buried; oil and gas can form and accumulate. Failed rifts are

Figure 7.9. A mature continental rift is a shallow basin near sealevel where evaporites form, together with volcanic sands and material washed down from the slopes. With the onset of drifting, the rift widens, and the margins cool and subside. The sea invades, coral reefs grow, organic matter accumulates in the sediments, and beaches form along the shores. Ultimately the margins sink enough to reverse the direction of continental drainage, and large quantities of silt and clay bury the reefs and the black organic shales, creating the continental shelves of a passive margin.

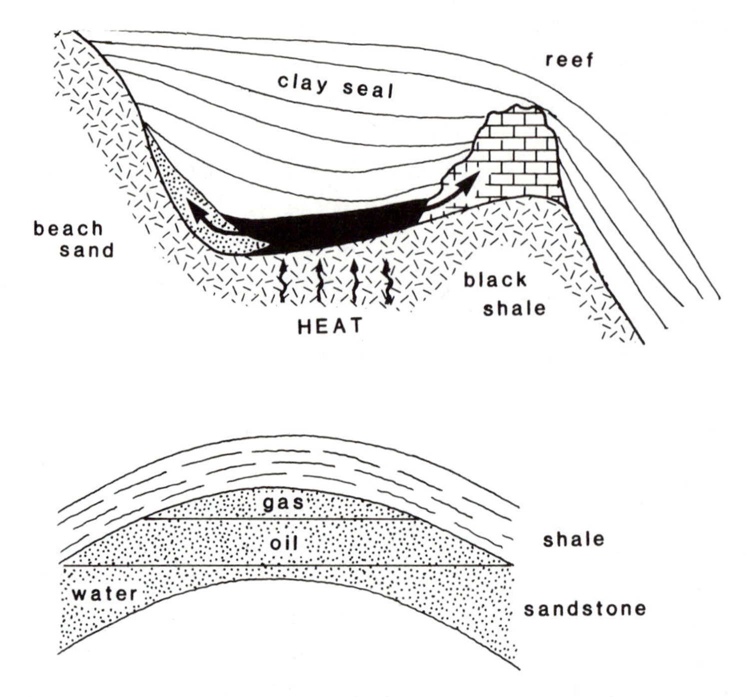

Figure 7.10. To accumulate oil and gas in such a way that they can later be extracted with profit, we need a source bed rich in organic matter. The organic matter must mature with prolonged gentle heat, and the oil and gas must be able to migrate into permeable and porous reservoirs, reefs for example, or beach sands. In a reservoir (bottom), the oil separates from the omnipresent water and floats to the top, and gas accumulates above the oil. A seal of clay and silt is required to prevent the oil from escaping to the surface.

even better. They stay near the source of the heat instead of drifting away from it. They do not subside so deeply later, and there is less risk that their oil will be destroyed by excessive pressure. Their reservoirs are not so deeply buried and hence less costly to drill. Most of them are now on land, where recovery is easier and cheaper. On successful margins, on the other hand, the earliest and most prolific reservoirs are likely to be quite deep, and if their contents have not already been converted into useless carbon dioxide, they will be expensive to find and exploit. The overlying shelf wedge of silts and clays is not so rich in organic matter, has few good reservoirs, and has not benefited from adequate heating. Oil accumulations do occur there, but they are not nearly as worthwhile as those of the failed rifts of the North Sea, Venezuela, Nigeria, or the Persian Gulf.

8

Converging plates and colliding continents

Where plates collide, subduction destroys the oceanic crust and removes the evidence for the existence of old oceans. At the same time, it is only in former subduction and collision zones, known also as *mobile belts,* that the few rocks remaining from ancient ocean floors are caught up and preserved. These rocks, together with paleomagnetic and some kinds of paleontological data, are our sole sources of information regarding vanished oceans.

Ancient mobile belts, long and narrow zones of deformed and often deeply eroded rocks, are common on all continents, where they sweep around stable cores of great antiquity, called *cratons* (Figure 8.1). The oldest mobile belts are Precambrian, and their significance is not yet fully clear, but those of the Phanerozoic certainly represent converging plate boundaries, either extinct ones like the Appalachian or Alpine ranges or active ones like the Andes or the Himalayas. All contain andesitic volcanic rocks, granite intrusions, and intensely folded and metamorphosed sedimentary rocks.

Plate-tectonic theory says fairly simple things about what must happen at converging plate boundaries. However, as we have gone on to apply it to ever more complicated mobile belts, the difficulties have increased, and the fit has become less satisfactory.

SCENARIOS OF SUBDUCTION

The simplest case of convergence is a collision between two oceanic plates. Either plate may be subducted, and a volcanic arc will form on the other one. On the other hand, if an oceanic plate meets a plate bearing a continent, the denser oceanic plate will be forced down, and volcanoes will grow on the continental side. Of course, if the plate boundary should happen to lie a little seaward of the continental edge,

Figure 8.1. Subduction zones and sutures mark the places where plates converge or once converged. The oldest sutures are of Precambrian age and lie in the interiors of continents, Africa for example. The youngest ones border the Pacific and are still active.

once again either plate may be subducted, but if that happens to be the slab attached to the continent, the end will come as soon as all oceanic crust has been consumed (Figure 8.2). The direction of subduction must then be reversed, and a new chain of volcanoes will form, this time on the continental side. The old slab will become detached and be absorbed in the mantle, and because it no longer drags the plate down, the region above it will rise. Such a reversal, involving abandonment of a slab of lithosphere 100 km thick and perhaps 700 km long, with the simultaneous creation of a new one, is rather mind-boggling, but theory demands it, because continents cannot be subducted, and the evidence is there that it actually does happen.

Let us take a closer look at the subduction zone (Figure 8.3). As one approaches the trench, a series of fault blocks step down into it, the result of bending and stretching of the crust. On the opposite side, where the plates meet, one finds a wedge of deformed rocks, the forearc. The forearc builds from slivers of the upper part of the subducting crust that are sheared off the way a carpenter's plane pulls woodcurls off a piece of lumber. Each successive sliver is thrust under the wedge that is already there. Therefore, the oldest slivers are on top,

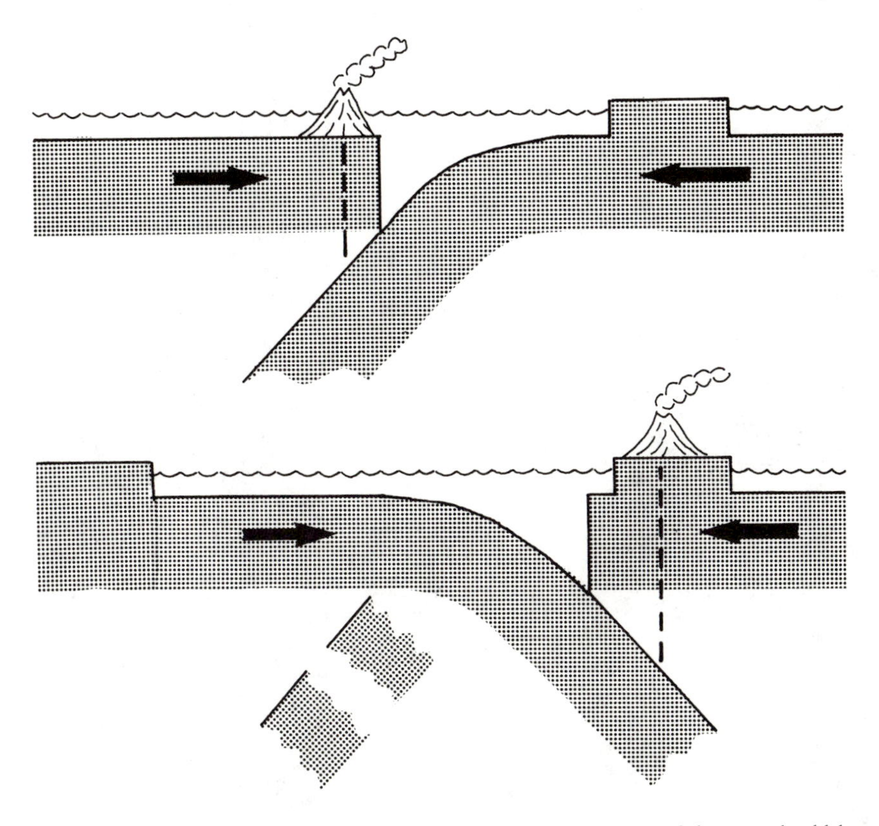

Figure 8.2. When plates collide, subduction consumes one of them. If that one should happen to carry a continent, it will in time find itself at the edge of the subduction zone, but it cannot go down because it is too light. The subduction then reverses direction, and the other plate descends.

and each new one will raise the pile a little more. Eventually, parts of the forearc appear above sealevel to form islands of deformed and metamorphosed sediments sometimes mingled with fragments of the oceanic crust. Behind the islands lies a forearc basin filled with sediment washed in from both sides, and then comes the volcanic arc. Often the far end of the collision zone is occupied by yet another basin, the backarc basin. This basin has a young crust, a high heatflow, and a set of magnetic anomalies, implying in every way that it is being formed by seafloor spreading. The cause and manner of this spreading are still matters of some speculation.

The volcanic arc consists of andesite lavas with prodigious amounts of volcanic ash. Unlike mid-oceanic basalt volcanoes, which flow quietly, andesitic volcanoes are explosive, sometimes violently so, as Krakatoa, Vesuvius, and Mount St. Helens have shown. The volcanic arc

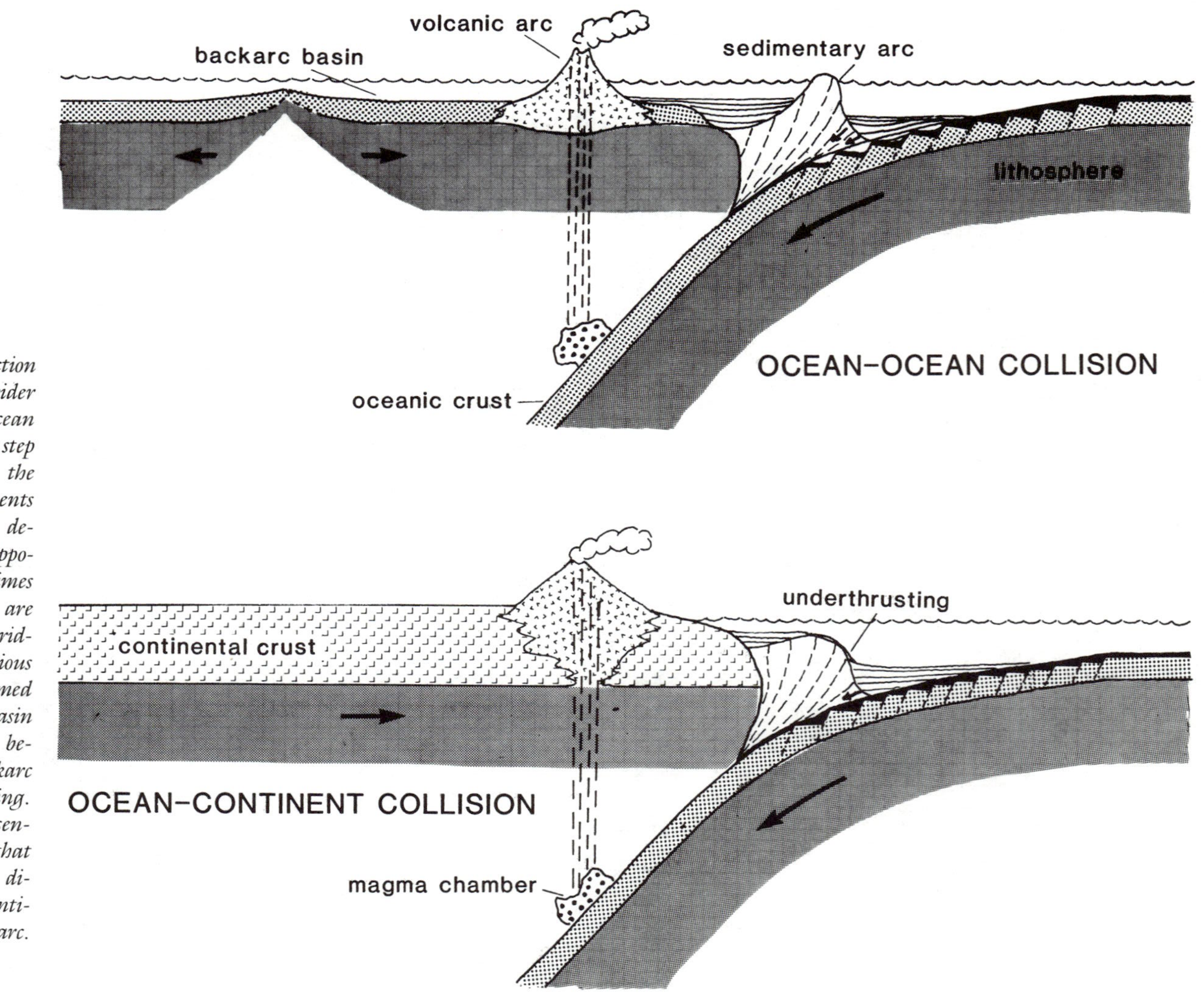

Figure 8.3. In detail, a subduction zone is complex and much wider than just the trench. An ocean-ocean collision zone (top) begins with step faults on the ocean side. In the trench, oceanic crust and sediments (black) are buried under trench deposits derived mainly from the opposite side. Sediments, and sometimes parts of the oceanic crust as well, are scraped off by the edge of the overriding plate and thrust under previous wedges to create an arc of deformed sediments. A sediment-filled basin and a volcanic arc follow, and behind the volcanic arc lies a backarc basin, at times a zone of spreading. A continent-ocean collision is essentially the same (bottom), except that the thrust complex generally lies directly against the edge of the continent without forming an island arc.

is found where the subducted slab reaches a depth of about 100–150 km. There its more volatile components, the sediments and their enclosed seawater, and the altered basalts melt in a tubelike magma chamber. This magma is rich in water vapor and in carbon dioxide (from oceanic calcareous oozes), and it rises and melts its way upward in plumes, each crowned at the surface by a volcano. Simple in principle, but there are some puzzling aspects. Why, for instance, do the eruptions tend to be episodic rather than continuous, as subduction itself is?

As we noted, the forearc wedge consists of sediments and some basalt scraped from the surface of the subducting plate by the edge of the overriding one. This is plausible enough, because the sediments are soft, and the deeply altered top of the basalt is not very strong either. On the other hand, in order to produce the magma for the andesitic volcanoes, not all of this soft crust can be scraped off; some must be subducted to be melted at greater depth. How much is dragged down and how much remains at the surface to build the forearc wedge is thus a key question. Under the west coast of South America, about 200 km^3 of sediment per kilometer of trench length were subducted during the Cenozoic alone. That would be enough to form a forearc wedge 50–75 km wide, but the existing one is narrow. Most of the sediment must therefore have been subducted. Implausible as it seems to anyone who has ever held a handful of oceanic ooze, the larger part of this soft and watery material commonly seems to have been stuffed down into the mantle.

Sometimes there is no wedge at all, or one so small that it must be temporary. Sometimes we observe that the edge of the overriding plate seems to be sinking rather than being pushed up. The best explanation for such observations is that the subducting plate somehow scrapes material off the bottom of the overriding plate and carries it down. If this happens to a continental plate, it thins the crust, reduces its buoyancy, and causes it to subside. Continental crust would thus be lost because of "tectonic erosion," in contrast to the normal subduction process, which results in growth of the continental edge by accretion. The evidence for tectonic erosion is still meager, but given the importance we attach to the indestructibility of the continental crust, it is a troubling suggestion.

Trenches with volcanic arcs as well as forearcs are common in the western and southwestern Pacific, and Japan, Indonesia, and the Philippines can be fitted into the same scheme. Continent-ocean collisions occur along the west side of the Americas from Chile to Mexico, at a slow rate along the coasts of Oregon and Washington, and from southern Alaska to Kamchatka along the Aleutian chain.

OROGENY AND GEOSYNCLINES: CLOSURE OF THE THIRD ATLANTIC OCEAN

Orogeny is the geological term for the folding, thrusting, metamorphism, intrusion of igneous rocks, and uplift that build most mountain ranges (*orogens*). Orogens contain large volumes of sediment that appear to have been laid down originally to a great thickness in deep, narrow, elongated troughs. Such troughs are called *geosynclines,* and scholars have argued for almost a century whether or not trenches can be considered to be the early stages of geosynclines, before deformation, intrusion, and uplift. Unfortunately, the definition of the word "geosyncline" includes eventual conversion to a folded mountain range, and this prevents us from applying the principle that the present is the key to the past. Who, after all, can say with confidence what will happen later to a trench and its sediments? Today we recognize that in plate-tectonic terms, a geosyncline is no more than a trench, its sediment fill, and its forearc wedge combined. Subsidence, sedimentation, and deformation occur simultaneously rather than, as once thought, subsequent to each other. Consequently, the concept of the geosyncline, treasured by generations of geologists, has virtually lost its usefulness.

To assemble Pangaea, North America and many pieces of Europe and Asia had to be joined to Gondwana, thereby closing the Paleozoic Atlantic Ocean (Figure 8.4). The resultant collisions raised mountains from Texas to northwestern Canada and all over western and northern Europe. It was here that John Dewey, in the late 1960s, pioneered the application of plate tectonics to major orogens. It has turned out to be considerably more complicated than his pioneering work suggested.

The first evidence for a subduction zone in eastern North America dates to about 500 million years ago, and a phase of mountain build-

Figure 8.4. Early in the Paleozoic, an Atlantic Ocean, probably the third of its kind, separated North America from Europe and Gondwana (top). Initially it had passive margins, but subduction soon set in. During the Ordovician there was a phase of intense mountain building, perhaps because of a reversal in the direction of subduction, or because of collision with a small continental sliver offshore. After another period of uneventful subduction, northeastern North America collided with the small continent Baltica (northwestern Europe) during the Caledonian-Acadian orogeny. A vast mountain range rose and shed its sediments across two continents.

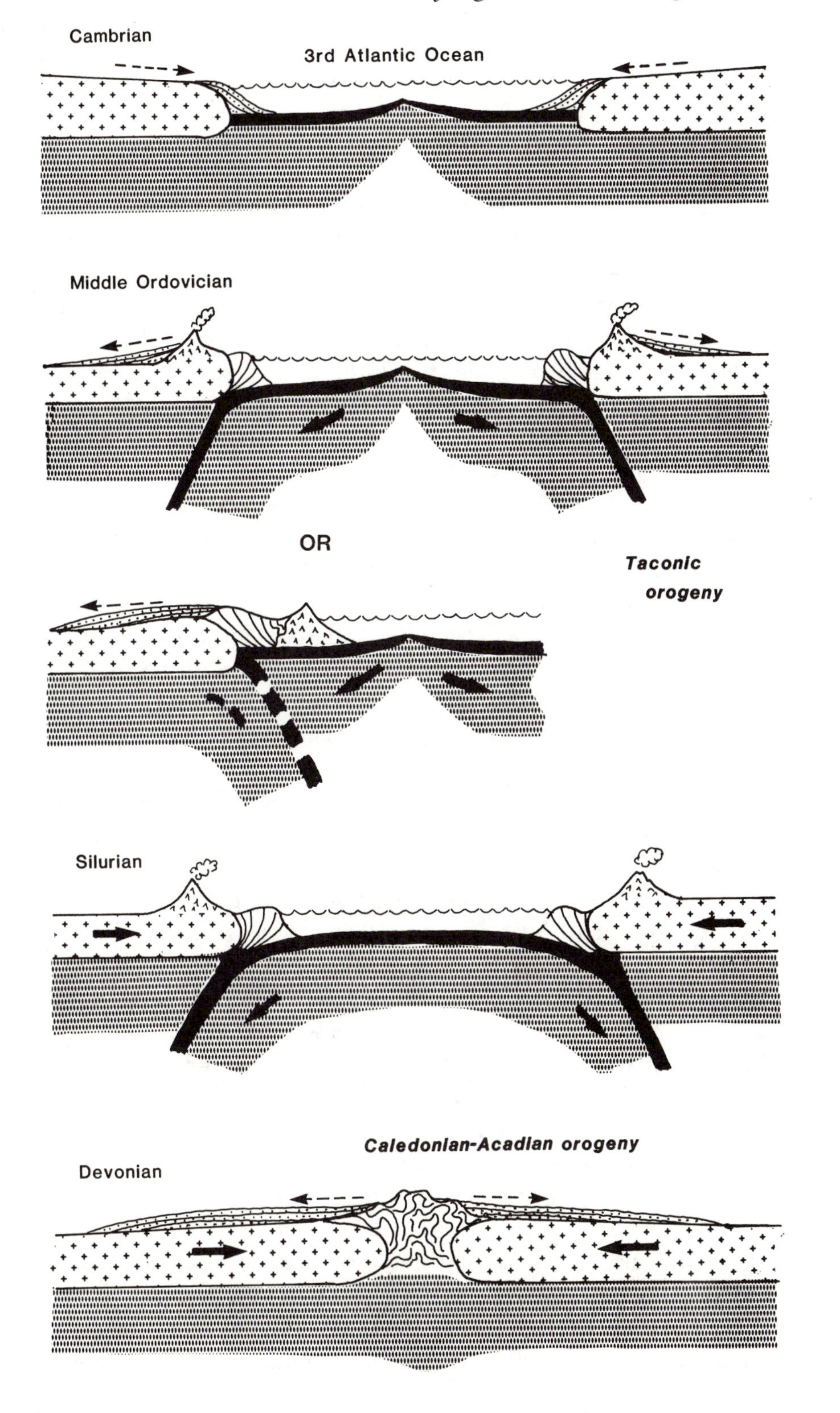
Cambrian
3rd Atlantic Ocean
Middle Ordovician
OR
Taconic orogeny
Silurian
Caledonian-Acadian orogeny
Devonian

ing, the Taconic orogeny, followed soon after. It was accompanied by volcanic activity, by considerable folding, and by the uplift of a land area to the east of the present Atlantic seaboard. Prior to the Taconic orogeny, the sediments that filled the trench and formed the deposits of the so-called "Appalachian geosyncline" had come from the mainland in the west, but afterward a huge delta spread from new eastern lands and laid down a thick wedge of redbeds in a desert climate.

The Taconic orogeny was just the beginning, and it closed no ocean. Volcanic activity and sedimentation continued, and so did subduction, until about 400 million years ago the northeastern tip of North America collided with the small continent of Baltica, today's northern Europe. This phase, the Acadian-Caledonian orogeny, produced another mountainous land, whose remains we find today in New England, eastern Canada, Scotland, and Norway. These Acadian and Caledonian mountains shed sediments to both sides: westward across much of eastern North America to form the Catskill delta, and southeastward over Europe as the Old Red Sandstone.

The final closure of the Atlantic Ocean required a third, or Appalachian orogeny. It began when the South American corner of Gondwana ran into Texas and Oklahoma and raised the Ouachita Mountains (Figure 8.5). Next, the southern Appalachians were thrust up by a collision with northwestern Africa. Finally, a number of small pieces were squeezed together between Baltica and Africa; they now constitute the core of central and western Europe. The time was about 300 million years ago.

It is a fascinating story, but there are problems with it. Consider, for instance, the southern Appalachians. They experienced both the Taconic and the Acadian orogenies, but afterward there was still an ocean to the east, and the true continent-to-continent collision was yet to come. Nevertheless, the two first orogenies raised land areas large enough to feed vast deltas from the east. Island arcs do not usually produce large volumes of sediment, nor do they support big rivers. Moreover, the intense volcanic activity, the widespread intrusions of granite, and the deformation that came with both orogenies seem excessive if what happened was merely intensified subduction or a collision with an island arc.

Seeking the more substantial bang that seems to be required, a current proposal favors two successive collisions with minor continents assumed to have been lounging conveniently offshore. Intensive orogenies creating large sediment sources seem more plausible if one is

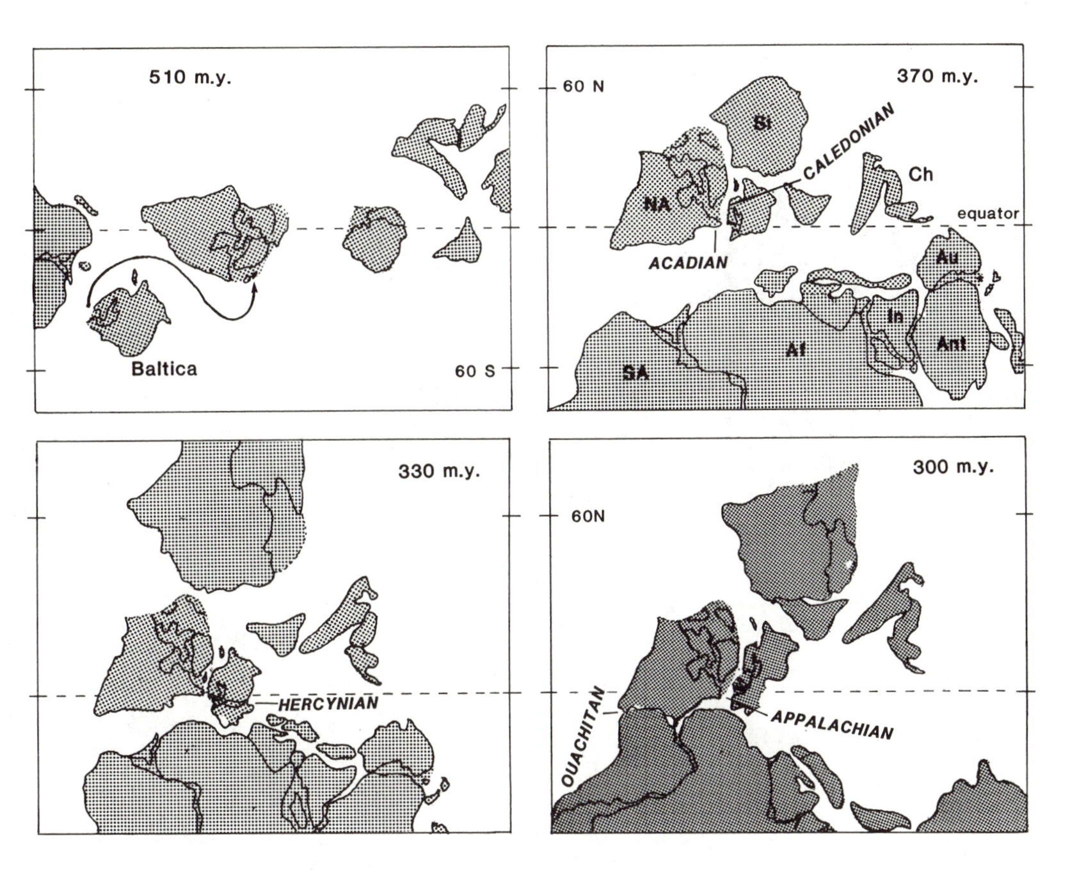

Figure 8.5. These maps help visualize the complicated series of events that accompanied the closure of the third Atlantic ocean. First Baltica, after swinging around North America, collided with that continent during the Caledonian-Acadian orogeny. Then central North Africa (Af) pushed into Baltica, and finally, beginning in the southwest, Gondwana was joined to North America in the Appalachian orogeny.

dealing with continent-to-continent collisions, even though one of the continents was rather small and has more or less to be pulled out of a hat. The assumption would explain the facts, but the need to draw on such ad hoc elements to keep things straight is not very elegant, and rather perturbing.

CONTINENTS COLLIDING

If less new crust is generated in an ocean basin than is subducted, any continents on opposite sides of the basin must eventually collide. Australia, for example, will some day push into southeastern Asia. A classic case of continental collision began about 40 million

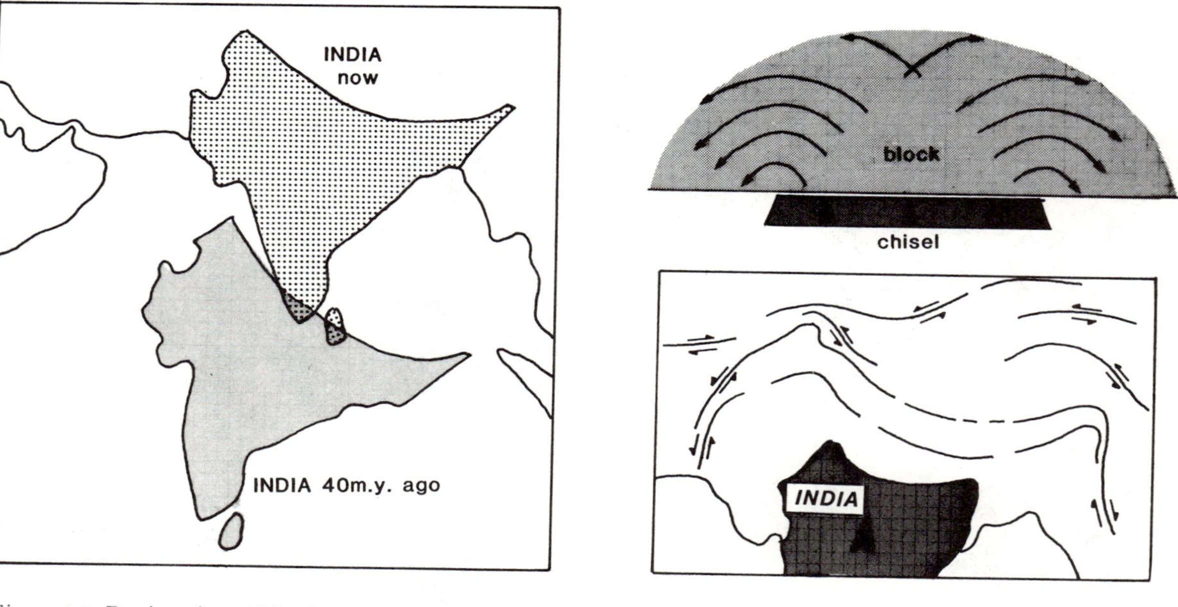

Figure 8.6. During the middle Cenozoic, India made contact with central Asia. Since then, the Indian wedge has continued to penetrate into the larger continent, producing a set of strike-slip faults very similar to the cracks that form if a rigid block is hit with a chisel (right). The many devastating earthquakes from Iran to China are the result of India's inability to stop suddenly.

years ago when India touched the underbelly of Asia, and the subduction of ocean crust came to an end (Figure 8.6). Surprisingly, the suture of this collision is not the Himalayan range, but lies well north of those lofty mountains, which are the products of later events and consist of slivers of the Indian plate that have been thrust southward.

Since that first encounter, India has moved 2,000 km farther north. Half of that distance can be accounted for by horizontal compression and by the overthrusting that created the Himalayas, but about 1,000 km must be ascribed to India's penetration into Asia, which still continues at about 5 cm per year. The process is analogous to pressing a chisel into a block of metal. The chisel drives a triangle of metal inward and forces aside parts of the block along shear zones that behave like the horizontal strike-slip faults explained in Figure 6.6. The lithosphere is rather too inhomogeneous to produce a perfect facsimile of this experiment, but with India the chisel and the Tibetan plateau the triangle, we can easily find the strike-slip faults along which the crust of central Asia is getting out of the way. The often calamitous earthquakes as far away as central China and Afghanistan are direct consequences of the collision between India and Tibet.

Continental collision first involves consumption of all oceanic crust, followed by abandonment of the subduction zone. The shallow sea that remains from the former ocean fills with marine sediments, and later with river sediments as the mountain ranges begin to rise. Eventually these deposits are caught up in the folding, and granites in profusion are intruded into them. In addition, the crust underneath appears to increase greatly in thickness to form a kind of root. The heat source for these granites and the existence of the root are not required by plate-tectonic first principles and demand an explanation. An ingenious suggestion by Peter Bird goes as follows: The lower lithosphere, below the Moho, is relatively cool and heavy and, if not kept afloat by the light continental crust above, might sink in the asthenosphere. When the subducted slab ceases to function and becomes detached (Figure 8.7), it tears at the lower lithosphere and causes part of it to come away and sink. This opens a path for the asthenosphere to rise and heat the crust, creating granitic magmas that invade the crust above. At the same time, space is provided where, as the collision continues, the continental crust of one plate will slide under that of the other, thus forming a thick root.

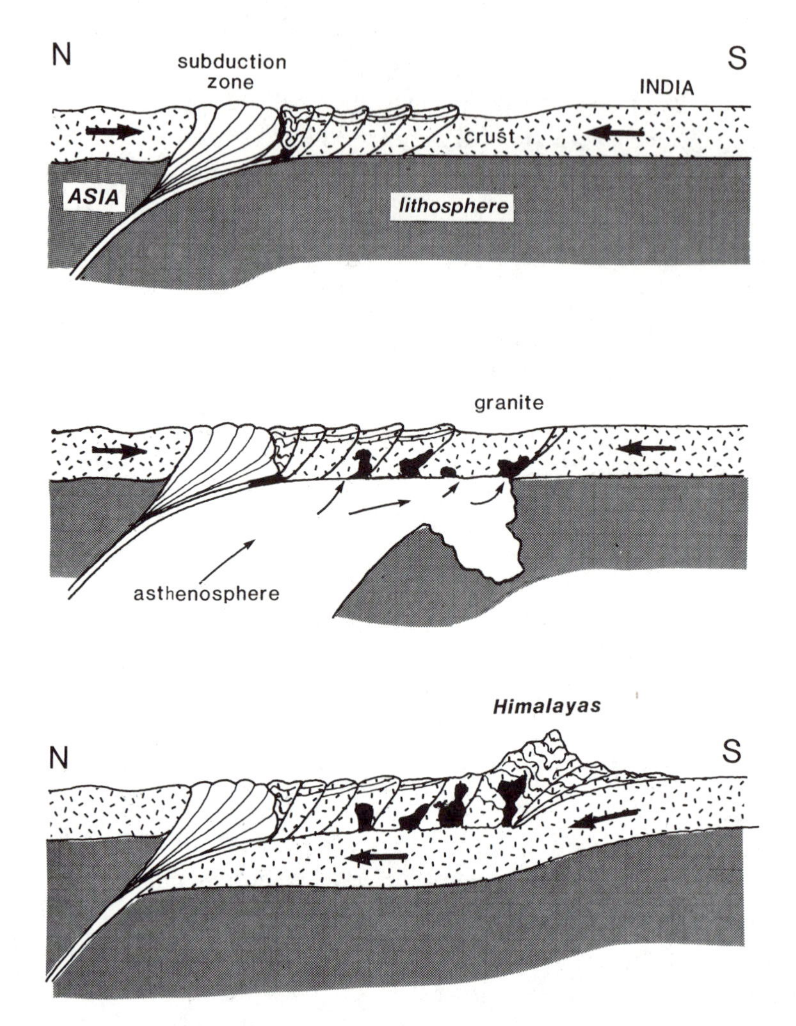

Figure 8.7. When the last remnant of oceanic crust in front of India had been subducted under Asia (top), the compressed subduction zone became the suture between the two continents. The drag of the then useless subducted slab may have torn away the lower part of the lithosphere, providing access for enough heat to produce vast intrusions of granite. It also thinned the slab, and its southern part became wedged under the northern half, doubling the thickness and raising the Himalayas as much by isostatic compensation as by compression.

FLOTSAM AND JETSAM

Consider the west coast of North America. Once upon a time, in the Precambrian, it was a rifted margin, its opposite continent long gone, perhaps to Asia. Located in eastern Nevada, Idaho, and British Columbia, this margin remained passive for hundreds of millions of years. An island arc may have existed offshore, but not until the late Cambrian are we sure that subduction had begun. Since then, the Pacific margin of North America has steadily grown westward, overriding the Pacific Ocean floor in several orogenetic steps, two during the Paleozoic, others in the Jurassic and Cretaceous.

The Pacific Ocean is all that remains of Panthalassa, but the entire floor of that superocean and most of the crust generated during the Mesozoic have been subducted. The process is well illustrated by the history of coastal California. Early in the Mesozoic, an island arc was driven against the continental margin, leaving a suture that can still be seen in the foothills of the Sierra Nevada. Steady subduction followed and brought considerable accretion in the form of several forearc wedges that added successive strips of rock to the continent. A deep forearc basin occupied the position of the Great Valley. On the continent behind it grew andesitic volcanoes above large intrusions of granite deep in the crust. Today, after long erosion, these granites form the backbone of the Sierra Nevada. In the early Cenozoic, with a final intense orogeny, came widespread emergence that brought the forearc out of the water to form the present Coast Ranges. Shortly afterward, a reorganization of plate movements replaced the subduction with a transform fault, the redoubtable San Andreas. Along this fault, which connects the subduction zone off Central America with a mid-ocean ridge off Oregon, a coastal sliver of California, part of the Pacific plate, is steadily moving north.

Once again we encounter the same sort of difficulties that bedevil the application of plate tectonics to the history of the Appalachians. There are many orogenies; yet as far as we can tell, subduction was continuous. The orogenies are marked by major deformation, large-scale intrusion of granites, and other events not so easily reconciled with the simple model of forearc tectonics and volcanic arcs.

Various explanations have been offered, each with its merits, but none fully adequate. One of the more intriguing suggestions rests on the fact that all over the Pacific Ocean, and especially in its older parts, we find pieces of oceanic crust that lie too shallow and are too thick for their age. Some may be old island arcs, others parts of fracture zones, still others plateaus of unknown origin, perhaps similar to Iceland. What happens if such anomalously thick pieces of lithosphere arrive at the trench? One might assume that they could not be subducted readily. It might be easier for the subduction boundary to jump seaward beyond the obstacle and start anew, abandoning the slab to the margin of the other plate (Figure 8.8). There is some evidence that supports this, because in the few places where such plateaus impinge

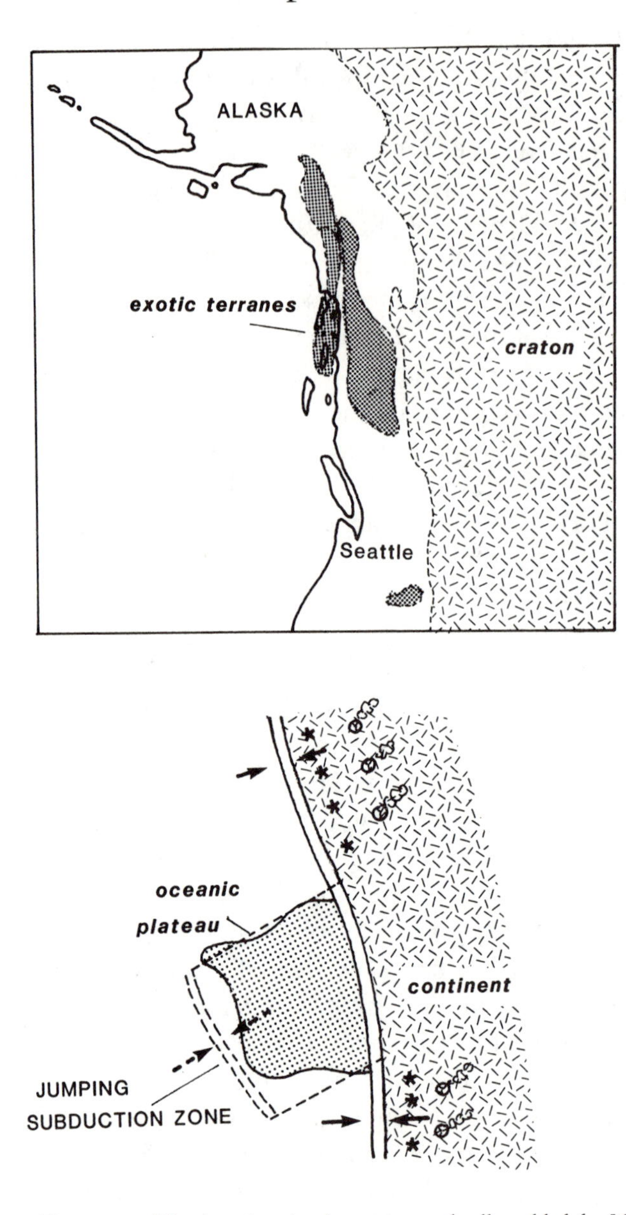

Figure 8.8. The broad strip that was gradually added by Mesozoic subduction to western North America contains slivers of crust that seem alien compared with the surrounding rocks. These exotic terranes (top) are currently regarded as having traveled from afar, perhaps even from south of the equator, before they collided with western North America and produced paroxysms of mountain building. They resemble anomalously thick oceanic crust, plateaus perhaps, or island arcs. When such a thick piece arrives at a subduction zone, it may be more efficient for the subduction to jump seaward (bottom) than to attempt to swallow it.

on present subduction zones, we see a break in the belt of volcanoes and earthquakes behind the trench, showing that subduction is not active there. Thus, it seems reasonable to postulate that such odd pieces of Pacific floor might play the same role in orogenesis that has been assigned to convenient, although unproven, continental slivers in the Atlantic.

Do we encounter such alien chunks of crust in the subduction complexes of the past? The answer seems to be yes. In British Columbia and southeastern Alaska there are rock formations quite different from those surrounding them that, by the nature of their rocks and fossils, appear to have come from far away, from beyond the subduction zone. The earliest of these *exotic terranes* may have arrived in northwestern Nevada during the Permian and Triassic orogeny. Then there is a long and narrow block that now forms the coastal islands of British Columbia but may have originated as much as 2,000 km farther south, judging from its paleomagnetic properties. The fragments of another, named Wrangellia, are scattered from Idaho to Alaska. Paleomagnetic data indicate that Wrangellia began its existence at 15° latitude, but whether it was north or south of the equator we cannot say.

Where did these and other pieces come from? Some have talked of the lost continent of Pacifica, once located in the southwestern Panthalassa, its fragments now scattered around the western and northern Pacific. The notion does not suit; few (perhaps none) of the anomalous plateaus and exotic terranes appear to consist of continental crust. Most, in fact, are like chunks of island arcs, torn from their roots. On the other hand, one should remember that western California, complete with Los Angeles and parts of San Francisco, will become a thin continent out at sea, somewhat like New Zealand, in a few tens of millions of years. Eventually it might run into a subduction zone and be attached to an alien continent as a truly exotic terrane.

The idea of exotic terranes has caught the fancy of many geologists concerned with the margins of the Pacific, and examples have multiplied like rabbits, in a manner reminiscent of the proliferation of hotspots when that concept first became fashionable. It is not clear just how exotic many of the proposed terranes really are, nor can it be confirmed that they all came from far away. To the interested bystander, the current rush to find exotic terranes appears a bit overdone and not likely to lead to a better global synthesis of the history of coastal orogenies.

If all this seems complicated and not helpful to an observer interested in a broad overview of the history of the earth, it is nothing compared with the confusion concerning the Tethys, that great eastern embayment of Pangaea, of which only the Mediterranean remains today. A swarm of microcontinents spalled early off Gondwana's eastern and northern shores, drifted north and westward along various routes, and eventually became parts of Spain, Italy, Greece, Anatolia, Iran, and Arabia. Other pieces were torn from southern Asia and traveled west to end up in the Balkans and Rumania. Along their way, these fragments rotated, collided with each other, broke apart again, were separated by short-lived mid-ocean ridges, and finally produced the tangle of mountain ranges that extends from the Pyrenees, the Alps, and the Apennines to the Balkans, the Caucasus, and eastern Iran.

The process continues. Arabia now pushes against Iran, a rift is opening from the Gulf of Aden to the Dead Sea, and the seafloor north of Egypt and Libya is being subducted under Crete and the Aegean Sea. Evidently the geological evolution of the Mediterranean has been hardly less turbulent than its human history. It is not surprising that, so far, no two proposed historical accounts of this complex sequence of events are similar.

9

Postscript to a revolution

Revolutions are often largely affairs of the young, and the revolution in the earth sciences was no exception. This seems natural when the overthrow of conventional wisdom is the issue, but graduate students, young scholars, and beginning professors require courage and persistence to face the considerable resistance from the older generation. The names of the young protagonists became scientific household words overnight, sometimes even before they outgrew student status, something not always appreciated by their seniors, who had needed decades to make themselves known, and then often only in small and specialized circles. Such instant fame is most gratifying, but not an unmixed blessing. After setting an entire discipline on its ear with a single lecture, what does one do for an encore? How, having just proposed subduction, does one spend the remaining 40 years of one's career? This anxious desire to live up to impossible expectations has led to some extravagance, but more often to a sense of profound boredom as geologists returned to the long and often tedious task of checking and consolidation.

Where does plate tectonics stand, now that revolution and revolutionaries have attained middle age? It has answered many questions and will continue to influence geological thinking for the foreseeable future. It is not a comfortable concept, and we are still learning to live with it. It is difficult enough to visualize how the Alps might have looked 60 million years ago. Now we must also visualize where they might have been. First the almost exclusive turf of marine geologists, plate tectonics has now shifted to the continents and to the history of the earth. There its impact is far from over.

The revolution had several stages. First there was the gathering of overwhelming evidence in favor of large horizontal movements of the earth's crust. Then came models that explained the crustal generation and crustal consumption taking place at boundaries later recognized as those of plates. These, in turn, provided us with a geometry for plate motions resting on the assumption that plates are not internally deformed. This assumption, almost an axiom, not required by the geological evidence, but necessary if we are to deal with plate motions by rigorous geometric methods, has enabled us to reconstruct the geography of the past 150 million years with impressive exactness. The true breakthrough, however, the real revolutionary moment, came with the simultaneous realization by several investigators that plates are large and thick and that continents and ocean basins are secondary features superimposed upon them. Only so has continental drift become acceptable; only so does the evidence for horizontal movement make sense.

Naturally, alternatives should be and were considered. Many suggestions were downright silly, but a much debated one was the expanding-earth hypothesis, still defended strongly by S. Warren Carey of Tasmania. An earth, small 250 million years ago, when the continents were one, and having expanded since then, thereby fragmenting and scattering them, makes some sense at first sight. It does not, of course, need subduction, nor does it account for the compression of the crust that builds mountains, but modifications can accommodate such things. Much more seriously, however, it cannot be reconciled with the dispersed continents of the Paleozoic and with the well-substantiated events that welded them together into Pangaea. Other conscientious objectors, unable to generate explanations better than plate tectonics, yet reluctant to accept it, have served their science well by forcing the big thinkers to maintain standards of critical scrutiny and intellectual honesty that might otherwise have been temporarily relaxed.

Yet, after 20 years, the edifice is beginning to show a few cracks, the paint is peeling, and it seems in danger of losing its simplicity and elegance. Violations of what the late Norman D. Watkins used to call the "principle of minimum astonishment" are becoming quite common. Attempts to reconstruct the plate configurations of the past 100 million years by strict geometric methods have run into serious difficulties that demand the invention of undocumented plates, or abandonment of the rule that plates are not deformed internally. Even the lithosphere is constantly tinkered with, some seeing the continental lithosphere as very much thicker than the oceanic one, whereas others peel its bottom off or speak of subcrustal erosion that thins the lithosphere in subduction zones. These views are taking the direction of restoring a fundamental difference between continents and ocean basins, a concept so fruitfully abolished by plate tectonics, and nibbling away at the integrity and permanence of the continental crust. Finally, we have had to call repeatedly for special events and ad hoc assumptions to deal with the episodic intensifications of orogenies that do not follow from the stately progress of crustal creation and destruction. The spreading edge has thus far remained rather simple, but there, also, some recent work is beginning to call for greater complexity.

Such developments are not really surprising to geologists, who learned long ago that the earth is always more complicated than it first appears. They are, however, disconcerting to others who prefer their world to be predictable, and especially to some plate-tectonics enthusiasts who insist that the entire world and all of its history must fit precisely into the new scheme, and never mind that this scheme rests on a mere 150 million years of observable seafloor-spreading history. Attempts to accommodate the growing number of difficulties have loosened some of the useful constraints that plate-tectonic models once possessed. As plates and plate boundaries multiply, small platelets appear and disappear for little reason other than that without them a postulated set of plate movements is not feasible. Reconstructions of plate history have become much more uncertain, diverse, and above all idiosyncratic.

Perhaps the time has come to acknowledge that there may be a limit to the usefulness of the geometric plate model and that we are rapidly approaching it. Applied on a global scale, plate tectonics has clarified much that was obscure, but it has unquestionably worked much better in the oceans, for which it was invented, than on the continents and continental margins, where it is causing some trouble. It is not clear whether this is because the continental lithosphere is more heterogeneous or whether it remembers its long and complex history in its tectonic responses. Whatever the cause, fitting orogenies in the standard scheme is beginning to produce so many ad hoc assumptions that we are rapidly returning to the prerevolutionary days when each corner of the earth had its own history, only loosely tied to others by general principles.

If this view is not too pessimistic, the way to proceed is to return to fundamentals. Plate tectonics and continental drift are often seen as synonymous, bound together as "the new global tectonics." That is not really so. Our acceptance of continental drift constituted the great advance that has allowed so many breakthroughs to be made, whereas plate tectonics is merely one way in which continents might move. It may not be the complete model, perhaps not even the right one, and we should abandon the idea that we possess a "global" tectonics at this time. If the geometry of plate motions will help us no further, we ought to pursue with new vigor the study of the processes that drive them across the surface of the earth: plate dynamics, a subject that has seen relatively little progress in recent years. It is interesting that Wegener's principal weakness, his inability to propose a plausible driving mechanism for continental drift, is also that of plate tectonics, but has not hindered its acceptance, because the evidence for horizontal motion is so strong.

The evidence for vertical movements, which are not demanded by plate tectonics, is not trivial either, as we shall see in the next chapter. It suggests to us that we do not yet fully understand how the earth works, and hence that the model is inadequate. Whether it is the push of upwelling magma at a divergent plate edge, the pull of the hanging slab in subduction, or convection deep in the mantle that drives the mobile earth, it seems that the ball is in the court of geophysics. It can, with some reason, be argued that we must first understand how plates move before we can hope to grasp why they do so, but we seem to have reached a limit in what we are able to learn from the behavior of plates at the surface. It might, therefore, be time to focus more on the geophysics of the deeper parts of our planet by various means, some not yet invented. Once we acquire a better understanding of what drives the plates, we shall be ready for another attempt to deal with how they behave. In the interim, and it may be a long interim, we can expect the simple image of the early, heady days gradually to lose its clarity.

These defects notwithstanding, plate tectonics has proved capable of settling many ancient controversies and casting light on numerous issues, including the most fundamental of all: the evolution of life. It has even been of some small help in such practical matters as finding oil and minerals. It seems improbable that we should ever return to fixed continents, but few, indeed, are the scientific theories that survive time intact. Our concepts of the dynamics of the earth and the behavior of its surface are sure to change, but the continents will continue to drift.

PERSPECTIVE

Like many of its kind, the geological revolution was slow in coming and not particularly novel in its most memorable achievement: the acceptance of continental drift. Yet the insights gained from this acceptance have been far-reaching, and we still have not fully assessed its consequences. It now seems amazing how we managed so long with a world of fixed continents.

Oceans and continents are fundamentally different. New ocean crust, formed on mid-ocean ridges, is destroyed by subduction, thereby ensuring that no record of the ancient oceans remains, but continents endure forever. Mountains rise as oceanic plates collide, and subduction explains their features, though not completely. Episodicity is evident in mountain building, but it does not seem to be demanded by the often steady subduction of oceanic crust. The collision of continents is also a more complex process than plate tectonics suggests.

Thus, notwithstanding the successes of plate tectonics, two major issues foreshadow its ultimate limits as a theory: the apparent pulse-like behavior of at least some of the earth's internal processes, and the growing need for ad hoc hypotheses, a sign that the theory may not be able to keep its promise to be the new "global" tectonics.

If it therefore seems unlikely that plate tectonics will be the last of the great geological theories, continental drift is sure to stay. The reconstructions of past arrangements of land and sea, which the theory permits us to make, will occupy geologists for decades to come. Some of their consequences will be explored in the next few chapters.

There is, however, another and more neglected aspect to plate tectonics. In the form in which it was conceived and is most often applied, it is a phenomenon of the later years of the earth. There is no surety, not even a likelihood, that these last half billion years properly represent the dynamic behavior of our planet during the 4 billion years that went before. We shall encounter this problem as a major obstacle when, several chapters hence, we consider the childhood and adolescence of the world.

FOR FURTHER READING

It is not surprising that there is a vast literature on the geological revolution, scientific as well as popular. I must choose here, and my choices are not likely to satisfy everyone. Nevertheless, I believe that the following works provide a reasonable selection for further pursuit of the subject.

ON THE HISTORY OF THE GEOLOGICAL REVOLUTION

Hallam, A., *A Revolution in the Earth Sciences,* Oxford University Press, 1973, 127 pp.
Sullivan, W., *Continents in Motion, the New Earth Debate,* McGraw-Hill, New York, 1974, 399 pp.

ON PLATE TECTONICS

The following book combines a comprehensive but pleasant overview of the theory of plate tectonics and the relevant evidence with an enchanting personal touch:
Uyeda, S., *The New View of the Earth,* Freeman, San Francisco, 1978, 217 pp.

Excellent at the beginning college level, though slightly older, is:
Wyllie, P. J., *The Way the Earth Works,* Wiley, New York, 1976, 296 pp.

The magazine *Scientific American* has carried numerous articles chronicling advances and new insights gained in the last quarter century. The reader should be warned, however, that the level of difficulty of this magazine, supposedly intended for the intelligent lay audience, has risen to the point that even professionals have to strain.

In a more reportorial style, the two journals *Nature* and *Science* carry reviews of progress in the earth sciences that often are both excellent and accessible.

Ancient oceans, ancient climates

If we are to believe Wegener's hypothesis,
we must forget everything that has been learned
in the past 70 years and start all over again.

R. T. Chamberlin,
1926 meeting of the American
Association of Petroleum Geologists

It is the clarity of the radically new and
absolutely simple idea which catches us as if it
were an intuition. By it, problems we considered
insoluble will resolve themselves or, rather, be
dissolved, either to disappear definitively or
to present themselves in a new way.

Henri Bergson, *The Creative Mind*

CONTINENTAL DRIFT AND ANCIENT ENVIRONMENTS

Attempts to describe the landscapes and seascapes of the past are as old as geology itself. One component of this arcane art, paleoclimatology, has grown into a separate science with its own practitioners; none other than Alfred Wegener wrote, with his father-in-law, W. Koppen, an early work on this subject. Today it seems curious that the paleoclimatologists who came after him managed to erect climate histories that for decades satisfied everyone, even though they did not consider continental drift. Oddly, also, reconstructing the ancient oceans has not been a popular pastime, even though oceans and climates are closely coupled.

Reconstruction of the environmental history of the earth rests on the nature of the sediments and on the fossils contained in them, with liberal additions of climatic and environmental theory, the whole seasoned with courage and creative imagination. Neither the sedimentary nor the fossil record can truly be regarded as what physical scientists are pleased to call "hard evidence." Fossils, especially, are prone to lead us on the treacherous path of circular reasoning. For the recent past we can infer their preferred environments by comparing fossils with their close living relatives, and we are permitted some confidence in the interpretation. But what about a group that has been extinct 100 million years or so? How shall we deduce its environmental dependences and determine what its members required from nature, except in the broadest of terms: marine or terrestrial, shallow or deep, forest or desert? It is hardly surprising that we have, unknowingly and sometimes even knowingly, often used the environment deduced from the presence of certain fossils to conclude why they lived there. This dubious procedure becomes doubly undesirable when we inquire what influence a changing environment might have had on the evolution of life.

Obviously I have put this a little sharply to make my point. Nonetheless, it is the beauty of the continental-drift theory, and a solace to earth historians, that it enables us to reconstruct the geography of the past without immediate recourse to sediments or fossils. This, combined with the principles of oceanography and climatology, leads us to infer many things about the physical environment independently and with a rigor hitherto not possible. This permits us to trace the interactions between environment and life with greater confidence and diminished concern about circular reasoning.

All this is still in its infancy. The paleoclimatologist and the new paleoceanographer have just begun to hone their tools, to formulate their questions, and to make the first attempts at solving them. Even these hesitant steps, however, illustrate the enormous impact of plate tectonics on our understanding of the earth in all its aspects.

10

The sea comes in, the sea goes out

On an open coast, twice daily the water's edge moves in, then out; the land expands slightly, then shrinks again. During an ice age, several times each 100,000 years the silver line between land and sea travels a much greater distance. Over the eons, the shore is never still, the land growing and diminishing with the tides of time. Much of geology is about this continuous movement of the coast; much of the geological record exists only because of it.

For many years we have contemplated this spectacle and wondered whether we saw the land rising or the level of the sea falling. Direct information can be found on geological maps depicting ancient shorelines; from them we obtain the proportions of land and sea as they have varied with time. Many such estimates exist (Figure 10.1), each a bit different from the others, and all marred by a lack of data in unexplored regions, by the ravages of time, and by the uncertainties of the timescale. Nevertheless, a coherent picture has emerged from a century of effort. It shows a gradual emergence of the continents over the last 600 million years, interrupted by a sharp drop in sealevel during the days of Pangaea and a big rise in the later Mesozoic. There have also been many shorter and less well documented fluctuations.

TRANSGRESSIONS AND REGRESSIONS

The sea covers and uncovers the land either because sealevel itself changes, inevitably a worldwide event, or because the land sinks or rises. No reference point exists to calibrate the vertical motions of past seas; they must be inferred from shifts in sedimentation patterns. A rise or fall of land or sea displaces the shore and leaves a characteristic trail of coastal deposits. We learn about vertical sealevel changes from the horizontal migrations of the shore: *transgression* as the sea en-

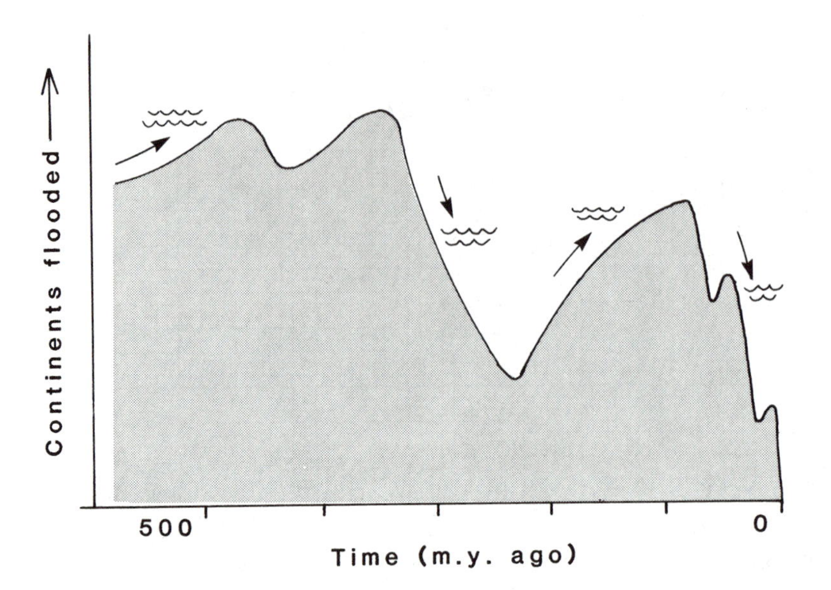

Figure 10.1. This curve of Phanerozoic sealevel change is based on the amount of land covered with marine sediments of various ages. During most of the Paleozoic the sea was high; then it fell sharply in the early Mesozoic. Another high occurred in the Cretaceous, but the Cenozoic was mainly a time of falling sealevel. The present stand is the deepest ever.

croaches, and *regression* as it withdraws. On a nearly level coastal plain, a small rise of the sea may move the beach a long way, but on a steep coast the same rise will hardly displace it at all. Therefore, unless we know the slope of the seafloor and the adjacent land, we cannot estimate the vertical change in sealevel merely by knowing the horizontal distance over which the shore migrates.

It is only natural that one should intuitively equate transgression with a rising sea or a sinking land, and regression with the reverse. We also easily acknowledge that, even when sealevel remains stationary, erosion may cause the coast to recede, or a delta may move it seaward. The point is not trivial: Sedimentation is a powerful force, and we must take it into account when we consider whether a given change in the level of land or sea will result in a transgression or a regression. Stated formally:

if sealevel rises or the land sinks
- faster than deposition = transgression
- slower than deposition = regression

The inverse is, of course, true for erosion, but erosion is seldom important, because waves do not cut deep and soon create a shallow platform that breaks their strength and thus protects the coast. Even though roads and houses do get swept into the sea, the loss of land by erosion is invariably modest.

Local transgressions and regressions are tediously common in the geological record, but those that were global in extent raise very interesting questions. We have already encountered the eustatic sealevel changes of the Pleistocene Ice Age, but they are by no means the only ones that had a worldwide extent. Superimposed on the general trend of Figure 10.1 are many fluctuations in the boundaries between land and sea, of which some appear to be synchronous across the world and are therefore the result of eustatic sealevel changes, though not obviously connected with glacials and interglacials.

During the Cretaceous, the inundation of the continents reached an extent not seen since the early Paleozoic, and less than 60 percent of the present land area remained above water. In North America, successive transgressions created a shallow sea from the Canadian Arctic to the Gulf of Mexico (Figure 10.2). In the west, this sea lapped onto a mountain range studded with volcanoes, whereas in the east it was bordered by a lower, more gentle land. Abundant sediments, dumped in from both sides, lined the shores with lagoons, swamps, deltas, and beaches, much like those of the present Gulf Coast, and provided a favored territory for the last dinosaurs. These coastal wedges carry the imprint of numerous small transgressions and regressions of the high Cretaceous sea. Similarly, Europe and Africa were covered with wide, shallow seas that have no counterpart in our time.

Of the cycles of transgressions and regressions that composed the Cretaceous flood, each lasting from 2 to 4 million years, at least five were probably worldwide. The highest Cretaceous sea may have been no more than 200 m above present sealevel, and the entire range of sealevels probably was only about 200–300 m. Today, a similar rise would inundate less than half the area that was flooded in the Cretaceous, because our continents stand high above the sea, whereas the Mesozoic lands were low and flat.

The Cretaceous transgressions and regressions were not unique. The Jurassic, too, had many such wiggles, superimposed on a slow general rise toward the Cretaceous maximum. Several of these were probably also eustatic and produced vast shallow seas. Other transgressions and regressions marked the Cenozoic, even before the Ice Age imposed its own rhythm on the level of the sea.

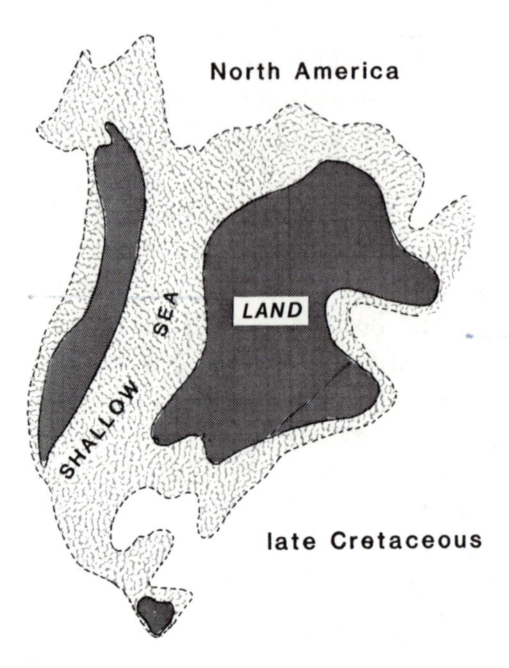

Figure 10.2. This map of late Cretaceous North America shows how large a difference even a small sealevel change of a few hundred meters makes in the configuration of land and sea.

THE CONTINENTAL MARGIN AS A SEALEVEL GAUGE

Whereas changes in sealevel with time can be inferred from global maps of ancient shorelines, it would be nice to study a more complete series of transgressions and regressions without having to hunt far and wide for the pieces of the puzzle. Fortunately, what we seek exists in the sediments of the passive continental margins. There, as the shelves slowly subside under the weight of the sediments accumulating with time, a record is kept of transgressions and regressions that is far more complete than that of the deposits of shallow seas now stranded on land.

We can read this record because of an exploration technique based on the old method of echosounding at sea. If we send a sound signal to bounce off the seafloor and measure the time until its return, we can calculate the water depth, provided we know the velocity of sound in seawater. If we use a more powerful source of energy with a lower frequency, part of the energy will penetrate the bottom and be reflected from boundaries between sediment layers. On this principle rests the method of marine seismic profiling (Figure 10.3), first devel-

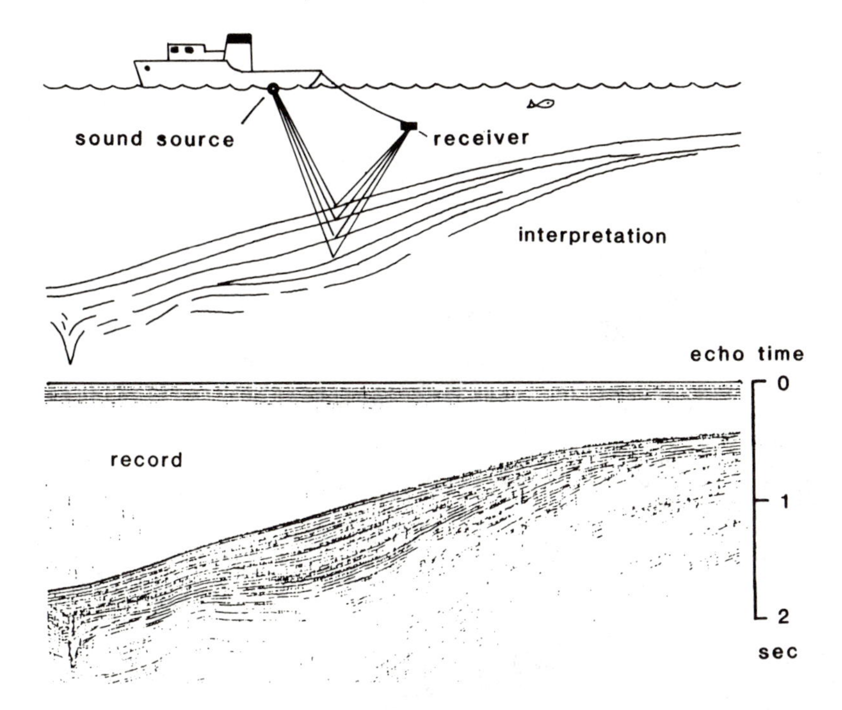

Figure 10.3. The deposits beneath the seafloor can be studied with seismic profiling, sending out sound pulses that are reflected by the seafloor and by various reflecting surfaces in the underlying strata. The record (bottom) shows return times plotted against the travel time of the ship and can be interpreted (top) almost as if it were a geological cross section.

oped by oceanographers and later perfected by the oil industry. A seismic profile resembles a geological cross section, but although it can, with caution, be so used, it is not. The horizontal dimension records time as the ship travels, not distance, and the vertical scale times the echo rather than measuring the depth. We can obtain the distance if we keep track of the ship's position, but to convert echo time into depth we must know the velocity of sound in rocks. That tends to be difficult and expensive, and we often use estimates instead. If we wish to know the nature and age of the reflectors, we must, of course, drill and core.

Most echoes return from surfaces that were once the bottom of the sea; each reflector therefore usually represents a single instant in time, a time line. Consequently, the course of past transgressions and regressions can be inferred from seismic profiles (Figure 10.4). Sometimes the character of a reflector tells us something about the sediments on that old seafloor; coastal deposits look different on a seismic profile than those laid down in deep water.

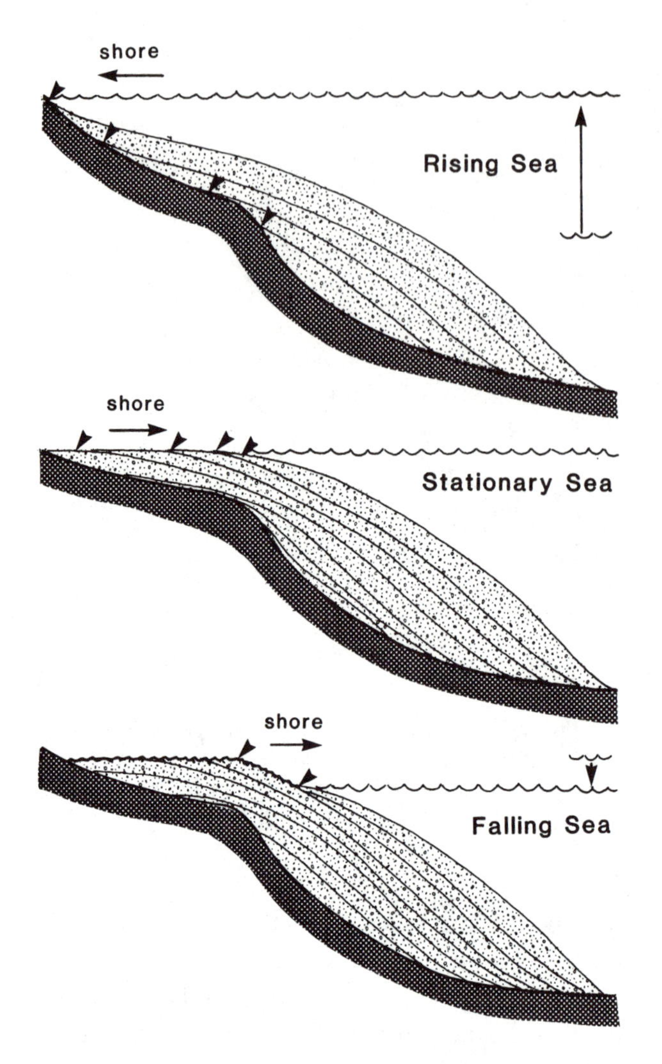

Figure 10.4. The reflectors of a seismic profile usually are ancient seafloors. The highest point of each reflector (arrows) will therefore mark sealevel at the time it was formed. When sealevel rose, each termination formed a little higher than its predecessor (top); when it was stationary, subsequent reflectors were displaced seaward, with their tops at a constant level (center). A falling sealevel caused each termination to be lower than the previous one (bottom). Seismic reflection records from continental margins thus enable us to determine sealevel changes.

Seismic reflection studies of continental margins have shown that they often consist of the deposits of numerous cycles of transgressions and regressions stacked on top of and against each other. If we knew the ages of these cycles, we could unravel the history of local sealevel changes, and if we had a large and widely distributed collection of local histories, correlation would tell us which of the cycles were worldwide. A eustatic origin rather than a rise or fall of the land would then be most probable. Naturally, such information is not easily come by; the oil companies have by far the best seismic reflection data and are generally the only ones who, through drilling, have acquired age information as well. The first and thus far only global compilation of sealevel changes based on seismic reflection studies of continental margins has come from the Exxon Corporation.

The primary trend of the Exxon sealevel curve (Figure 10.5) is much like those obtained in a more conventional manner, with high sealevels early in the Paleozoic and during the Cretaceous, but without the gradual emergence of the continents that appears in Figure 10.1. Superimposed on this general trend are large transgressions and regressions, supercycles, which show a remarkable episodicity, an earth rhythm, a "pulse of the earth." Like the teeth of a somewhat abused sawblade, each supercycle rises rapidly, slows to a standstill, then drops precipitously, almost instantaneously. In fact, the authors of the curve claim that they have never seen a gradual sealevel drop, in contrast to the Cretaceous and Jurassic outcrops on land, which show much more symmetric behavior. There were 14 supercycles, each lasting from 10 to 80 million years in the Paleozoic and Mesozoic, but much shorter in the Cenozoic. The supercycles are themselves composed of cycles, one to six for each, and more than 80 in all. Cycles may be shorter than supercycles, but they are quite like them in shape and in amplitude.

CAUSES OF EUSTATIC SEALEVEL CHANGE

Unfortunately, although the method on which the Exxon curve is based has been published and seems to be sound, few actual data, especially those for age and synchroneity, have been released. That may be understandable for economic reasons, but until that sort of proof is forthcoming, the Exxon curve is no more than a rather intriguing and somewhat startling proposal. Still, whether it is a complete or only a partial picture of true sealevel change, it presents a major problem. Even if we ignore the details for the moment, we have to deal with supercycles, because their existence has been established beyond doubt on independent evidence from the Jurassic or the Cretaceous. Why has sealevel risen and fallen episodically throughout much, perhaps all, of geological history? Can something be learned from the durations of the oscillations, from the rates of rise and fall, or from their remarkably small and uniform vertical range?

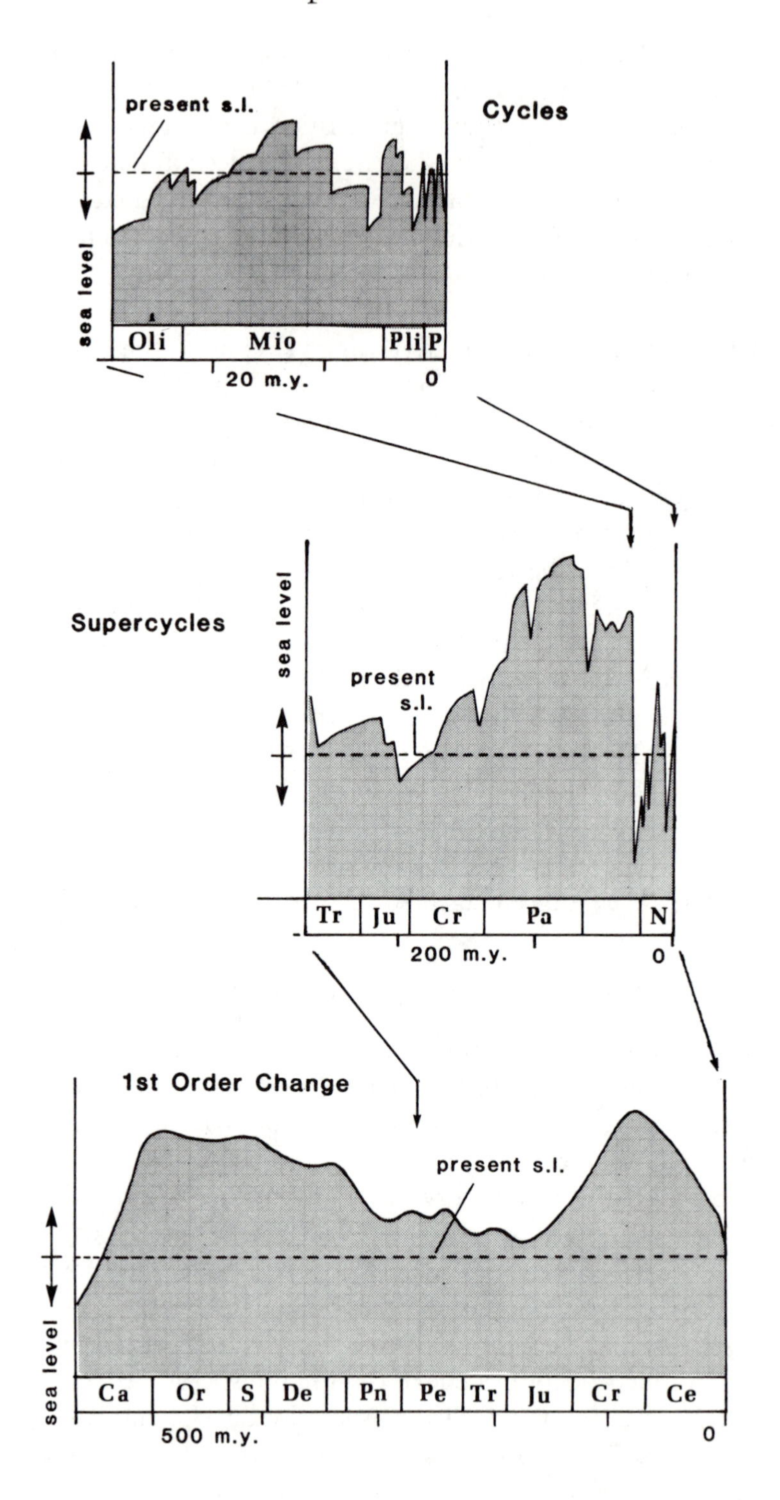
present s.l.
Cycles
sea level
Oli
Mio
Pli
P
20 m.y.
0
Supercycles
sea level
present
s.l.
Tr
Ju
Cr
Pa
N
200 m.y.
0
1st Order Change
present s.l.
sea level
Ca
Or
S
De
Pn
Pe
Tr
Ju
Cr
Ce
500 m.y.
0

The level of the sea can be changed by altering the volume of water in the oceans or by changing the sizes of the containers, the ocean basins. Each can be accomplished in several ways, some capable of causing a large change in sealevel, others trivial in their consequences.

Most of the water on earth was released from the interior very early (see Chapter 14), but smaller amounts are still being added by volcanoes and hotsprings, although much of that is merely recycled groundwater. This ongoing addition is much too small to explain a sealevel rise of about 1 cm per 1,000 years, the average for both supercycles and cycles. Moreover, there is no known way to dispose of the excess when the time comes for the sea to fall once more. Rapid changes in both directions can, of course, be achieved by removing water from the sea, storing it elsewhere for a while, and then releasing it again. Glaciations and deglaciations are the only examples of sufficient scale that come to mind, because the capacities of streams, lakes, and groundwater are much too small. The same is true for other processes, such as a change in atmospheric moisture content or ocean temperature, which can at most alter sealevel by centimeters or meters. The rate of sealevel change for glaciations and deglaciations is measured in meters per 1,000 years, much too fast, and we are quite certain that there were no ice ages during the Mesozoic. Clearly, nonglacial eustatic changes cannot be explained by changing the volume of water.

What about the containers, the ocean basins? Can we change their volume enough to produce the right effect? Some proposals can be easily discarded: Closing and evaporating some small ocean like the Mediterranean, something that indeed happened about 6 million years ago, would raise the global sealevel by no more than 15 m. Might we fill the oceans partly with continental sediments, which are in abundant supply? This would indeed raise sealevel, but too slowly and not enough, and timely speedups of subduction would be required to dispose of the sediment layer each time the sea was to fall again. That seems far-fetched.

There is a different path, though it also involves plate tectonics. The floor of the ocean is not flat like a dish but convex upward, like the bottom of a wine bottle, because of the mid-ocean ridge. The ridge crest, where the crust is young and hot, is shallow, but as the crust

Figure 10.5. Sealevel changes determined on seismic reflection profiles from many parts of the world show change on three scales. The first-order change (bottom) resembles that of Figure 10.1, although with less extreme range. Superimposed are variations of similar amplitude but lasting only 20–50 million years, called supercycles (center). Supercycles are composed of cycles, again of similar amplitude but of much shorter duration (top).

moves away and ages, it cools, contracts, and therefore subsides. It is a simple process, and the increase in depth with age is proportional to the square root of the age of the crust (Figure 10.6). When spreading is rapid, young and hence shallow ocean crust is found farther from the ridge axis than when spreading is slow. Therefore, the volume of the mid-ocean ridge depends on its rate of spreading. This volume will be increased, thereby reducing the capacity of the ocean basin and raising sealevel, if spreading is accelerated, whereas sealevel will drop if spreading slows down. Obviously, the same effect can be achieved by increasing or decreasing the number or the total length of the mid-ocean ridges.

Will this be adequate? If we take a ridge 10,000 km long, about one-eighth of the present system, and abruptly change the spreading rate from 2 cm to 6 cm per year, both perfectly reasonable values, the new equilibrium profile will be complete in 70 million years (Figure 10.6), no sooner, no later. The sea will rise, rapidly at first, then more slowly, and reach a new height of 120 m above the old level, certainly in the right ball park. If we decrease the spreading rate back to 2 cm per year, sealevel will recede to its former position in another 70 million years.

Are such modest changes in the rate of spreading or in the number and length of mid-ocean ridges in accord with observations? The answer is yes: A decrease in the average spreading rate has occurred since the Cretaceous, and its magnitude is sufficient to account for the overall drop in sealevel during the early Cenozoic. We also know that some mid-ocean ridges have disappeared during that interval and others have formed, but our data lack the precision necessary to calculate what the accompanying volume changes of the ocean basins might have been.

If this appears to account for the primary, very long term change, the supercycles and cycles of the Exxon sealevel curve are not so easily explained in this manner. For one thing, they tend to be brief compared with the many tens of millions of years that it takes for a spreading rate change to have full effect. For another, we have no evidence at all that there have been so many and such large changes in spreading rate or in the number or length of the mid-ocean ridges that they could account for 14 supercycles and more than 80 cycles. Still, we need not reject the hypothesis out of hand, because we have not yet examined the role of another variable: the rate of sedimentation.

We saw earlier that whether a shore moves landward or seaward depends on the balance between the rate of sealevel rise and the rate of

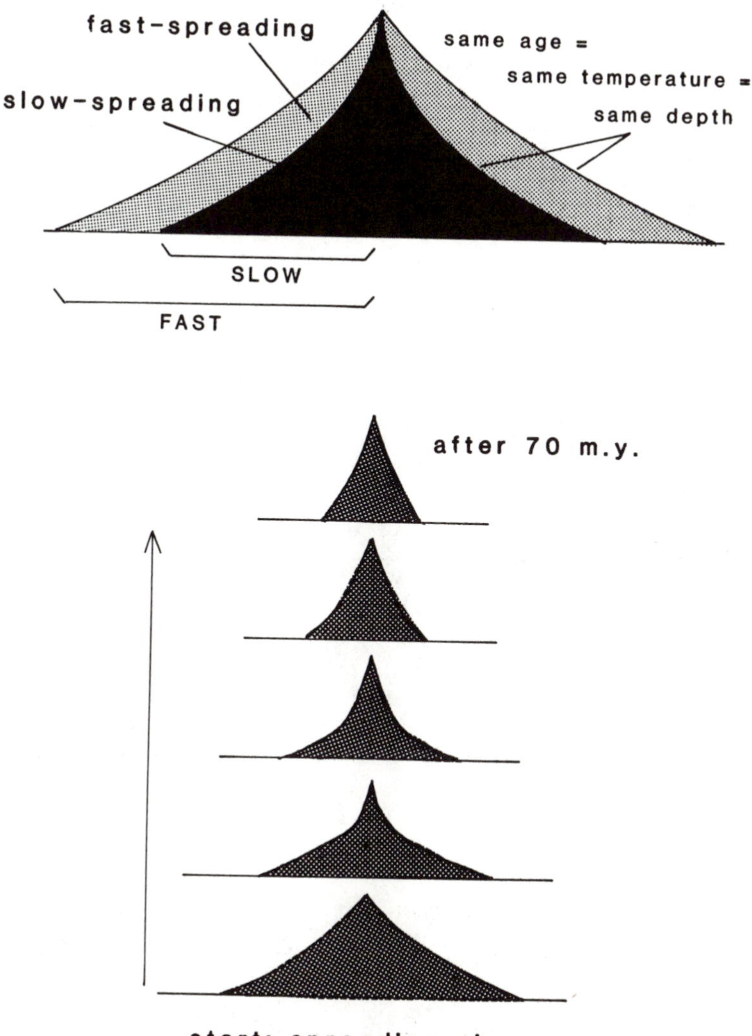

Figure 11.6. The cross-sectional area of a mid-ocean ridge depends on its spreading rate. The ridge is due to expansion by heating, and because cooling is a function of time, crustal segments of equal age have the same temperatures and therefore the same elevations. On a fast-spreading ridge, points of a given age are farther from the axis than on one that spreads more slowly. Consequently, such a fast ridge has a larger cross section and displaces more water (top). If one cuts the spreading rate to half, for example from 6 to 3 cm/year, the ridge profile will become narrower, taking 70 million years to reach equilibrium (bottom).

deposition. In the long run, the shore will be placed where the rate of sedimentation is equal to the rate of sealevel rise. If we increase the rate of rise of the sea, sedimentation will fail to keep pace, and the shore will migrate landward. If, on the other hand, the rise of the sea

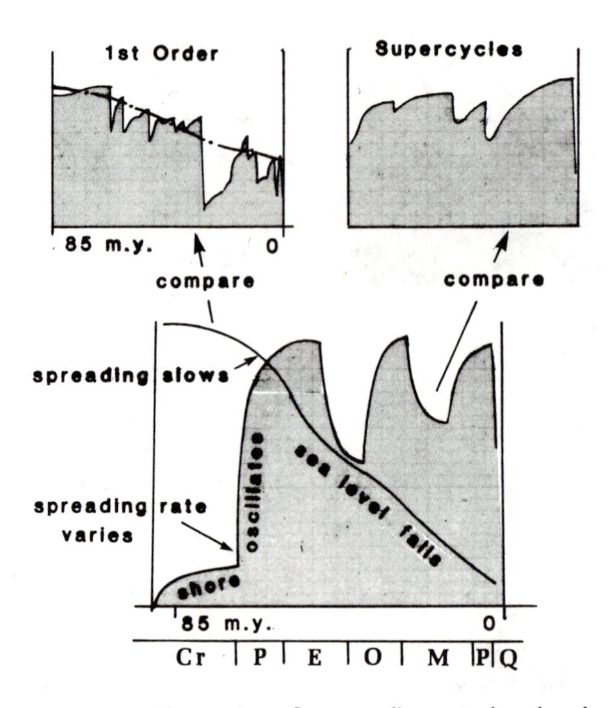

Figure 10.7. Changes in seafloor spreading rate alter the volume of the ocean basins and cause sealevel to rise or fall, thus producing a trend like the 1st order change eustatic sealevel curve (upper left). If the rate of spreading varies (without reversing) the interaction with sedimentation at the shore will produce transgressions and regressions that resemble the supercycles of the Exxon curve (upper right).

merely slows down, without reversing to a fall, too much sediment will arrive at the shore, and the coast will build seaward, thus starting a regression. A quantitative model can easily be computed (Figure 10.7); we find that small changes in the rate at which the sea is rising or falling will cause transgressions and regressions that exhibit the sawtooth shape of the Exxon curve, without requiring actual reversals of the rise or fall of the sea. Perhaps all we need is some variation in the rate of seafloor spreading, in combination with the effect of sedimentation rates, to explain complex sawtooth curves of sealevel changes.

Alas, we are a long way from being able to reproduce the Exxon curve accurately with known changes in seafloor spreading and sedimentation rates. In fact, it appears that we are at liberty to play with so many variables that we may never be able to prove that the idea does *not* work. The time seems right to abandon this line of inquiry and try another.

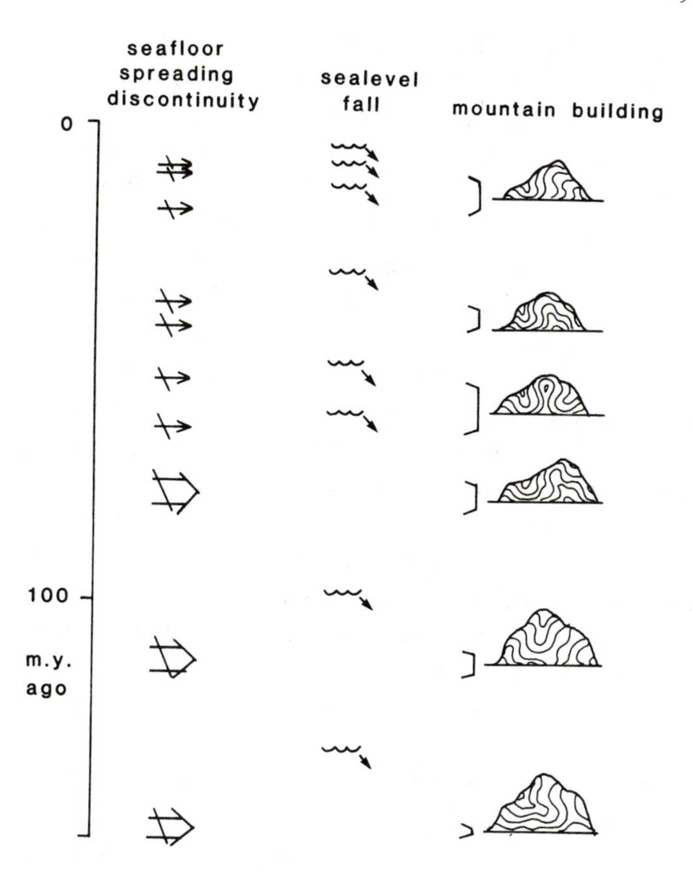

Figure 10.8. The record of the last 200 million years shows some correspondence between times when seafloor-spreading rates changed or sealevel fell and periods of mountain building. This might imply that all three have the same cause(s), except that the chronology is so flexible that one can suit the correlations to a large extent to one's preference.

VIEW FROM THE CRATON

Having tried, with promising but inconclusive results, to change the capacity of the container, is there perhaps something we can do about its rim? That is, Can we change the level or size of the continental blocks? The thought is an old one, inspired by the frequently made observation that many major regressions seem to coincide in time with orogenic phases. The folding and emergence of mountain ranges during plate collisions, the uplift of continents above a dying subduction zone, or the formation of a root where two continents collide, such as we noted for the Himalayas, offer various means to raise the continents up out of the ocean. Especially the last method is attractive, because roots of that kind eventually vanish again, so that the continents sink once more to accommodate a transgression.

There are other possibilities. Since its birth, the earth has gradually cooled, and that may have slowly thickened the continental crust. Thicker continents are more buoyant, and a gradual emergence would thus be expected, as, indeed, some of the evidence suggests. On the other hand, the process of crustal thickening, which is anyhow not universally accepted, cannot deal with transgressive-regressive cycles.

All continental cores, the *cratons,* contain basins with shallow marine deposits that display a long history of alternating submergence and emergence. The eustatic sealevel changes of the oceans are insufficient, or are improperly timed, to account for all of those, and vertical movements of the continents themselves seem necessary. It is curious that many, though certainly not all, episodes of subsidence appear to be synchronous and may coincide with plate-tectonic events such as spreading rate changes (Figure 10.8). This implies a global mechanism that causes plates not only to move horizontally, as plate tectonics proposes, but also to rise and sink. Should we perhaps consider a rhythmic flow in the asthenosphere, back and forth between continental and oceanic regions? Is the evidence strong enough (many think not) to force us to admit that the theory of plate tectonics is incomplete?

Before we get carried away, however, I note that the game of correlating orogenies and sealevel changes on a global basis has been played before and has never yielded much more than controversy in the long run. One should also remember that the durations of many of the events and the uncertainties of their timing are often of the same order as their spacing in time, so that anyone possessed of firm resolve can usually find a correlation that suits.

Be that as it may, the simple story of the ups and downs of the sea evidently leads to weighty questions about how the earth works. Not only are they far from being answered, there is not even unanimity about which questions are the right ones. Contemplating these issues has led us into a new realm of thought that we shall explore further in the following chapters: how to deal with major problems by thought experiments rather than by straightforward reasoning from sparse evidence to conclusion. How not to lose oneself in fascinating but ultimately useless creativity is a problem that will concern us more and more in the remainder of this book.

11

Searoad to an ice age

Continental drift alters climate and oceans in many ways, raising mountains, which cast rainshadows and make deserts, and enlarging or shrinking the continents. More important are the shifts of land masses to other latitudes and climate zones, and especially the opening and closing of seaways that control the circulation of ocean water. The ocean, like the atmosphere, is a great engine driven by solar energy, that carries heat from low to high latitudes. Ocean currents do this much more effectively than the atmosphere, because water holds more heat than air and holds it longer. Changes in ocean circulation induced by plate motions are therefore in principle capable of significantly altering the global climate. To examine 100 million years of ocean history, however, we must understand the ocean's present behavior. The detour is somewhat lengthy, but the reasons for it will be obvious later.

HOW THE OCEAN WORKS

We can think of the present ocean as a rather irregular bowl filled with two fluids: a thick layer of dense liquid at the bottom, a lighter thin one on top. Both, of course, are seawater; the density difference is due mainly to a sharp change in temperature at a shallow boundary called the *thermocline* (Figure 11.1); salinity plays only a minor part. Near the surface, to as great a depth as waves can stir, the water is usually quite uniformly warm. Below the thermocline the water is cold, cooling only a little more toward the bottom. The density difference between the two layers is small, less than 0.05 g/cm^3, but that is enough to make it difficult to raise deep water to the surface. The stratification of the present ocean is therefore quite stable, and that severely limits its fertility.

The biological productivity of the ocean, often regarded as virtually limitless and a guarantee of a vast and mostly untapped supply of food,

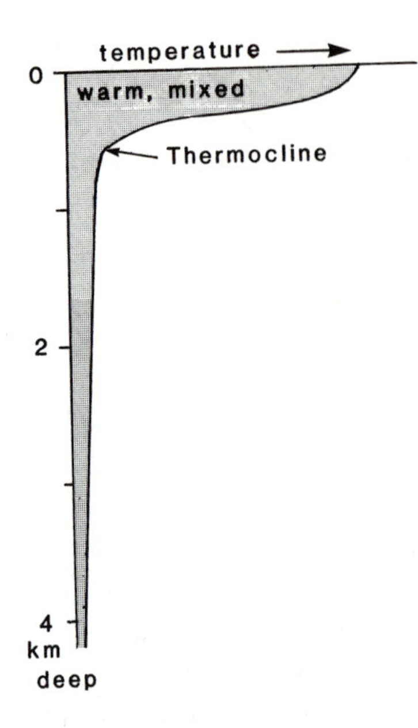

Figure 11.1. The upper part of the ocean is warmed by the sun and mixed by waves and currents. A sharp drop in temperature, the thermocline, separates it from the cold abyss. The water above the thermocline is much less dense than that below, creating a stable stratification that impedes the exchange of water between the surface and the deep.

is actually just marginally greater than that of a desert. In only a few areas, mostly in oceanic currents and along some coasts, is the ocean really fertile. An astonishing 90 percent of all organic matter in the ocean is produced over less than 10 percent of its surface. It is produced by phytoplankton, which need not only sunlight, water, and carbon dioxide for photosynthesis but also nutrients such as nitrogen and phosphorus. These nutrients are dissolved in seawater, but in such small amounts that one healthy plankton bloom can exhaust the supply in hours or days. Dead organisms sink and, in decaying, give up their nutrients, but most of this decay takes place below the thermocline. Unless deep water can be recycled to the surface, these nutrients are lost to the cycle of life that takes place at shallow depths in the sun.

Turbulence in currents can recycle some of the deeper water, but the truly important process is one called *upwelling* (Figure 11.2). Upwelling has various causes, but it is especially common where a coast lies to the left of a current in the Northern Hemisphere, or to its right south of the equator. Winds blowing from the land also cause upwelling. The rising water is also cold, and where upwelling occurs along a warm coast, off California for example, it is accompanied by frequent and dense coastal fogs. Upwelling thus provides the fertility that is respon-

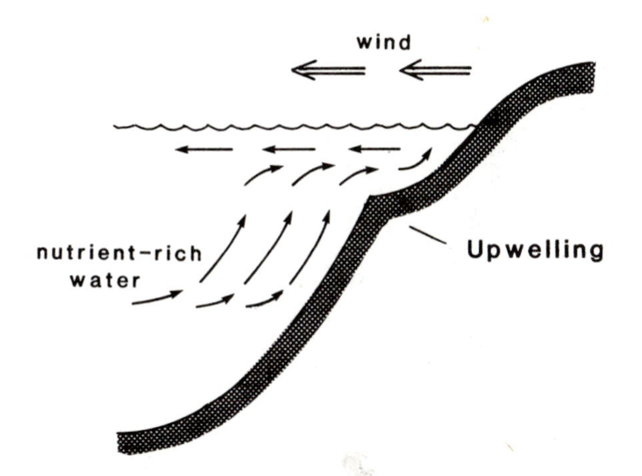

Figure 11.2. The stable stratification of the ocean limits the supply of nutrients to the surface. Currents stir the sea to limited depth and so fertilize the surface zone, but the principal agent in bringing up nutrients from the deep is upwelling. A seaward wind, or a current along the coast deflected seaward by the Coriolis force, drives the surface water away from the shore. It is replaced by water welling up from a depth between a few hundred and a thousand meters.

sible for many of the major fisheries of the world, such as those of Peru and Ecuador, Japan, and West Africa.

The main ocean currents are driven by the planetary winds (see Chapter 3), the trades on either side of the equator and the westerlies at about 45° latitude, and are, like those winds, modified by the Coriolis force (see Chapter 3). A simple ocean extending from pole to pole and bordered by land would have equatorial currents driven westward on either side of the equator by the trades (Figure 11.3). At the western shore, part of each current would be reflected back along the equator, and the remainder would form large gyres in the Northern and Southern Hemispheres under the influence of the westerlies. Another gyre, this one flowing counterclockwise, would exist in each subpolar zone. The currents of the real ocean are, of course, somewhat different, because our continents are not so simply arranged, but the basic pattern is the same (Figure 11.4).

The equatorial currents are heated by the sun, and the water is warmest where the equatorial return flow meets the east coast. The two gyres, on the other hand, bring back water cooled during its voyage in higher latitudes, so that the temperature distribution at the surface is asymmetric (Figure 11.3). A similar asymmetry exists in the salinity, which is low in the eastern equatorial region because of the tropical rainfall there, but increases westward as the warming water

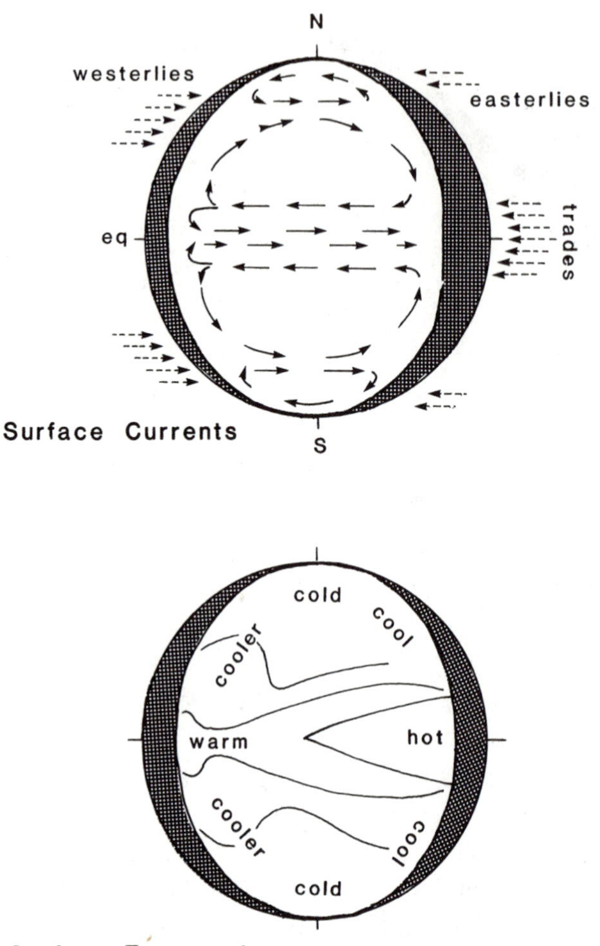

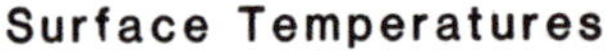

Figure 11.3. The surface currents of a simple ocean surrounded by land (top) are driven by the planetary winds. Two equatorial currents blown westward by the trades turn back when they encounter the opposite shore, producing an equatorial countercurrent and two large gyres in the Northern and Southern Hemispheres. The surface temperature distribution follows the current pattern (bottom), with the warmest waters at the end of the equatorial countercurrent.

evaporates. It is highest in the subtropical centers of the gyres, where rain rarely falls.

This is only part of the story, although the most visible part. By far the larger volume of water lies below the thermocline and below the wind-driven surface circulation. This deep water is not stagnant; it flows in predictable patterns driven not by the wind but by density differences that cause heavier water to sink and lighter water to rise and take its place. In the present ocean, density differences are due

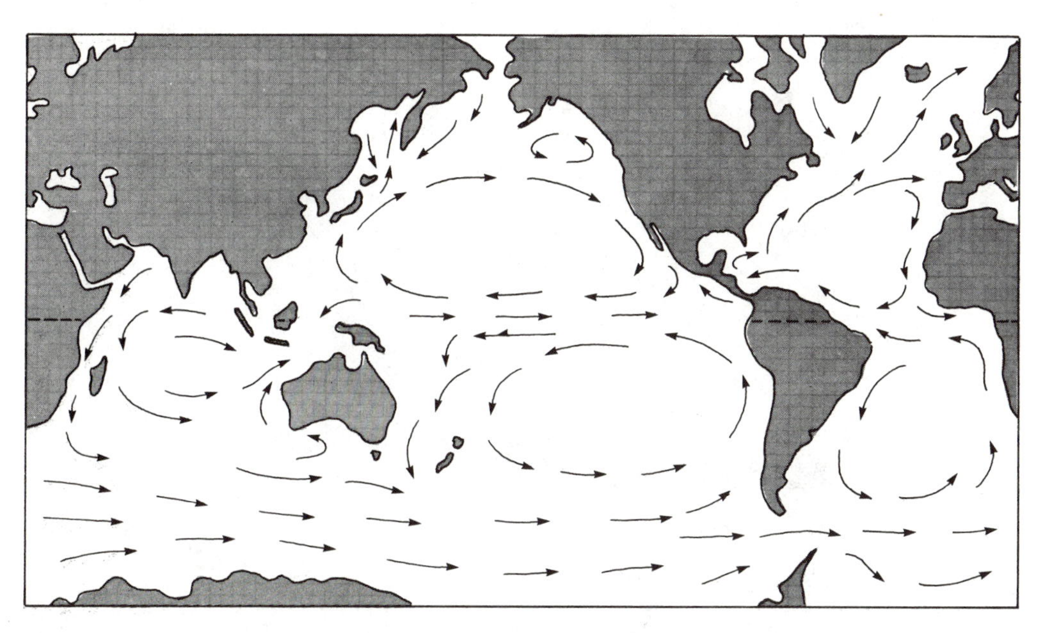

Figure 11.4. The current pattern of the ideal ocean (Figure 11.3) corresponds well with that of the topographically more complex real ones. The map brings out the importance of seaways connecting oceans with each other, such as the passages through the maze of the Indonesian Archipelago, or the Drake Passage between South America and Antarctica.

mainly to temperature, which has a wider range and a larger effect than the salinity. Unlike freshwater, which reaches its maximum density at 4°C, seawater is densest at its freezing point, which depends on the salinity but always lies below 0°C. Around Antarctica, very cold water flows in the circum-Antarctic current (Figure 11.5). There, ice forms in the southern winter; as it freezes, it excludes most of the salt. The cold water below the ice increases in density because of the added salt. Driven by the density difference, it travels north along the bottom as abyssal currents in all oceans. Slightly warmer water of ordinary salinity rises from intermediate depths in a vigorous upwelling that makes the circum-Antarctic region one of the most fertile oceans in the world. In spring and summer, ice floes and icebergs break off from the icecap, float north, and melt along the way. Their meltwater is hardly salty, but so cold that it also sinks and travels north at an intermediate depth.

In the North Atlantic, conditions are similar, but on a much smaller scale. Cold, saline water is formed near Labrador and Greenland by the freezing of sea ice. This water is neither as dense nor as copious as the Antarctic bottom water, but it possesses enough density contrast to traverse the Atlantic southward at a depth of about 2 km (Figure 11.5)

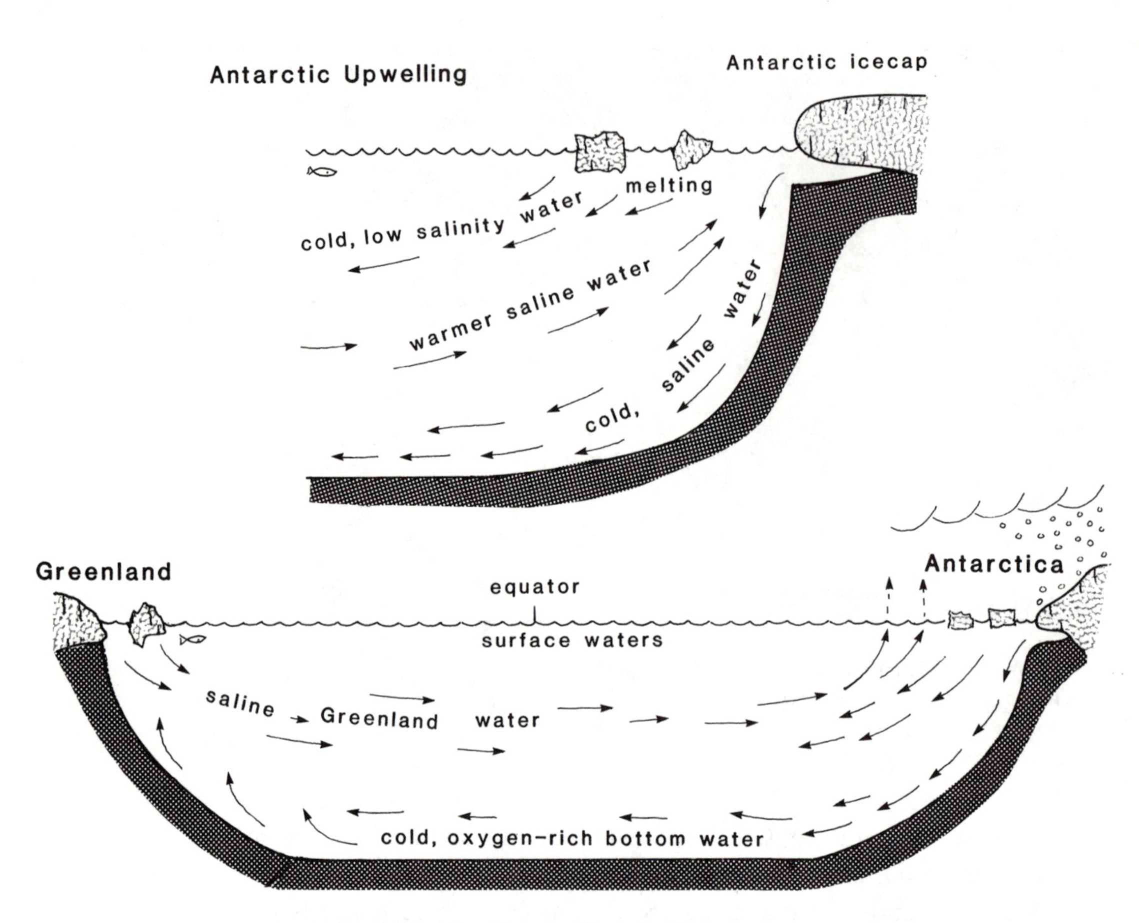

Figure 11.5. The deep circulation is driven by differences in temperature and salinity that produce density contrasts. As a result of the freezing of sea ice, very cold and saline water forms at the edge of Antarctica. It sinks and travels north in all oceans as an abyssal current. Farther away, icebergs melt, and another water mass is formed, also cold, but less saline. It travels at much shallower depth and not so far. Near Greenland, a similar process leads to the formation of Atlantic intermediate water, which, after traveling south, surfaces near Antarctica. Being somewhat warmer, it brings moisture to the air and snow to the southern continent.

and to contribute to the upwelling in the circum-Antarctic seas. A similar but much smaller effect can be observed in the northwestern Pacific.

Because all abyssal water has a single source, it begins with a uniform temperature near 2°C, in contrast to a range of surface temperatures from about 27°C on the equator to 2°C near the poles. Taking about four centuries for its voyage north through the Atlantic, and 1,500 years in the Pacific, the abyssal water is slightly warmed by the ocean floor below and the warmer water above, so that when it arrives at its northern destination, it has become light enough to rise and begin the return trip. On its way north, narrow passages, such as a single deep channel east of the Falkland Islands in the southwestern Atlantic, or another one near Samoa in the western Pacific, provide the only entrances to more northern basins and so control the directions and patterns of abyssal flow.

CONTINENTAL DRIFT AND OCEAN CIRCULATION

The abyssal gorges of Samoa, the deep gap east of the Falklands, the Straits of Gibraltar, or the Drake Passage which allows the circum-Antarctic current to flow between South America and Antarctica, illustrate the special importance of key points called gateways for the ocean circulation, abyssal as well as at the surface. The opening or closing of gateways can be achieved by minor tectonic events, which can thereby have an impact on ocean circulation and climate as large as that of the stately drift of continents, and often much more abruptly.

To appreciate the role of gateways fully, let us return for a moment to our simple ocean circulation model and modify it slightly to arrange four continents in such a way that the equatorial currents flow around the world unimpeded (Figure 11.6). This should heat their water more than in the original model; some of it may circumnavigate the earth more than once and so warm up even more. Such an ocean would be warmer overall, the effect being especially noticeable at high latitude, and the latitudinal temperature contrast would be reduced. A warmer ocean would yield more moisture to the atmosphere, and so it should rain more on the continents.

Conversely, we might set up gateways at high latitude that would allow circum-polar currents to pass. These currents would isolate the polar continents from warmer seas, cause the temperature there to drop, increase the temperature contrast between lower and higher latitudes, and might even induce an icecap to form if an adequate supply of moisture were available. The return flow to the equator would be colder and the whole earth cooler.

The breakup of Pangaea provides us with examples (Figure 11.7). Panthalassa was an enormous ocean. Reaching from pole to pole, it extended at the equator over about 80 percent of the circumference of the earth. It must have had a simple circulation with one main gyre in each hemisphere, the subpolar ones being much reduced. Water much

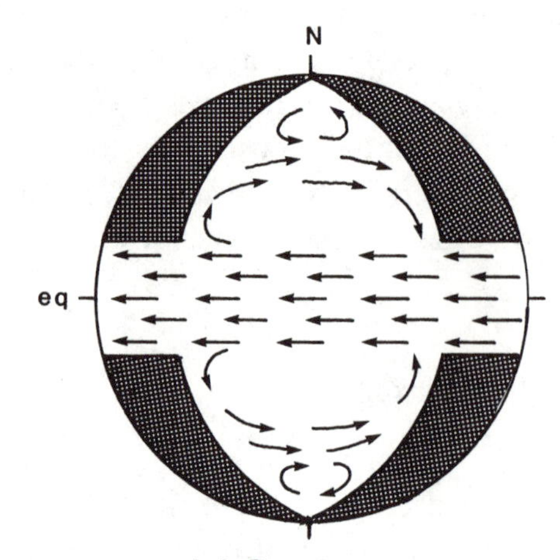

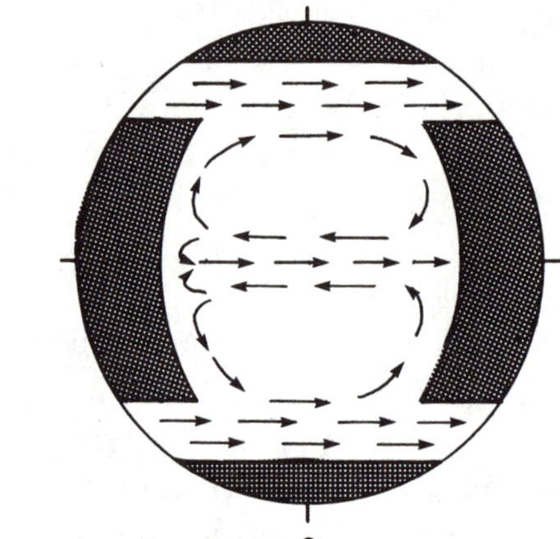

Figure 11.6. These diagrams illustrate the role of barriers and seaways in ocean circulation. If an equatorial current is allowed to pass completely around the earth, perhaps even several times, before being deflected north and south, a more even distribution of warmth across the latitudes will be the result (top). If, on the other hand, the equatorial current is restricted, but circum-polar flow is possible (bottom), the polar continents will be isolated from the warmer waters at lower latitudes, and a steeper temperature gradient from equator to poles will be produced.

warmer than that of the western Pacific today bathed the east coast of Pangaea. The latitudinal temperature gradient was surely small, and the circulation of the ocean, and hence of the atmosphere, was more sluggish than now. About 200 million years ago, the dominance of the equatorial current was enhanced further when a seaway broke through from the Tethys embayment westward and, assisted by the early opening of the southern North Atlantic, wedged the northern and southern

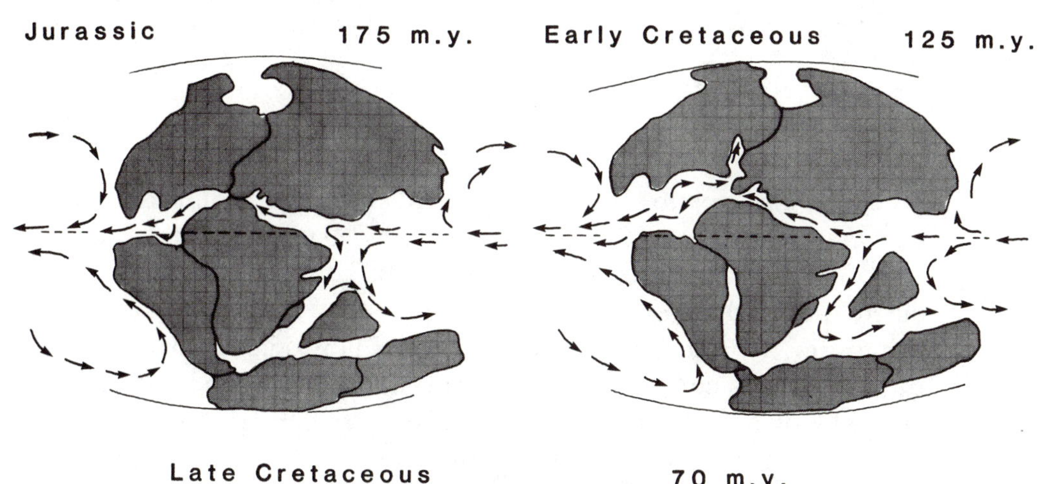

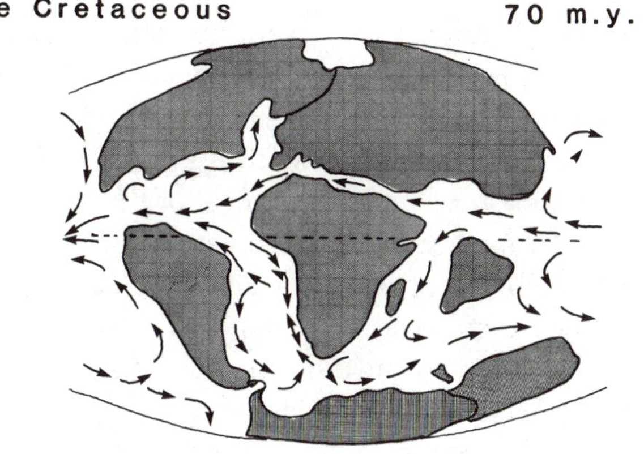

Mesozoic Ocean Circulation

Figure 11.7. During the Mesozoic, the ocean circulation evolved from a simple pattern in a single ocean with a single continent (Figure 11.3) to a more complex situation in the new oceans of the Cretaceous, but throughout the open circum-equatorial path and the absence of polar currents made for a distribution of temperature more even than that of today.

continents apart. The circum-global equatorial current further warmed the earth and reduced the latitudinal temperature gradient even more.

As the breakup of Pangaea continued, with separation of Africa from Antarctica about 125 million years ago, the circulation became somewhat more complicated, and even more so when the South Atlantic opened. Nevertheless, at the end of the Cretaceous the oceans must still have been quite warm at high latitudes. Even the early Cenozoic earth resembled its Mesozoic ancestor more than its modern descendant. We shall return to the question of the warm Mesozoic seas in the next chapter, but let us first trace the evolution of the ocean to its present state.

ONWARD TO THE ICE AGE

When the Cenozoic began, there were Atlantic and Indian oceans, as well as that remnant of Panthalassa, the Pacific, but the arrangement of the continents and hence the ocean current patterns were still very different from those of today.

Early in the Cenozoic, circum-equatorial flow still dominated the current pattern, but the drifting of the continents had produced quite modern-looking low-latitude and subpolar gyres in every ocean except the North Atlantic (Figure 11.8). As the southern continents continued to push away from Antarctica, however, the Tethys began to close in the Near East and Middle East during the Oligocene. The Mediterranean formed a little later as Spain approached North Africa. In the western Pacific, ridges and islands were crowded together by the march of Australia toward southeast Asia, blocking most of the flow from the Pacific into the Indian Ocean. A last equatorial gate was closed about 5 million years ago when the Isthmus of Panama rose to link North America and South America. Matters went in the opposite direction in the Southern Hemisphere. Australia's northward drift cleared a seaway south of Tasmania before the end of the Eocene, and the opening of the Drake Passage south of Chile completed the circum-polar path between 30 and 25 million years ago.

The ocean circulation changed greatly as a result. The equatorial flow weakened, and the waters turning north and south were cooler. Inevitably, ever-cooler water returned toward the equator along the west coasts of the continents, reducing evaporation and hence rainfall there. The desert of western South America may date from this time. Compared with the Mesozoic, we are in a far better position to check such assertions, mainly because of the large amount of paleontological data that has come from more than 600 holes drilled in the ocean floor by the research drillship *Glomar Challenger* during the past 15 years. Because we find in these cores planktonic microfossils that once lived near the surface, as well as their bottom-dwelling, benthic relatives, the paleontological record simultaneously illuminates surface and abyssal conditions.

The paleontologist's deductions about past oceanographic conditions can be checked against temperature information contained in the oxygen isotope ratios of the calcareous shells of planktonic and benthic microfossils. For the first half of the Cenozoic we need not correct

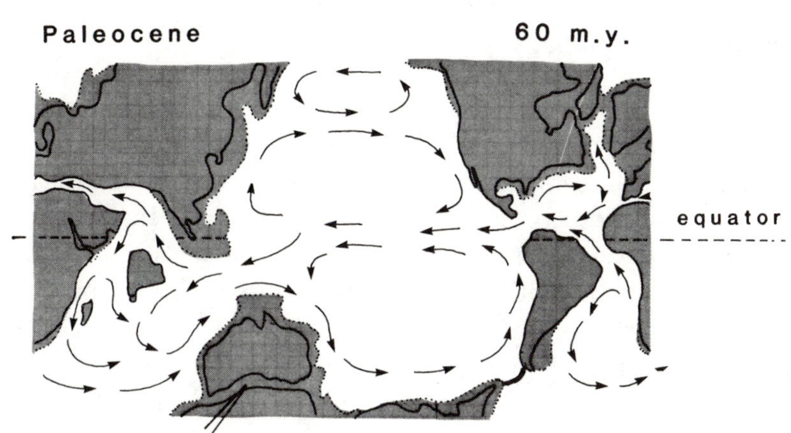

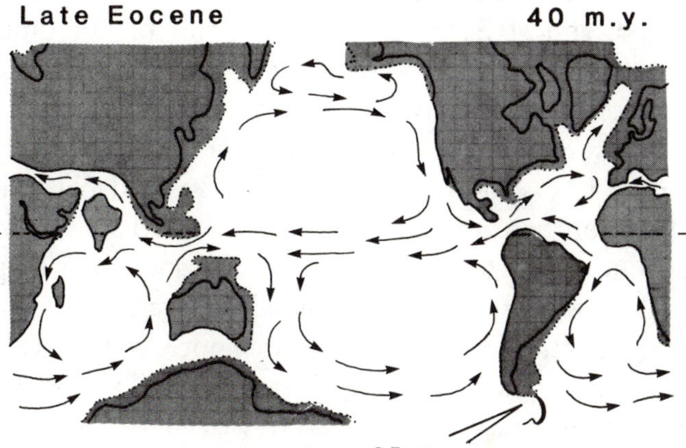

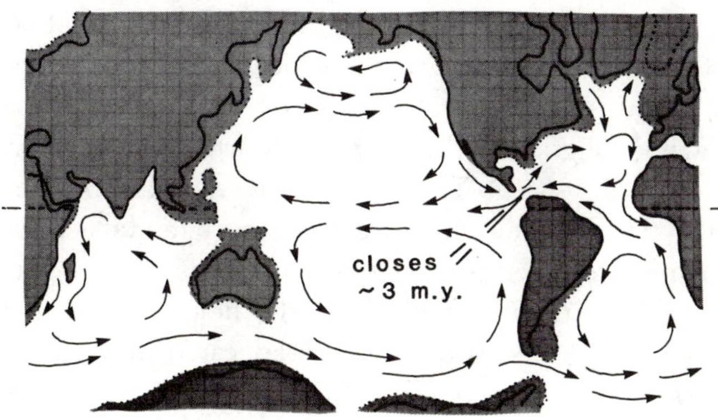

Cenozoic Ocean Circulation

Figure 11.8. The Cenozoic history of ocean circulation is dominated by two events: the gradual closure of the circum-equatorial seaway completed in the Pliocene, when the Isthmus of Panama emerged, and the opening of the Antarctic circum-polar seaway. The sequence of events eventually led to the late Cenozoic Ice Age (Table 1.1).

those ratios for water locked up in icecaps on continents, because there is no evidence for even minor glaciation anywhere. Not before the latest Eocene did mountain glaciers appear in Antarctica, though it had been in a polar position for scores of millions of years. Instead, the southern continent was covered with a temperate broadleaf forest not unlike that of southern Chile today. Ocean temperatures at middle latitudes resembled those of the present, but the sub-Antarctic region was still much warmer (Figure 11.9). The bottom water everywhere was at the very most 10–15°C cooler than that at the surface. A slight cooling toward the end of the Eocene had little apparent effect.

At the boundary between the Eocene and the Oligocene, however, a major event happened, almost instantaneously in terms of our time-scale. It was the first of several step-wise changes in conditions that ultimately led to the Ice Age. The surface water in the far south cooled suddenly and drastically, and deep ocean basins everywhere quickly filled with water some 10°C colder than before. Understandably, this had an unhappy effect on bottom-dwelling organisms accustomed to more comfortable conditions. Many became extinct, and for a long time afterward the abyssal benthic fauna remained impoverished. Cold water covered the Antarctic shelf and penetrated into embayments where sea ice could form, and glaciers grew on land. The cooling coincided with a global regression of the sea, but we cannot yet blame an icecap for the lowering of sealevel, because as far as we can tell, none existed. The Northern Hemisphere does not seem to have been much affected by all this, although the warmth-loving forests that grew in northern Canada during the Eocene made way for more temperate ones.

During the following Oligocene, the supply of warm water to high latitudes diminished further as the Tethys gradually closed. About 25 million years ago the circum-Antarctic current was born, and its encircling waters further insulated the polar continent from the warmer seas to the north. It is at this time that we have the first evidence for Antarctic glaciers at sealevel: Ice-rafted cobbles appeared in deep-sea sediments far offshore. Still, the continent was far from completely covered with ice, and forests, although stunted, grew here and there.

A true Antarctic icecap did not form until the Middle Miocene, about 25 million years ago, rather curiously arriving during a relatively mild time. This event is heralded loud and clear by another stepwise change in the oxygen isotope ratios (Figure 11.9), due this time not to a drop in temperature but to the removal of a large volume of water to Antarctica to form an icecap. From that moment on, oxygen isotope

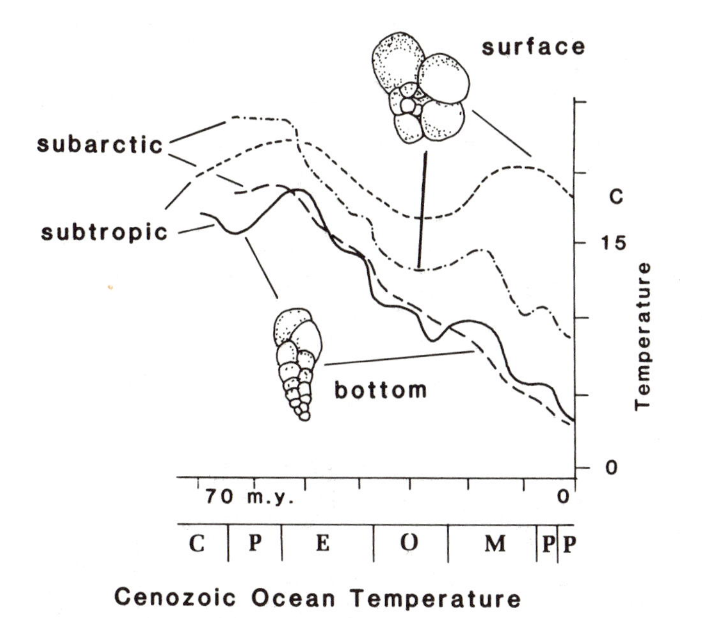

Figure 11.9. Oxygen isotope measurements on planktonic and benthic foraminifers show that the temperatures of the surface and deep ocean waters were dropping throughout the Cenozoic, generally slowly but sometimes suddenly. Major cooling occurred, for example, at the Eocene/Oligocene (E/O) boundary, and during the Middle Miocene (M). Two curves each are shown for the subtropics and for higher latitudes, one for deep water (with a benthic foraminifer) and one for the surface (with a planktonic one).

ratios no longer reflect temperature alone, but changes in water volume due to glaciations and deglaciations as well.

The formation of the Antarctic icecap accelerated the production of very cold water, and the modern abyssal circulation dates from that moment. The world climate itself was not yet especially cold, but once that first icecap had formed, further global cooling became inevitable.

Why was the formation of the Antarctic icecap delayed some 10 million years after the start of the circum-Antarctic current, and why did it occur during a mild rather than a cold phase? The cause may lie in the far northern Atlantic, where the separation of Greenland and Norway finally produced a major seaway early in the Miocene. During the Middle Miocene, the last barrier, the Iceland-Faroe Ridge, sank deep enough to permit normally saline, not very cold water from the northern North Atlantic to travel south. When this water began to share in the upwelling around Antarctica, its temperature, higher than that of the chilly local waters, increased the evaporation and thereby augmented the snowfall on the continent. Thus, the tardy arrival of the

Antarctic icecap may not have been a matter of temperature as much as of insufficient precipitation.

From that moment on, the world was in the Ice Age. Another episode of drastic cooling occurred about 6 million years ago, at the end of the Miocene. It was accompanied by a drop in sealevel, this time surely due to a large increase in ice volume. This must have taken place in the Southern Hemisphere, because there is still no evidence for northern icecaps. The drop in sealevel temporarily isolated the Mediterranean which in short order evaporated to dryness, leaving a thick deposit of salt.

The drying of the Mediterranean removed enough salt to lower the salinity of the oceans by 6 percent. This raised the freezing point of seawater slightly, so that sea ice could form at somewhat lower latitudes than before. This, in turn, may have increased the albedo and so caused a small drop in temperature. Some scholars even propose that the Mediterranean salinity crisis was the main culprit in the late Miocene cooling, but it is just as likely that it was the other way around.

The late Miocene cooling, especially at middle and high latitudes, and the increasing equator-to-poles temperature contrast had many consequences. One of these was a colder ocean and thus a reduction in continental rainfall. In East Africa and southern Asia, the drier climate led to extensive replacement of forest with savanna. It has been suggested that this change persuaded some forest-dwelling primates to come out of the trees into the open, where they ultimately evolved into *Homo sapiens*.

Finally, after a briefly warmer Pliocene, northern icecaps formed 2–3 million years ago, their birth revealed by yet another abrupt shift in the oxygen isotope ratios, by glacial deposits on the northern continents, and by ice-rafted gravels in the northern Pacific and Atlantic oceans. Although the first northern icecaps arrived during the closure of the Panamanian isthmus, we cannot hold that gradual event responsible for the sudden onset of the northern Ice Age. The blocking of the equatorial circulation did, however, enhance the Gulf Stream in the North Atlantic, which carried warm Caribbean water north and supplied more moisture for rain and snow in northeastern North America and northwestern Europe, so contributing to the growth of the ice.

A GOOD AND SUFFICIENT EXPLANATION?

Three major oceanographic changes accompanied the Cenozoic evolution toward the Ice Age: the segmentation of the equatorial current,

Table 11.1. *Summary of the evolution of the Ice Age*

Time (million years ago)	Events
>50	Ocean flows freely around the world at the equator Uniform climate and a warm ocean, even near the poles Deep water much warmer than now; circulates slowly No icecap or even glaciers on Antarctica
48–45	Slight cooling
38	Sharp cooling of surface water in the south and of deep water everywhere Deep-water circulation speeds up; cold benthic fauna First glaciers in Antarctica, but no sea ice yet Seaway between Antarctica and Australia fully open
35–30	Circum-equatorial circulation broken by closures of Tethys in Middle East and Near East and by barriers in the western Pacific Partial circulation around Antarctica
30–25	Opening of Drake Passage Circum-Antarctic current established Antarctica much colder; sea ice, but no icecap yet
15	Iceland-Faroe Ridge sinks; North Atlantic water wells up around Antarctica; more moisture and copious snowpack Antarctic icecap develops Present abyssal circulation established
6	Sharp cooling; Mediterranean salinity crisis
5–3	Isthmus of Panama closes Gulf Stream intensified; more precipitation on northeastern North America and western Europe
3	Icecaps form in Northern Hemisphere Ice Age is under way

the initiation of circum-Antarctic flow, and the development of the modern deep circulation (Table 11.1). Do these three events together furnish an adequate explanation for the global cooling, mainly at high latitudes, for the increased north-south temperature gradients,and for the sudden shifts from one condition to the next? Is the interaction between continental drift and ocean circulation by itself enough to do the job? The story just presented certainly sounds persuasive, but doubts remain. The abrupt changes, for example, do not seem to correspond well with the more gradual pace of continental drift. Moreover, we cannot yet evaluate quantitatively the various processes and the climate changes they were able to produce. It would thus be unwise to disregard other possible explanations.

One factor we have not yet considered is the possibility of albedo

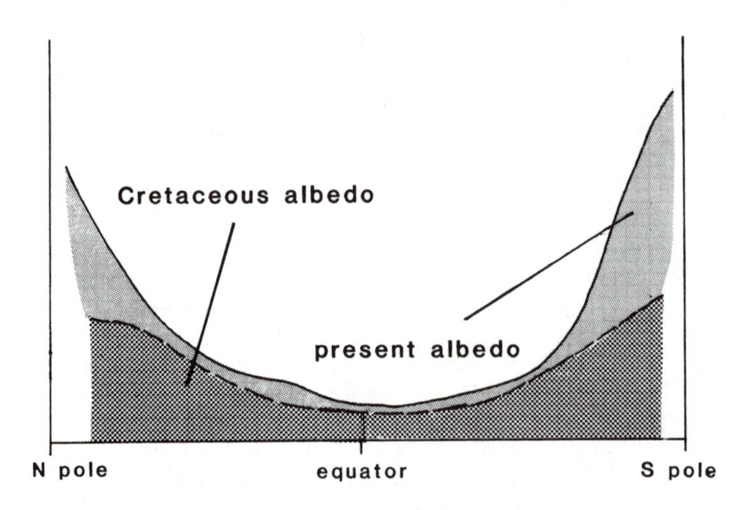

Figure 11.10. The albedo is a major factor in the surface temperature distribution of the earth (see Chapter 3). Because there were no Cretaceous icecaps, and because the configuration of land and sea was then very different, the north-south profile of Cretaceous albedo was much lower than that of the present earth. The consequence must have been a warmer earth and a smaller temperature gradient from equator to poles.

changes and their effect on the planetary heat budget. Albedo changes are not restricted to variations in the cover of ice and snow. Land has a higher albedo than water, so that transgressions and regressions are accompanied by albedo changes. If we have land rather than sea at the equator, much less heat will be retained. Replace sea with land at a high latitude and not only will the summers be cooler, but the seasonal contrast will also be greater, because land loses more heat in winter than water does. The semi-arid and arid lands of the subtropics reflect much heat that would be retained by equatorial and temperate forests.

As the continents drifted, the land areas at different latitudes changed, decreasing in the southern temperate zone and increasing drastically in the subtropical belt of the Northern Hemisphere, which was largely desert. With the help of many simplifying assumptions, one can calculate the albedo changes that resulted (Figure 11.10). Evidently the effect was far from trivial, and it may well have been an important factor in the climate changes of the last 100 million years.

That, however, is about as much as we can say. Our present knowledge does not allow us to translate albedo changes quantitatively into climatic effects. Other factors need to be considered, about which we know even less, such as the climatic consequences (other than albedo changes) of large transgressions and regressions and of the numerous mountain ranges that rose during the Cenozoic, or subtler matters such as evolutionary changes in vegetation. Grass was a successful innovation of the early Cenozoic, and the spread of pastureland and steppe, not seen in the world before, altered not only the albedo but also patterns of evaporation and precipitation.

To be reasonable, we should therefore assume that plate tectonics and paleoceanography present us with only part of the Cenozoic climatic story and should remain eager in the search for other causes. Remembering that very nearly everything in this chapter was unknown 20 years ago, there seems to be ample cause for optimism that another two decades will bring us key answers to the puzzles of climatic evolution and the onset of ice ages.

12

Other times and other oceans

At the moment we are still far from being able to discuss, even with the limited confidence displayed in the previous chapter, oceans older than those of the Cenozoic. About the Precambrian ocean we can speak only in generalities, and we are mostly forced to guess at the geography, sediments, and pelagic life of the Paleozoic and Mesozoic oceans. Nevertheless, new tools have been developed, and our knowledge is growing fast; another decade will much improve the picture. Let us at least sample what is being done.

A TALE OF TWO OCEANS

In the previous chapter, a paleogeographic reconstruction of the Cretaceous ocean circulation indicated that whereas the equator was as warm as it is today, the higher latitudes were then much warmer. This deduction agrees with paleontological data (e.g., the occurrence of palm trees at subarctic latitude), and it is borne out by a complete lack of evidence for any polar ice and snow.

Oxygen isotope data, although not yet abundant, confirm that Cretaceous surface temperatures ranged from 30°C at the equator to 15°C as far north and south as 55° latitude. Ratios from bottom-dwelling organisms give a rather astonishing value of about 15°C for deep water, much warmer than the 2°C of today, although we do not have values for depths greater than about 2,000 m. Therefore, we do not know the temperature of the deeper abyssal waters directly, but unless the temperature change with depth in the Cretaceous was fundamentally different from that of today, the bottom waters cannot have been much colder. This means that the water of the polar regions, the coldest surface water on earth and hence the most plausible source of deep water because of its high density, was equally warm.

Obviously, in a world with such small temperature differences between equator and pole and between surface and bottom, the driving force for the deep-ocean circulation must also have been small, and the flow sluggish. It is even conceivable that sometimes, and in some areas, the Cretaceous deepsea was as stagnant as the Black Sea is today.

We have direct evidence for unusual deepsea conditions in the form of black shales in the deepsea sediment record of the Cretaceous, especially between 115 and 125 million years ago, and again around 90 million years (Figure 12.1). Black shales also occur in abundance in marginal Cretaceous seas, where they are the sources for some of the world's most prolific oil fields, such as those of Venezuela. The high content of organic carbon of these shales, up to 30 percent, implies that insufficient oxygen was dissolved in the water to decompose all organic matter that sank to the ocean bottom, before it was buried.

Oxygen is dissolved in seawater when it is in contact with the atmosphere. When the surface water sinks and travels along the bottom, its oxygen is consumed, some by the breathing of benthic organisms, much more by the decay of organic matter. If the flow of deep water is relatively rapid, as in the modern ocean, then enough oxygen is supplied at all times, most organic matter is decomposed, and its nutrients are returned to the water. A sizable bottom fauna can exist, and very little organic matter is buried in the sediments. However, should the rate of flow decrease, as it likely did in the Cretaceous, then the time since the deep water last saw the sky will be extended, oxygen will be consumed more completely, and farther downstream there may be little or none left. The excess organic matter will be buried in the sediments, and black shales may form.

Another factor that certainly contributed to the accumulation of organic matter in Cretaceous deep-sea sediments was the high temperature of the deep water. Because much less oxygen can dissolve in seawater at 15°C than in water at 2°C, it is a priori to be expected that the warm abyssal waters of the Cretaceous contained less oxygen.

In the long run, a negative-feedback process prevents the deep water from remaining permanently free of oxygen. When the supply runs out, organic matter is no longer decomposed, and the recycling of nutrients is sharply reduced. This lowers the fertility of the surface waters; biological productivity slows down, less organic matter sinks to the bottom, and the oxygen content of the water recovers because there is less demand. For this reason, deposition of black shales tends to be intermittent; very black layers alternate with others less rich in organic matter. We find this fine lamination preserved in Cretaceous

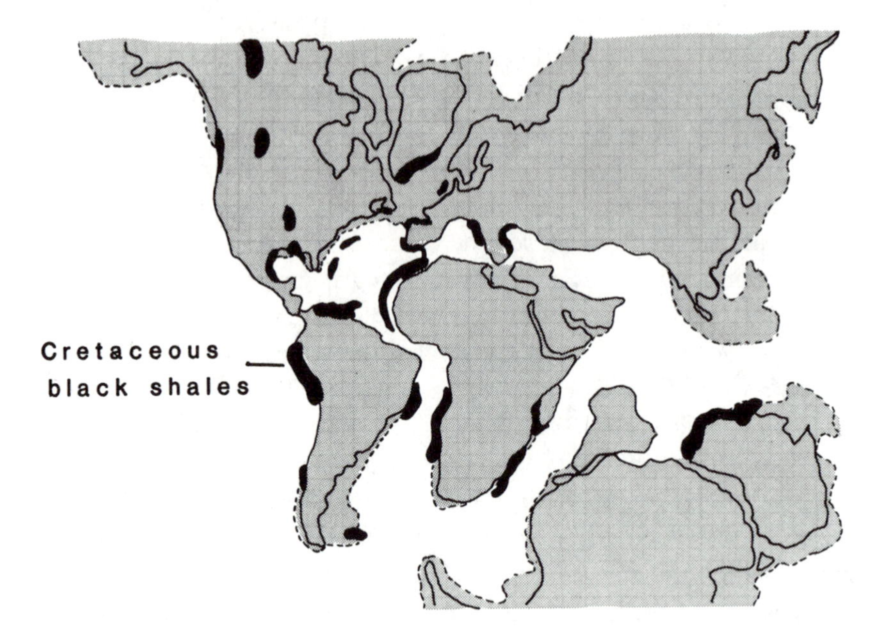

Figure 12.1. Black marine shales rich in organic matter were formed in abundance at various times during the Cretaceous. This occurred at depths as great as 3,000 m, as well as in coastal seas. These black shales are now important source beds for oil.

black shales, evidence in itself that the oxygen content was low enough to prevent burrowing organisms from destroying it.

The oxygen-free zone need not have extended down to the deepest parts of the Cretaceous oceans, as it does in the modern Black Sea. Sometimes the rain of organic matter consumes virtually all oxygen in the water column. This produces an oxygen-minimum zone at intermediate depth, something that occurs in the present Pacific, but it leaves the abyss with enough oxygen to support a benthic fauna. It is probable that most of the Cretaceous black shales were deposited on continental slopes in an oxygen-minimum zone, rather than in a totally anoxic abyss.

The burial of such large amounts of organic matter meant that equally vast amounts of nutrients were permanently removed from circulation and hence no longer available to plant life near the surface. Ironically, it appears therefore that the Cretaceous ocean, producer of such rich oil source beds, was actually not particularly fertile, much less so than ours.

The unusual warmth of the Cretaceous deep water has some other interesting consequences. The density contrast between water at 25°C and that at 15°C is so small that differences in salinity become important. We must therefore consider the possibility that at least some of the Cretaceous deep-water circulation may have been driven by salinity differences rather than by temperature as in the present ocean.

Let us consider an example. At low latitudes, insolation and evaporation easily produce a salinity of 3.8 percent or more, which, at a temperature of 15°C, will make the surface water as dense as normal seawater (salinity of 3.6 percent) at 2°C. Mediterranean deep water at 15°C and with a salinity of 3.8 percent flows out of the Straits of Gibraltar and spreads clear across the central North Atlantic, being only very slightly less dense than the cold but less saline Atlantic deep water (Figure 12.2). It therefore is quite conceivable that supersaline water, formed by evaporation in marginal seas, was an important source of deep water during the Cretaceous. One can easily imagine the Tethys or the shallow early South Atlantic spilling dense salty waters into adjacent ocean basins, where they would have sunk to the bottom. From there they might have flowed north and south to rise in the polar regions, in a reversal of the present sense of flow. Fed in this manner by waters of more local origin, the Cretaceous deepsea may have lacked the simple circulation pattern and uniformity of the abyssal flow of later Cenozoic oceans.

Why should the Cretaceous climate have been so uniformly warm from equator to pole? Theoretical studies of the Cretaceous climate that have considered continental drift and albedo changes have suggested that these factors do not suffice to explain the small temperature range or the warmth of the polar regions. Even if we allow for such rather arbitrary assumptions as a more evenly distributed cloud cover, we still need something else, a greenhouse effect perhaps, or some external cause. But before we embark upon fascinating speculations about orbital changes, cosmic clouds, or variations in the output from the sun, it would be well to remember that our information is slim, that it may have been misinterpreted, and that in not fully understanding the role of the ocean, we may have underestimated it. Moreover, if a warm state of the earth is normal, it would make more sense to call on extraterrestrial events to explain the climatic patterns of ice ages, which are rare.

In summary, the Cretaceous and the late Cenozoic oceans represent two extreme states. The warm Cretaceous ocean was weakly stratified, lacked a vigorous circulation, and was such an effective sink for organic matter and nutrients that its overall fertility was low, although its oil potential was high. The cold late Cenozoic ocean had steep temperature gradients with latitude and depth and therefore possessed

Figure 12.2. The present Mediterranean illustrates how, in the Cretaceous, abyssal water may have originated in shallow equatorial seas rather than in cold, high-latitude regions of the ocean. Evaporation is higher in the Mediterranean than the influx of fresh water, and a saline brine forms and sinks. This dense water of about 15°C flows across the sill at the Straits of Gibraltar and spreads far across the central Atlantic, just above the much colder but less saline Antarctic abyssal water.

a vigorous circulation. Nearly complete decay of organic matter recycled all nutrients and ensured high fertility, notwithstanding the difficulty of transporting the nutrients across the barrier of the rather stable density stratification.

If we compare the comings and goings of warm and cold ocean states with other events that vary with time, some vague but intriguing correlations can be seen. Warm oceans correspond to transgressions, to much solution of calcium carbonate, and to deposition of black shale; cold oceans correlate with regressions, with deposition of calcareous oozes and full recycling of organic matter. Warm oceans contain a very diverse life in complex biological communities, but the number of individuals in each ecological niche is small. A cold ocean possesses abundant life, but it is less specialized: The number of species is smaller and the community structure simpler. Extinctions accompanied each transition from a warm state to a cold state, and evolutionary explosions often occurred during the reverse. A rhythm is vaguely discernible: long periods of warmer oceans punctuated about every 32–36 million years by a colder state (Figure 12.3). Food for thought, but cautious thought, until we know much, much more.

VAST AND SHALLOW SEAS

In the present emerged state of the world, we tend to consider the shallow shelf seas as mere transitions between continent and ocean. During the Cretaceous and even more during the Paleozoic, enormous shallow seas, which have no modern counterpart, ranked rather more equal in environmental importance to the land and the deepsea. It was in those shallow seas that the abundant life of the Paleozoic evolved to leave deposits quite different from anything we find today. The sudden diversification and abundance of fossils at the beginning of the Paleozoic allow us to make fine temporal and environmental distinctions and to connect events that happened in distant places. Regarding the early Paleozoic in this new, bright light, we find it a wet world, the continents inundated more than they have ever been since then, and the rise of the sea continuing. Before this rise ended, much less than half of the area of the continents remained above water.

In North America there were several successive transgressions (Figure 12.4), each beginning with widespread sandstones laid down on beaches and in shallow seas. Offshore, on clear and sunlit shoals, which occupied most of the submerged parts of continents, algae and sponge-like organisms formed extensive limestones. As the sea advanced farther and the lands shrank, limestones became almost the only rocks, products of a sea filled with the largest reef complexes the world has ever seen. Needing sunlight, shallow depth, and plenty of food, the reefs on the North American continent, which was then just north of the equator, grew best on the windward edges of banks and shoals. There, wave turbulence furnished nutrients for a rich growth of plankton, which in turn fed a complex community of reef organisms. Altogether, the reefs, limey shoals, and channels formed a seascape rather like that of the present Bahama banks, but on a vast scale and in the middle of a continent instead of in the middle of the ocean.

Toward the later Devonian, this age of reefs came to an end as the

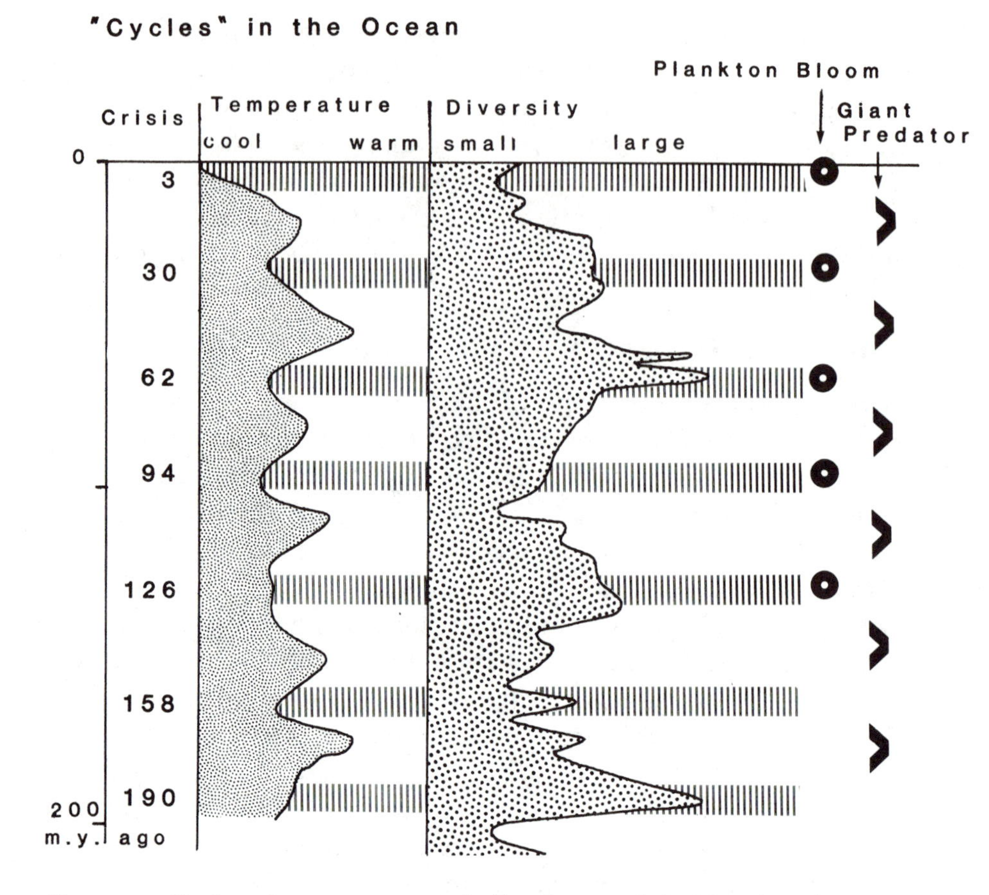

Figure 12.3. Fossils, paleotemperatures, and other features of the Mesozoic and Cenozoic suggest that the ocean may oscillate between long warm and brief cool states. During a warm interval, faunas and floras were diverse, and giant predators, such as huge sharks or marine reptiles, flourished. During the colder periods (shown here with vertical shading), life was more abundant but less diverse, and plankton blooms, often consisting of a single species, were common. There appears to be a faint periodicity of about 30 million years.

sea began a slow withdrawal in many stages. Eventually this regression led during the Permian to nearly complete emergence of the continents, which had by then drawn together in the supercontinent Pangaea. A little earlier, in the Pennsylvanian, a large part of eastern and central North America had emerged to form a vast, low plain across which many rivers meandered, bringing debris to the sea from the Appalachian mountains newly risen in the east. The low coast must have resembled that of the Amazon and the Guyanas today: extensive swamps, flat shores, and sluggish streams that built deltas in a warm and muddy sea. Land plants had just evolved, and for the first time in

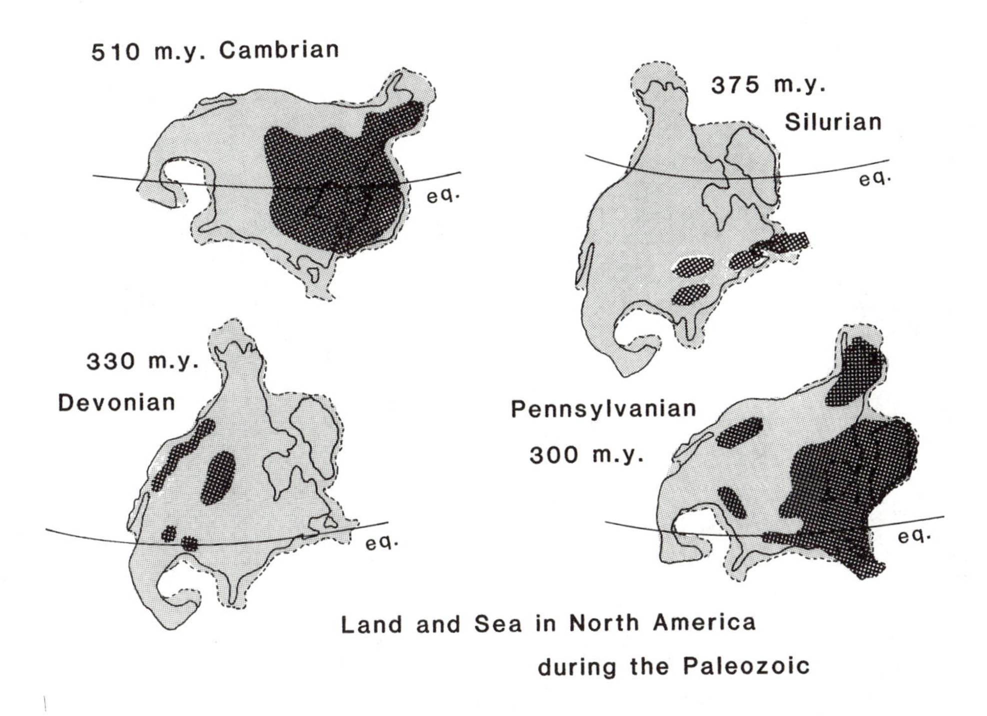

Figure 12.4. During the Paleozoic, the sea stood high most of the time, widely flooding the continents, as this set of maps of North America, properly oriented with respect to the equators of the past, clearly shows.

history forests covered the plains, laying down the coalbeds we now mine in eastern North America and in Europe and Siberia, where conditions were similar.

The Pennsylvanian sediments of the central United States consist of many curious repetitions of the same sediment types. Each cycle (Figure 12.5) begins with coastal and river sandstones resting on an old, eroded surface. They are followed by clay with tree roots and then by coal. Next come swamp clays, which suggest the beginning of an inundation, and then a shallow marine facies with shales and algal limestones. An unconformity announces the next cycle.

At least 50 such cycles can be recognized, each evidently representing a regression followed by a transgression. Such a rhythm comes naturally to many deltas as the river swings back and forth, having built seaward as far as possible in one place, then seeking a shorter path to the sea to one side or the other. A new delta is begun, and the old one subsides and is taken over by the sea. The Mississippi delta, with its active fingers near New Orleans, between many abandoned and now partly eroded subdeltas, is the classic example of this behavior.

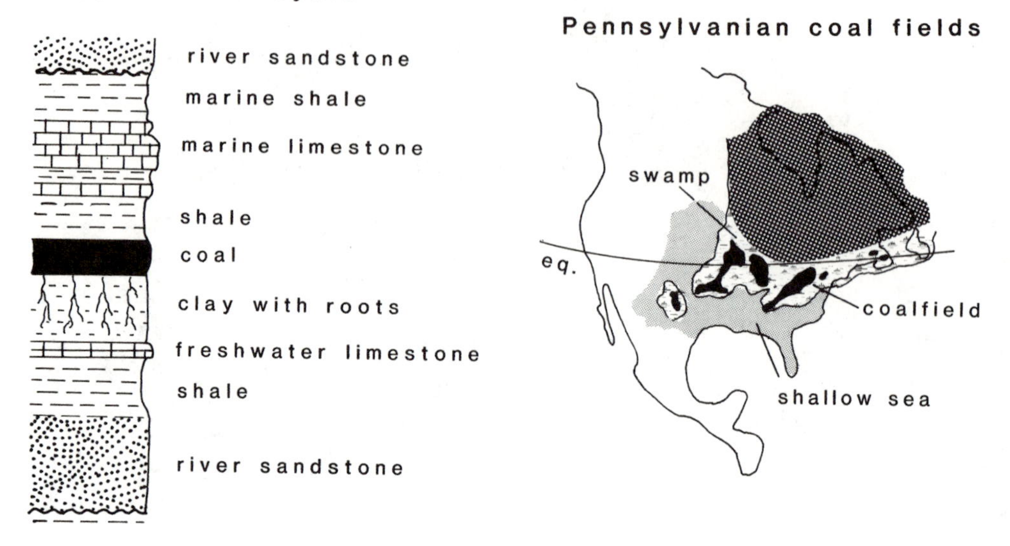

Figure 12.5. In the shallow Pennsylvanian (late Carboniferous) sea of the southeastern and central United States, sediments were deposited in curious cycles, each beginning with a river facies and evolving via a coastal swamp to very shallow marine conditions. These cycles, possibly related to glacial eustatic sealevel changes, produced major coal deposits.

The Pennsylvanian cycles are thin and seem brief and in this respect fit this model. They are, however, also very extensive, much too extensive for a single delta, no matter how big the river. Moreover, so far in the past, brevity is relative, and the tens or even hundreds of thousands of years of each cycle are much too long for a migrating delta. This has suggested a different explanation to some geologists. During the Pennsylvanian, an ice age was in progress in Gondwanaland, presumably accompanied by the eustatic sealevel changes that come with glaciations and deglaciations. Might it be that the Pennsylvanian cycles of North America faithfully recorded the waxing and waning of the icecaps of the remote south polar region of that time so long ago? It is an intriguing thought and worth pursuing, but there are problems. A good-sized glaciation ought to lower sealevel by more than these transgressions and regressions in a very shallow sea seem to imply. Moreover, we do not find such cycles, at least not as distinctly, in sediments of the same age elsewhere, in the Carboniferous of Europe, for example, where they ought to be if they were caused by eustatic and hence global sealevel changes. So the Pennsylvanian cycles remain a minor but fascinating reminder of our ignorance about even relatively simple things.

TOWARD A HISTORY OF SEAWATER

We have explored, if only in fragments, the history of sealevel changes, ocean circulation, and such properties of seawater as temperature and oxygen content, and we shall discuss the source of the water in Chapter 14. What can we say about its chemistry? Has the composition of this rather concentrated salt solution, containing a little, and sometimes a lot, of just about every element, also varied with time? Would not its very complexity suggest an eventful past?

The dominant ions in seawater are sodium, magnesium, and chlorine, with somewhat smaller amounts of calcium, silica, and potassium. Except for gases like oxygen, carbon dioxide, and sulfur dioxide, the dissolved components are supplied mainly by rivers that obtain them from the weathering of rocks. Because this has been going on for eons, there must also be processes that remove the salts, or the oceans would be much saltier than they are. For some elements we do understand the process of removal, but for many, including all of the principal components of seawater, the exit route is far from clear. Some salt is removed in the form of evaporites, but the 10 percent that became rock salt in the Permian and the 6 percent that was deposited during the Mediterranean salinity crisis are hopelessly inadequate to explain why the present salinity is much too low to account for billions of years of salt input by rivers.

Most geologists believe that the composition of seawater has remained essentially constant for at least the past 2 billion years. The reasons for this belief are rather intuitive: Life has inhabited the sea since the Precambrian, and direct descendants of the first inhabitants are still found there, implying that there has been no drastic change in environment. Furthermore, the body fluids of most organisms have the same concentrations of salts (although not always the same composition) as seawater. This would make eminent sense if life began in an ocean with the same salinity as the present one. On the other hand, life is viable under an astoundingly wide range of chemical conditions, and similarity of form between ancient and modern organisms does not necessarily mean similar physiology or biochemistry. Moreover, many plants and animals easily tolerate a wide salinity range. Clearly, this kind of reasoning is insufficient to settle the argument.

The geochemical evidence is sparse and not very satisfactory. Few minerals are good indicators of ocean chemistry and at the same time stable enough to be preserved in sedimentary rocks. What little we know is compatible with, but does not prove, a composition constant over time. We can show, however, that the average composition of the continental crust, which, through weathering on land, has supplied the dissolved substances has not changed much since the late Precambrian. If the input to the oceans has remained approximately constant over 1 or 2 billion years, why should not the same be true of its composition? This would, however, presume that whatever process removes elements from the ocean does it in the same proportions that weathering supplies it. Given the inevitably rather different nature of these processes, that hardly seems a plausible proposition. Furthermore, why should the processes that remove salts from the sea have remained constant over the years, even if the input has? These are uncomfortable points, but given our almost complete lack of evidence, it still seems simplest and safest to assume that except for such minor excursions as the loss of 10 percent of the salt during the Permian, the composition of seawater has not changed much over time. That way, at least, we avoid the arbitrariness inherent in any other approach.

Here the discussion has rested for many years, and we have accepted an ocean that could not have halved or doubled its salinity, was always at least somewhat oxygenated, and could not have altered its composition rapidly because the river input itself is incapable of rapid change. As a result, no one has ventured to call for a geochemical drive for the evolution of life in the ocean.

A chance discovery has recently changed this perspective. This discovery came from a mismatch between the heat that theoretically should be released as new crust forms on a mid-ocean ridge and the heatflow actually observed there. For crust older than about 5 million years, theory and observation agree, but for crust of zero age at the ridge crest there is a heat shortfall of at least 40 percent compared with the predicted value. The theory is impeccable, and the errors of measurement are small. Therefore, we must accept the fact that a very large amount of heat (10^{16} kilocalories per year) remains unaccounted for and is apparently eliminated by some means other than conduction through the rocks, the form of heat loss we conventionally measure. A different way to dispose of heat, the method most automobile engines use, is to circulate cold water through the young, hot rock. The water, once heated, will rise and escape to the seafloor through submarine hotsprings. During the past five years, many such hotsprings have indeed been found on mid-ocean ridges, where they properly and quite exactly account for the global deficit in heatflow.

Here, our interest in these springs stems from the fact that the circulating water loses some of its constituents to the hot crust,

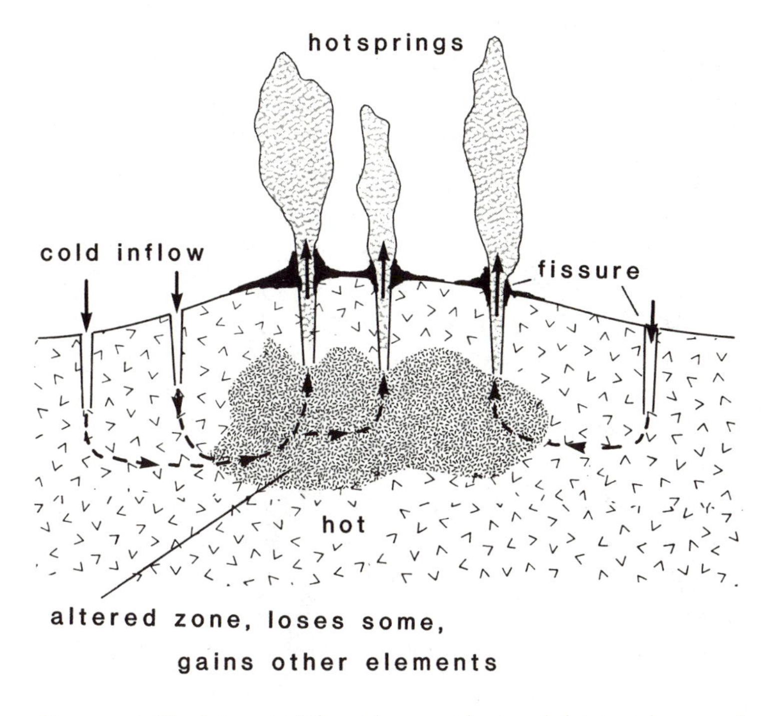

Figure 12.6. The hot crust of the mid-ocean ridge is cooled in part by upward conduction of the heat to the seafloor. In addition, cold water circulates through fissures in the crust. The water, once heated, returns to the ocean as hotsprings. During its passage through the crust, it leaves some of its constituents behind and extracts others, thereby altering the crust. At the vents, deposits of iron and manganese oxides (black) form that occasionally carry worthwhile amounts of such metals as silver and copper.

whereas elements from the rock dissolve in the brine and escape to the surface with it (Figure 12.6). For example, the entering seawater loses all of its magnesium and sulfate, and possibly part of its sodium and chlorine. It gains silica, iron, manganese, carbon dioxide, and many other components. The exiting water is therefore quite unlike the abyssal ocean water, not only in temperature but also in composition. Being so far out of equilibrium, it rapidly deposits crusts of manganese and iron that contain many other metals, including copper and silver. We currently estimate that submarine hotsprings furnish nearly all of the iron and manganese in the sea, the iron in its reduced, ferrous state (see Chapter 15), and as much as 45 percent of the silica and 15 percent of the calcium supplied by rivers. The total flow of water is also large: In 8–10 million years, the equivalent of the entire ocean volume circulates through the hot crust.

This discovery forces us to rethink the chemical history of seawater.

First of all, we may have found the long-sought sink that prevents magnesium and sulfate, and perhaps common sea salt as well, from building up to very high levels, beyond those observed. We also have located a major source for quite a few important elements, including ferrous iron, which, because it binds oxygen, limits the oxygen content of seawater. Equally important is that this *hydrothermal process* can speed up or slow down rapidly as the rates of seafloor spreading change. Doubling the spreading rate ought to approximately double the rate of circulation of the cooling water, reduce the magnesium and perhaps sea-salt concentrations by half, and increase silica and iron. We saw in Chapter 10 that such a doubling or halving of the spreading rate during the past 100 million years was quite definitely within the realm of possibilities. For a younger, hotter earth further in the past, an ocean with a much different and more variable composition, lower salinity, more silica, and less oxygen, must be considered likely.

Of course, we have no solid evidence that the composition of seawater actually ever was much different in the past from what it is today. However, now that we know of a process that might produce major change, the search for such evidence will become more worthwhile. The key may well lie in ancient oceanic basalts altered by the hot brines that passed through them on the crests of long-gone mid-ocean ridges. The serendipitous discovery of the chemical processes in submarine hotsprings, sought originally merely to solve a rather obscure geophysical puzzle, has revived interest in an old and very important problem.

PERSPECTIVE

The impression of antiquity and permanence conveyed by the vastness of the sea is false. Through the eons, the level of the sea has continually moved up and down, sometimes covering half the continents and creating wide shallow seas for which we have no counterpart today, then withdrawing to the very edges of the continents. Rise and fall occurred, as was the case with climate changes, on many timescales ranging from tens of millions to tens of thousands of years. Except for the brief ones produced by the freezing and thawing of icecaps, we can only guess at their causes. The processes of plate tectonics affect sealevel as mid-ocean ridges shrink and expand, but this explains only part of the phenomenon. The rest remains obscure, somewhat ironically, because the geological record is so much a record of sealevel change.

As continents drift, the configurations of the oceans also shift, and so do the ocean currents, the distributors of heat across the surface of the earth. An open equatorial path makes for a warm earth; an unimpeded circum-polar flow for a cold one. Though continental drift may not explain the advent of the Ice Age entirely, it certainly was a key factor, thus assigning plate tectonics a major role in climate control.

Other matters of the sea cannot be so easily related to the forces residing in the earth's interior. It appears, for instance, that the ocean oscillates between unstable warm and stable cold states, with appropriate large impact on marine life, but the timescale of tens of millions of years is too short for plate tectonics, and the cause is obscure.

Salinity is another problem. Salt is forever being added by rivers and must be eliminated somehow, or the sea would be far saltier than it is. The mechanism is unknown, but the presumption has long been that it maintains a constant salinity. Now it appears that hotsprings on mid-ocean ridges dispose of part of the excess. This implies that the composition of seawater must vary with the spreading rate and cannot have been forever constant.

This brings us to a reasonable end of our exploration of the consequences of the geological revolution for our understanding of the Phanerozoic history of the earth. We shall now turn to those long early years when everything began, years that must be probed with much less information and much more speculation.

FOR FURTHER READING

The fields of paleoceanography and paleoclimatology have not yet produced many overviews suitable for the edification of the interested observer. The technical literature is becoming voluminous but is quite inaccessible. Here are some slim pickings:

Turekian, K. K., *Oceans,* Prentice-Hall, Englewood Cliffs, N.J., 1976, 149 pp.

Menard, H. W., *Ocean Science: Readings from the Scientific American,* Freeman, San Francisco, 1977, 307 pp.

Wilson, J. T., *Continents Adrift and Continents Aground,* Freeman, San Francisco, 1976, 230 pp.

Turekian's book is another in the excellent series by Prentice-Hall mentioned at the end of the first section, and Wilson's book has a good selection of basic articles. Both, however, predate the development of modern paleoceanography.

The following is a rather personal but fascinating story about the use of deep-sea drilling in the study of oceans, with a first-hand account of the discovery of the Mediterranean salinity crisis:

Hsu, K. J., *The Mediterranean Was a Desert – A Voyage of the Glomar Challenger,* Princeton University Press, 1983, 197 pp.

Finally, these are two useful reference works at the college text level:

Frakes, L. A., *Climate through Geologic Time,* Elsevier, Amsterdam, 1979, 210 pp.

Kennett, J. P., *Marine Geology,* Prentice-Hall, Englewood Cliffs, N.J., 1981, 813 pp.

The four-billion-year childhood

The surf is brushing at my steps; I seek
An aged cliff that stands among the sleek
Young chargers of the sea.
Rounds of anemone
And areas held by sea urchins devise
The narrow range in which the tide will rise
And fall; though cliffs themselves
And all the earth's vast shelves
Crumble. And there the mode of permanence
Is framed in the sea-tide's changefull cadence.
Howard Baker, *Ode to the Sea*

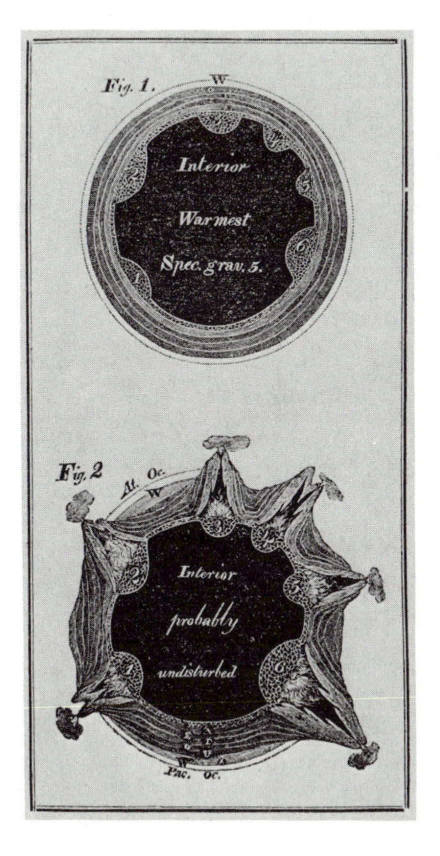

THE YEARS WHEN NEARLY EVERYTHING BEGAN

The early earth is alien, like and yet unlike the planets we have learned so much about in recent years: a little of Mars, some of the moon, a dash of Venus. From that remote planet Earth to the familiar world of the Phanerozoic leads a four-billion-year road, nine-tenths of the entire history of the earth.

When the earth began, it was a ball of cosmic debris: no land nor sea, no mountains nor valleys, no clouds, no winds, no life. When the Precambrian ended, blue seas fringed sandy shores, and myriad tiny green plants drifted in fertile seas, where they fed many and diverse small but already complex animals. The land was still bare, but the air was marginally breathable, and the temperature pleasant enough.

How, precisely, was the beginning – those first hundreds of millions of years for which no record remains? No one knows, and we use physics and chemistry, our knowledge of the cosmos seasoned with our imagination, to speculate, to dream. We can think of various ways in which the earth might have evolved to its present state, but so far proof and disproof have eluded us, leaving half a billion years to contemplation.

Another three billion years take us from the oldest rocks to the Phanerozoic. Along the way the earth was molded to nearly its present condition. We also encounter life, but the record is silent regarding its origin and not very eloquent about what followed afterward. The thread is thin and so faded that once again we are forced to speculate, testing only occasionally against rare facts rather than, as is the practice of geologists, reasoning from the evidence to the explanation. No firm answers have yet been found for even the most fundamental questions, but complex interactions between life, ocean, and atmosphere can be dimly discerned. Since I began to think about writing this book, an entire theory about the origin of the atmosphere, then widely accepted, has fallen into disfavor, and the age of the oldest fossils has been raised by almost half a billion years. The pages that follow are obviously a progress report; they can be no more.

Lest one believe that the Precambrian is not very important except to scholars who love deep, fundamental questions or puzzles without answers, let it be noted that it is economically one of the most important segments of earth history. Precambrian rocks contain an enormous share of all known mineral resources, and with the exception of coal and oil, the bulk of those that remain to be discovered will quite likely also be of Precambrian age. In this context, questions about the origin of everything are somewhat peripheral, but the chronology and record of events are of prime importance.

13

Birth of the solid earth

An array of isotopic data leads us to accept that the earth is approximately 4.5 billion years old. The oldest rocks, at Isua in Greenland and on the Limpopo River in South Africa, were formed at least 3.8 billion years ago. Distorted as they are by time and history, they indicate the presence of continental crust, of a sea in which sediments were laid down, and of weathering processes on land that made such sediments. In 700 million years the earth had evolved from a ball of cosmic dust to a planet with at least some of the properties of its present surface, albeit in different states and configurations.

Even after the first 700 million years, which are blank pages, deciphering the earth's early history remains exceedingly difficult. There are three main reasons for this, common to all of earth history but greatly magnified in the Precambrian: our inability to determine time with precision, our frequent failure to understand the rock record of events, and the defects of this record, the missing or mutilated pages. Fossils are rare in Precambrian rocks and, except toward the end, are useless for correlation and stratigraphy, and isotopic ages have uncertainties that can be as long as as the entire Cenozoic. A great deal of information has been lost to erosion, burial, subduction, and a metamorphism often so intense that we cannot remove its overprint. As we descend further into time, more and more often does the key of the present fail to unlock the door to the past. Frequently we are limited to theory and reasoning and must be content when the facts, failing to prove us right, at least do not tell us that we are wrong.

BEFORE HISTORY

The earth today is hot, its temperature at the center estimated at 3,000°C. It is not a homogeneous body; a core 7,000 km in diameter is contained within a 2,900-km-thick mantle, on which rests the thin skin of the crust. The core is very dense and probably consists mainly of iron and nickel. It is, in a manner not yet fully understood, the source of the earth's magnetic field.

Geochemical considerations suggest that the average composition of mantle and crust combined is like that of a class of meteorites called carbonaceous chondrites. If we melt such a chondrite, a scum rich in large, light elements and similar in composition to the continental crust rises to the top, whereas the residue underneath resembles the upper mantle (Figure 13.1). As a result, the upper mantle is depleted of light elements, but below a depth of 700 km the original chondritic material may still exist. Hotspot volcanoes, such as Hawaii, tap very deep layers of the mantle. Their lavas, quite different from those of the mid-ocean ridge that rise from much shallower levels, support the chondritic nature of the lower mantle.

A puzzle remains. If the upper mantle has lost much of its light material to the crust, it ought to be heavy, heavier than the unaltered chondritic lower mantle. Why, then, does the heavy upper mantle not sink and gather just above the core? Although it seems remotely thinkable that the shell structure of the earth is unstable, it is more likely that some kind of "lightener" has been added to the upper mantle. Water and carbon dioxide, subducted with altered oceanic crust and sediments, would be good candidates for this role.

The earth, quite certainly, formed as an aggregate from a dust cloud orbiting the sun. To permit the various layers to segregate, a period of melting was necessary. Various sources of energy may have brought this about: the gravitational energy released when the dust gathered and contracted into a ball, the impact energy of large meteorites and asteroids as they plunged into the growing planet, or the friction produced by tides. Combined, these sources of energy might raise the temperature of the new planet by at least 1,000°C, but that is not enough.

By far the largest source of energy, good for several thousand degrees, was radioactivity, mainly from an unstable isotope of potassium, from thorium and uranium, and from a number of now extinct, short-lived isotopes. Long-lived isotopes were also much more abundant then (Figure 13.2). Uranium-235, with a half-life of 0.71 billion years, for example, has already passed through six half-lives, and less than 1/64th of the original amount remains.

If we assume that the initial dust ball was homogeneous, we need a lot of melting to collect the nickel and iron in the core. Additional

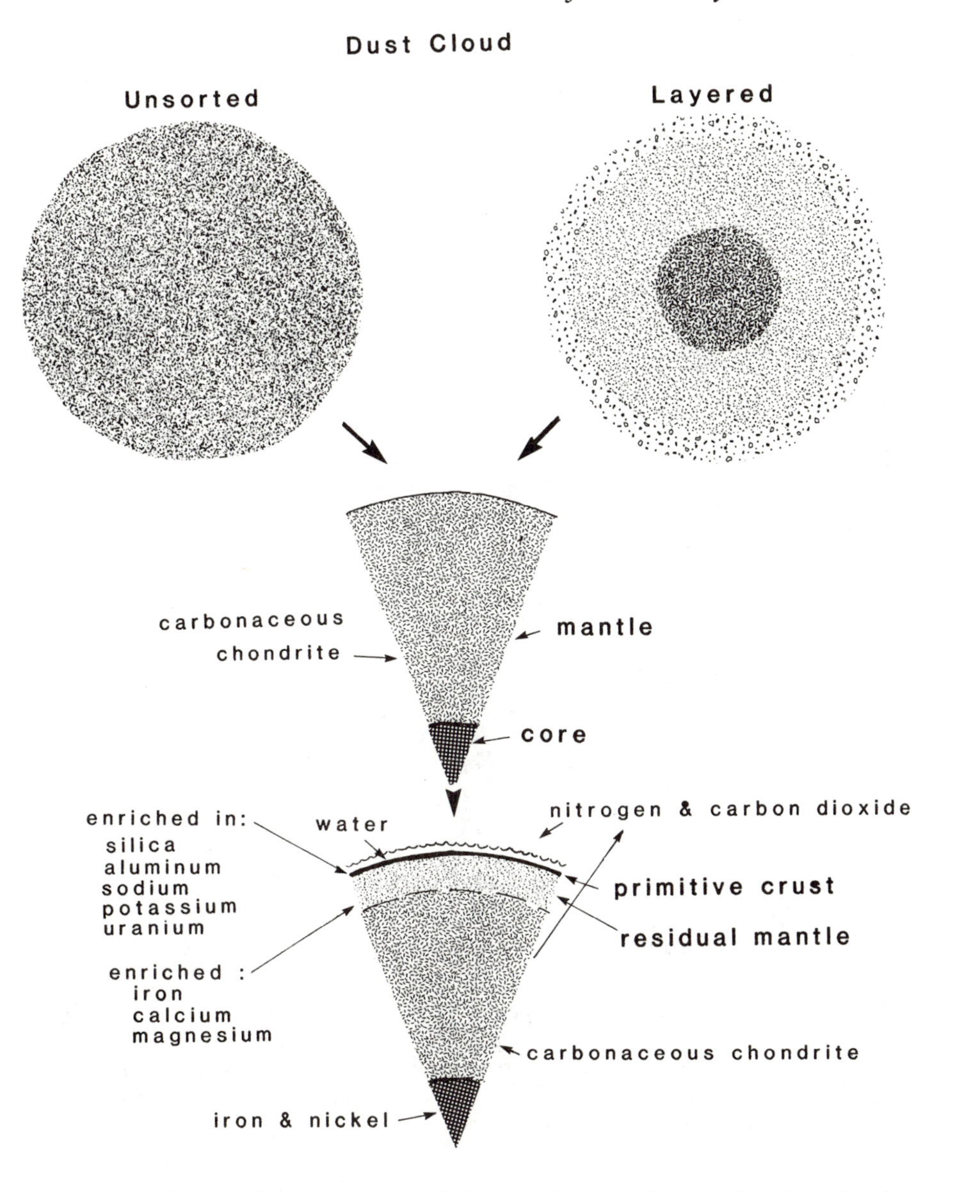

Figure 13.1. The earth formed from a cloud of cosmic dust (top) that may have been already stratified, with the heaviest particles in the center. From this came a planet with a dense iron and nickel core covered by a shell resembling in composition a carbonaceous chondrite. The outer part of the chondrite subsequently melted and separated into a heavy residue, the mantle, and a lighter primitive crust surrounded by an ocean and an atmosphere of some sort.

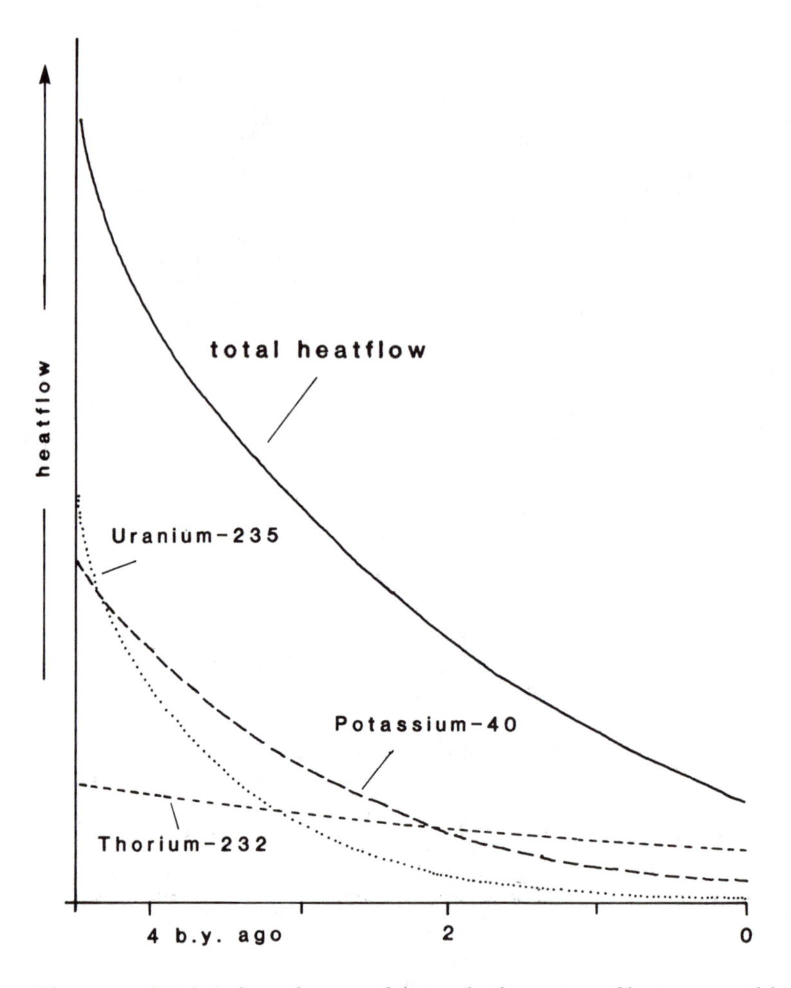

Figure 13.2. During the early years of the earth, the amount of heat generated by the decay of unstable isotopes was much larger than it is now. Many short-lived isotopes have vanished entirely, and others, with halflives of the same order as the age of the earth (thorium-232, for example) or shorter (such as uranium-235), have greatly decreased in abundance and hence produce much less heat.

melting would be needed to separate the chondritic mantle into a lower and an upper zone, put a crust on top, and sweat out the ocean and the atmosphere. During accretion, of course, the heat built up continuously, but there is room to argue about the exact moment when the planet became hot enough to melt. At present, the best chemical and physical reasoning suggests that this happened very early as a result of the formation of the core.

Of course, the dust cloud itself might have become sorted into layers under the influence of gravity, with the heaviest elements nearest the planet-to-be. The core could then have formed right away, and perhaps even the oceans and the atmosphere, and we would be at liberty to postulate a later phase of heating to make oceanic and continental crust out of the primitive mantle.

Whatever the nature of the dust cloud may have been, we may think of the primitive earth as a nickel-iron core wrapped in a chondritic mantle. When this mantle, or at least its upper part, began to melt and segregate a primitive crust is of particular importance for the history of the earth's surface. Was it very early and sudden, or might it have been later and perhaps gradual, continuing for billions of years? We shall have to turn to the Precambrian rock record itself for the answers, but we can learn something from what has been discovered in recent years about the other earth-like planets of the solar system.

The surfaces of Mercury, Mars, the moon, and perhaps Venus have not been altered by subsequent plate tectonics, remaining in a state long obliterated on earth. Their nature indicates that they were once covered with magma oceans, giant versions of the lava lakes of Hawaii. Light scum continuously rose to the surface, congealed to primitive crust, then was dragged down again by vigorous convection. When finally the surfaces of these oceans had cooled and solidified, a last bombardment with space debris about 4 billion years ago shattered the thin crust one more time. On the moon, the remains of the broken primitive crust are now the lunar highlands, whereas the maria, the dark blotches visible from earth, are the rejuvenated and then again congealed lava lakes, undisturbed ever since. The moon and Mercury, small and thus having cooled fast, ended their histories at this point, but Mars continued a while longer, acquiring a very thick crust loaded here and there without bending with giant volcanoes. Venus may even have reached an initial plate-tectonic stage. It is quite possible that the earth also was once covered with lava oceans, and almost certain that it must have suffered the same space bombardment. It has remained very active ever since, however, and 4 billion years of upheavals, including those of plate tectonics, have erased the record.

THE FIRST CONTINENTS

Thus, at an early time the earth had an upper mantle and a primitive crust, and probably also an ocean and an atmosphere. The separation into crust and mantle was not complete; enough light material remains in the upper mantle to raise the volume of the continents from the present 0.1 percent to 7 or 8 percent of the volume of the earth. When did the segregation of mantle and crust begin, how long did it continue, and how did the protocrust evolve into continents and ocean basins? It is tempting to imagine a surface dotted with giant volcanoes belching ash and noisome gases, rain pouring ceaselessly from dense reddish clouds, lightning glaring on wet black lavas. How do we go from this dramatic image to a reasonable facsimile of the present earth?

To make a typical continental granite out of primitive crust we must recycle it many times by melting, cooling, subsidence, burial, and renewed melting. Each time, some of the heavier components will remain behind while the lighter ones will gather and rise to the surface, there forming a kind of scum. Once there is land, weathering will help; it retains on the continents only the most resistant minerals, the oxides of silica and aluminum, the potash feldspars, and the quartz, and removes solubles to the ocean. By these processes the upper layers of the continents evolved to a very special chemical composition entirely different from that of the oceanic crust and from the presumed primitive crust (which no one has seen).

Some of the earliest Precambrian rocks, 3.8 billion years old, already possess an advanced continental composition and show evidence for the activity of weathering and erosion. How and when did these continents form and rise above the sea to make this possible? That the crust segregated from the mantle is not, by itself, enough to answer this question, because a patchy primitive crust evolving to separate continents in a sea of mantle is as likely as a thin continuous shell of crust completely covering the earth (Figure 13.3).

To clarify what was involved, let us examine what might have happened if the melting phase had ended with a continuous shell of primitive crust under a continuous ocean. Figure 13.2 told us that in those early days, say 4 billion years ago, much more heat was produced than today. We must therefore assume that the mantle was convecting vigorously under the thin crust, like a pan of soup on a stove turned up too high. The hot mantle should have been less viscous than it is today, its flow more rapid, and, one might guess, the convection cells smaller but more numerous. The image of a planet dotted with a dense scatter of hotspots is probably rather appropriate, more so than the linear patterns of modern seafloor spreading.

Above each rising hot limb (Figure 13.3), the thin crust would tend to fissure, and a volcano would build there from lavas separating out of the underlying mantle. Under the weight of this volcano, the crust would sink until it reached a zone hot enough for it to begin melting. This would allow further differentiation, and some light elements

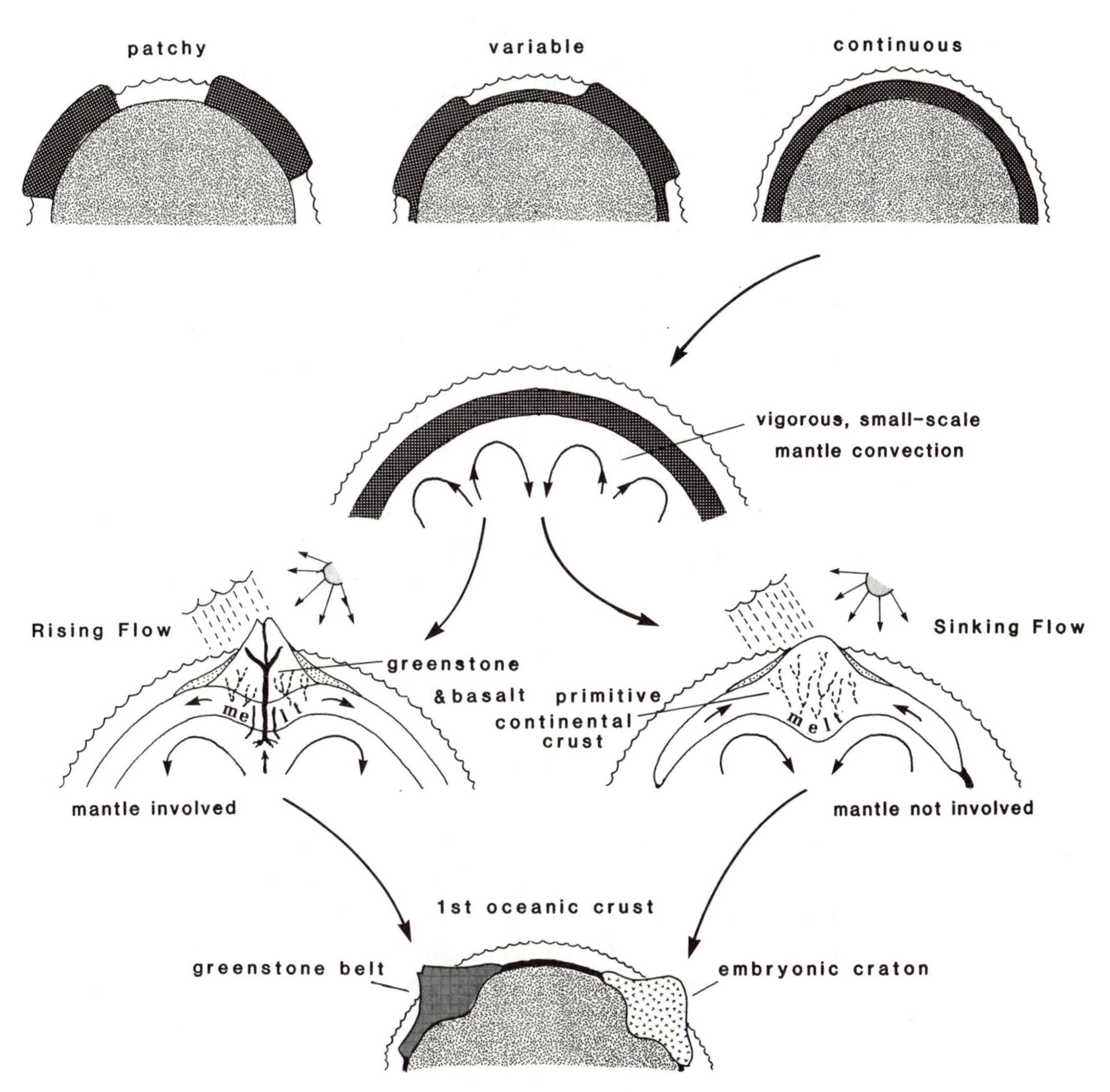

Figure 13.3. We can only guess whether the primitive crust was patchy, lumpy, or formed a continuous shell (top). Assuming, for the sake of simplicity, the latter, we speculate that embryonic continents were formed by vigorous but small-scale mantle convection (center). Above a rising current, melting of the crust would drive it toward a more continental composition, but progress would be delayed by intrusion of basalt from the mantle as the crust stretched and fractured. Above a sinking flow the crust would be compressed and thickened, no mantle material could interfere, and progress toward true continental crust would be faster, aided by weathering of the emerging land. In this way, small islands of greenstone and continental crust would be created, separated by oceanic crust as the primitive crust was consumed and the mantle stripped bare.

would distill off into the crust above. If the volcano grew large enough, it might break the sea surface; weathering and sedimentation could then contribute to further differentiation. The primitive crust would so evolve toward a more continental composition, but slowly, because mantle material would continue to be added. In the main, the rocks formed in this setting would be like the *greenstone belts* of the early Precambrian: deformed complexes of dark volcanic rocks, with shales and sandstones rich in volcanic fragments and known as *graywackes.*

Where the cool limbs of two convection cells met and foundered, the primitive crust, itself too hot and thus too light to be subducted, would be compressed and therefore thicken. A downward bulge would form that would melt at the base, producing a lighter magma that would invade the crust above. Because of the greater buoyancy of the thicker crust, land might emerge and give weathering an opportunity to speed up the process of differentiation. In this case, no new material from the mantle below the crust would be involved, and the crust should evolve more rapidly toward a continental composition than would be possible above a rising convective flow. In this manner the continental cores, the cratons, may have originated.

Obviously, as we continue to concentrate the primitive crust into thicker patches, we should eventually in some places run out of it, and the mantle between the embryonic continents would be stripped of its cover. There, true oceanic crust would then be generated in the same way as still happens on divergent plate edges.

This purely hypothetical model leads us to expect randomly distributed small continents of granitic composition wrapped in greenstone belts of less advanced material. The older Precambrian regions, such as those in South Africa and Australia, offer us fine examples of such configurations, but there are other ways to explain them. Rather than pursuing this exercise further, let us turn to the evidence instead.

A TIME OF GROWTH

We have already raised the question of the timing of the differentiation into oceanic and continental crust. Did this happen early or late, suddenly or gradually? The oldest rocks show that some continental crust existed 3.8 billion years ago, but on all continents the Precambrian rocks display evidence for a wide range of ages. Clearly, the process of making continental crust and continents was not completed all at once. In fact, continental crust and continents need not even have been

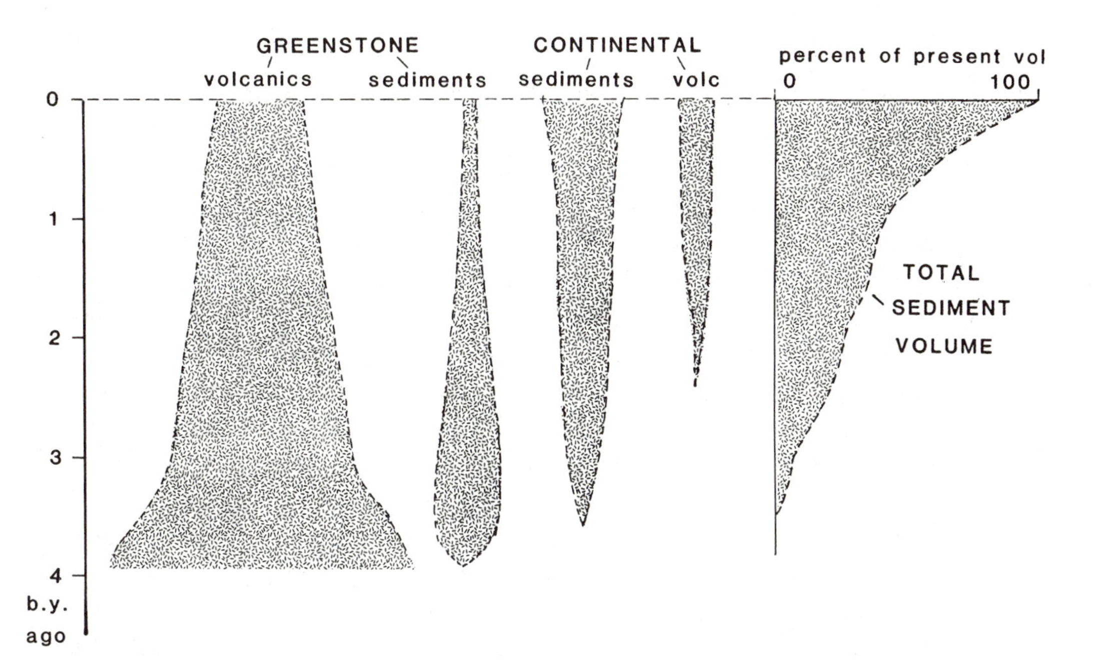

Figure 13.4. The history of Precambrian continental growth is reflected in the types and volumes of sediments generated on the continents and dumped in the sea. Initially, all sediments were derived from volcanic rocks and primitive crust or from greenstone belts. Continental deposits do not appear until later and do not dominate for another billion years. The growth of the sediment volume (right) indicates a slow, perhaps stepwise, increase in emerged land.

formed simultaneously. The process of melting and recycling of primitive crust might well at first produce large slabs so thin that they were mostly covered by sea. Only gradually might thickening have occurred so that true continents emerged. We are able to trace continental evolution with the aid of the changing composition of Precambrian sediments (Figure 13.4). If all or most continents rose above the sea early, we would expect little change in the composition of sediments since, let us say, 3 billion years ago (to allow generous time for weathering to do its work). Instead, we find that throughout the early part of the Precambrian, sediments were mainly derived from a primitive crust or from oceanic lavas. Truly continental sediment types did not become abundant until much later, appearing at different times on different continents. The same is true of lavas of purely continental origin.

Another argument can be drawn from an estimate of the volume of sediments produced over time (Figure 13.4). Although it is not easy to make this estimate, because of burial, recycling, or subduction, the increase over time in the amount of sediment shed from the continents has been so large that it dwarfs the uncertainties. Because very nearly all sediment originates through weathering and erosion on land, this increase with time can only mean that the area of the continents exposed above the sea must have similarly increased.

We conclude, therefore, that the volume of continental crust increased steadily throughout the early Precambrian, at least until 2.5 or 2 billion years ago. The area above sealevel expanded similarly, but more slowly, because the formation of continental blocks apparently lagged somewhat behind the formation of continental material.

Isotopic ages of Precambrian rocks support this long-lasting continental growth, if we avoid those rocks whose radioactive clocks were reset by later metamorphism. They also suggest growth in spurts (Figure 13.5). Around 3.5 billion years ago, only about 5–10 percent of the volume of continental crust appears to have been present. After a quiet spell of several hundred million years, a stage of major growth took place between 2.8 and 2.5 billion years ago to leave a continental crust close to half of the present volume. Subsequent construction phases at 1.9–1.7, 1.2–0.9, and 0.7-0.5 billion years ago added most of the rest. Very little continental growth can be attributed to the Phanerozoic. Thus, it seems that the first true large continents appeared about 2.5 billion years ago at the boundary between the two divisions of the Precambrian, the Archean and the Proterozoic, bringing large lands and shallow seas with their many environments. It was a momentous event and may have had large biological consequences, as we shall see in Chapter 18.

Why the evolution of the continents should have been episodic is an open question to be added to the growing list of earth processes that seem to march to some kind of beat.

TIME AND PLATE TECTONICS

Geologists are uniformitarians at heart, although they do not emphasize it so much nowadays. It is therefore only natural that when plate tectonics took the status of ruling theory, most of us assumed that it had operated from time immemorial, that ridges had rifted, continents drifted, and plates subducted as long as there had been a lithosphere and an asthenosphere. Lately, however, some have begun to wonder if this need be so. This heresy does not please every adherent of the ruling theory, and the discussion has suffered somewhat from the accusation that those who oppose classic plate tectonics for the Pre-

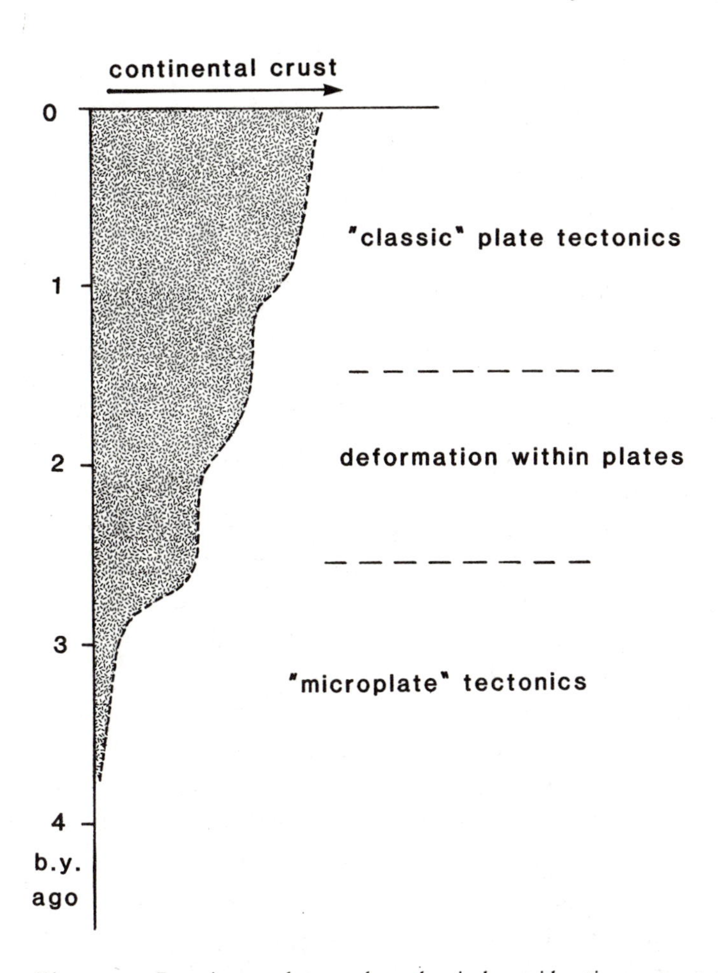

Figure 13.5. Isotopic age data and geochemical considerations suggest that the amount of continental crust increased greatly during the Precambrian, probably mainly during major growth phases between 3 and 2.5, just after 2, and around 1 billion years ago. As the heat production diminished, the plates thickened, and the continents grew, there may have been also a change in tectonic style from "microplate" tectonics to internal deformation of plates to the classic style complete with subduction.

cambrian simply demonstrate their ignorance of its principles. Such a reaction is not particularly useful, because it ought to be obvious that any energy-consuming process claimed to have been operating without change in pace since the earth began violates the prohibition against perpetual motion machines. Although James Hutton could see neither a beginning nor an end, we know now of the beginning and cannot doubt the end. The second law of thermodynamics demands, and the thermal history of the earth makes clear, that the engine has been slowing from the first and must come to a stop. It seems only reasonable that the dynamic behavior of the earth should vary with the speed of its motor. The question is, How much?

We have already speculated that the early Precambrian, the Archean, was a time of small drifting plates, jostling each other, colliding, and at times coalescing (Figure 13.5). Eventually, such "microplate" tectonics might produce larger plates, large enough to straddle convection cells. Simultaneously, the gradual decline in heat production would slow the convection and push the system toward larger cells. Oceanic crust must have formed, right away if the protocrust was patchy, or later if it was not.

The early Archean lithosphere probably was warmer than the present one, not only because of the higher heatflow of the time but also because of the smaller sizes of convection cells and plates. Today, oceanic lithosphere comes to the subduction zone cooled thoroughly by a voyage that may have taken more than 100 million years. With small-scale convection, however, the crust would sink while still young, and hence would be hot and light. Some geologists think, as I do, that this is a plausible model for the early days; others deny it. In terms of the available evidence, a standoff for the moment.

Let us just assume that, for a while, the Archean lithosphere was hot and light. This increased its buoyancy, perhaps so much that even oceanic lithosphere would have been too light to sink. Eventually, however, progressive cooling must have ended this state, and real subduction became possible. When this happened is wholly uncertain, and the entire presumption that there was once a lithosphere unable to subduct is open for debate. On the other hand, the first firm evidence for plate tectonics in the classic sense, that is, with rifting and subduction, comes no earlier than about 1.5 billion years ago, and we have reason to believe that a somewhat different style of plate tectonics prevailed early in the history of the earth.

Whatever this style may have been during the early and middle Precambrian, paleomagnetic evidence indicates that the continents were drifting with respect to each other around 1.5 billion years ago, and perhaps another billion years earlier. However, the same data also show that on some continents, such as those that later joined as Gondwanaland, the individual old nuclei changed their positions relative to each other very little even as late as the Proterozoic. Yet at the same time the greenstone belts that lie between them exhibit clear evidence of several periods of major deformation (Figure 13.6). In many cases, the paleomagnetic data leave little room for intervening large oceans, and so do not permit us to assume that these belts were produced by

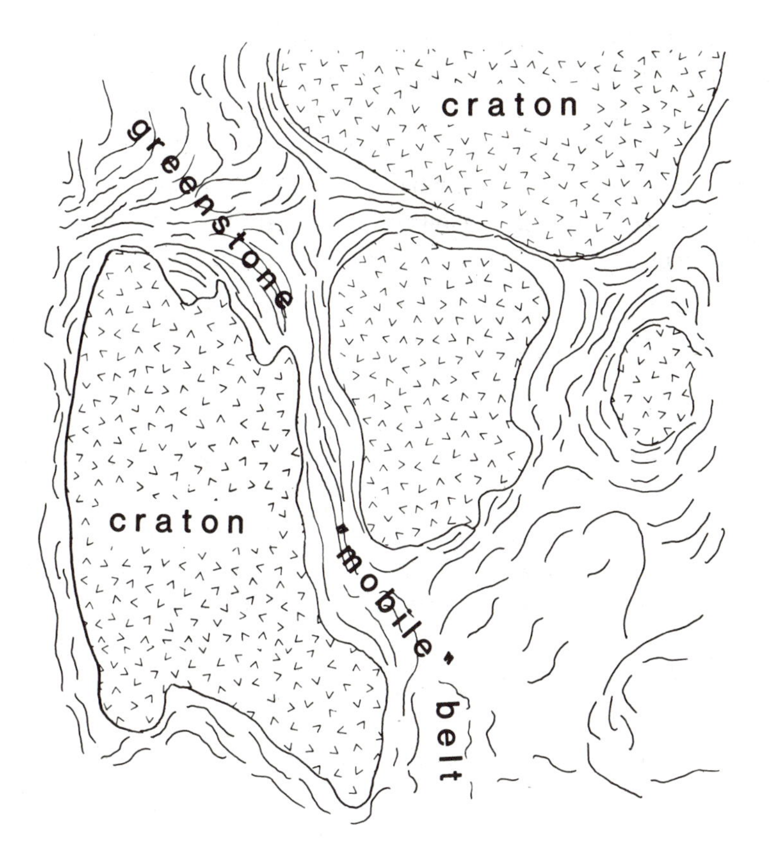

Figure 13.6. The oldest parts of the continents preserve the imprint of former tectonic styles. The terrain illustrated here after a satellite photo measures 100 km across and shows continental nuclei (cratons) wrapped in greenstone belts on a scale much smaller than that of the present plates.

continental collisions. Also, evidence is lacking in the rock record for subduction and even for oceanic crust. Therefore, the greenstone belts do not seem to be the Precambrian equivalents of the Appalachians or the Himalayas, but represent instead a tectonic style all their own, with major deformations not on plate boundaries but wholly within plates. A possible candidate for the process that formed them is the tearing off and sinking of the lower lithosphere that Peter Bird of the University of California at Los Angeles invoked in connection with the collision between India and Asia (see Chapter 8). This process is not necessarily limited to plate collisions, and in the Precambrian it might have provided volcanism, deep-water sedimentation, deformation, and metamorphism, of which the rocks of the greenstone belts bear eloquent witness.

The evolution of the continental crust with its various episodes of growth was thus probably accompanied by a gradual change in tectonic style. After an early stage of what might be called microplate tectonics, true plate tectonics with subduction appeared some time between 2 and 1 billion years ago, as far as we can tell at the moment. For a long time, however, it may not have been the dominant process it is today, and major deformation within plates, no longer common now, may have been an important feature of the middle and late Precambrian world.

14

Water for the sea, air for the atmosphere

Having introduced a semblance of order in the early evolution of the solid earth, we must now turn our attention to the oceans and the atmosphere. Let us begin with the oceans by observing that the problem of their origin, as distinct from the origin of ocean basins, has two parts: the source of the water and the source of the substances dissolved in it. We discussed the evolution of seawater in Chapter 12; its origin and initial composition remain to be considered here.

For the atmosphere, the main issues are its present composition and how that came to be. Judging from what we know about the rest of the solar system, our atmosphere, with its 78 percent nitrogen, 21 percent oxygen, and minor amounts of argon, carbon dioxide, and water vapor, is an odd one indeed. It is so odd that our growing knowledge of the very different atmospheres of Venus, Mars, Jupiter, Saturn, and some of their moons helps us little in understanding our own.

WHENCE THE WATER IN THE SEA?

Whatever model one espouses for the accretion of the earth, the composition of its parent material makes inevitable the early release of large amounts of water and gases. Therefore, around 4 billion years ago the earth had an ocean, although perhaps not a full one, and an atmosphere still quite unlike the atmosphere of the present. How ocean and atmosphere came into being and evolved during the early history of the earth has been the subject of much theorizing. The few examples to be discussed here represent mere possibilities and are, at least in part, mutually incompatible.

How much water there was on the early earth is still a matter of discussion. In 1951, the late W. W. Rubey proposed that the source of all water and gas was the mantle, that both were vented to the surface by volcanoes and volcanic springs, and that the volume of water increased steadily over time, although perhaps somewhat faster in the beginning (Figure 14.1). This thesis was supported by the fact that volcanic fluids contain salts in the right proportions to account, at least approximately, for the composition of seawater. Rubey's model was straightforward and consistent with the evidence of the day and was quickly accepted. It does, of course, imply a steady rise in sealevel that conflicts with the evidence cited in Chapter 10, but that evidence covers only the last few hundred million years. The conflict is thus not disastrous.

Obviously, it could also be the other way around, and the water released by volcanoes might be recycled seawater (Figure 14.2). That is indeed what more recent geochemical studies demonstrate; only a tiny fraction of volcanic steam can be shown to be "new" water, released from the mantle for the first time. In fact, some evidence suggests that the mantle, instead of releasing water, is gaining it through subduction of oceanic sediments and crust. How much of that water and of the oxygen and carbon dioxide that accompany it are later returned via volcanic eruptions we do not know, but it seems possible that the earth's surface is losing rather than gaining the precious fluid.

Karl K. Turekian of Yale, who came up with this surprising suggestion, has also pointed out that if the original mantle was indeed like the average carbonaceous chondrite, as most people believe, it should contain at least 20 times as much water as is now present at the surface of the earth. In his view, the problem is not where a suitable quantity of water might have come from to fill the oceans but rather what happened to the much larger amount that should have been released during the initial melting. If he is right, talking about how and when the early ocean filled should be rather futile, but time will tell.

During the last few decades, research in the geochemistry of isotopes and rare elements in Precambrian rocks has begun to place rather useful limits on our creativity, but it also occasionally comes up with such a startling reversal of what long seemed obvious. Clearly, it is premature to imagine that we have a firm grip on even so basic a problem as the source of ocean water. Let us therefore accept the thesis that the early Precambrian ocean was almost full, and proceed.

How did this early ocean acquire its salinity? Most dissolved substances in the sea were derived from weathering on the continents, and much smaller amounts, to judge from present conditions, from rock alteration, submarine "weathering," in the sea. If that is so, the rate of

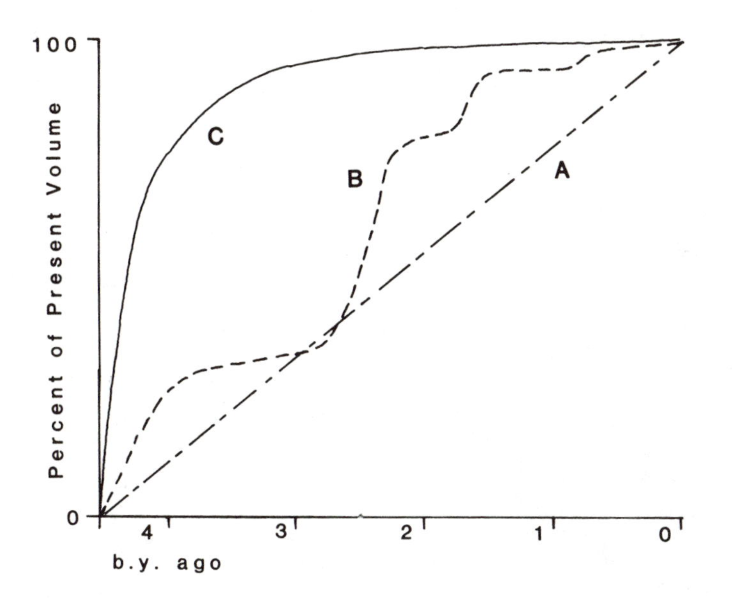

Figure 14.1. In principle, the ocean could have filled with water in three different ways, but one of those, a slow and steady increase (A), is improbable because of the evidence that the oceans were already full in the late Precambrian (Figure 10.1). Rapid filling at the time the primitive crust separated from the mantle (C) is more plausible, but the stepwise growth of the continents suggests that there may have been later additions of water as well. Unfortunately, we cannot say whether those episodic increases were large, as shown here (B), or trivial, or even if they took place at all.

salt supply should be roughly proportional to the area exposed above the sea. The previous chapter has made it clear that this area increased drastically during the early and middle Precambrian from 5–10 percent of the present surface in the earliest Archean to almost 75 percent, 2.5 billion years ago. The small initial land area would have meant an equally small influx of salt into the sea. Taking into account the expansion of the land area over the next billion years, the present salinity could not have been achieved much earlier than 2 or 1.5 billion years ago. After that, the situation is quite different: We need processes to dispose of excess salt so that the salinity can be kept within reasonable limits. The role of the newly discovered deepsea hotsprings in all of this may well have been very important (see Chapter 12), but we cannot evaluate it quantitatively at this time.

THE EARLY ATMOSPHERE

It was once thought that the primitive earth had an atmosphere composed mainly of methane, ammonia, hydrogen, and water vapor, as

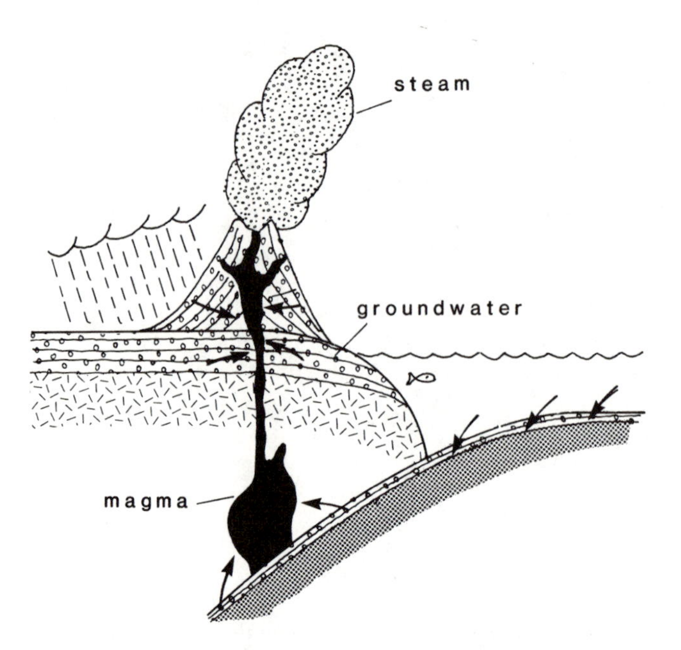

Figure 14.2. Volcanoes release a great deal of steam, but it consists mainly of groundwater recycled from the volcano body and from the underlying rocks. Another part may be seawater derived from subducted oceanic crust and sediments. Only a very small fraction is "new" water.

could be expected from a solar dust cloud. Scholars interested in the origin of life liked such an atmosphere, because it would have contained many elements essential to the building blocks of life and lacked the oxygen inimical to them.

If the earth ever possessed such an atmosphere, it blew away very early. It could not have evolved gradually into the present one, because the concentrations of the noble gases neon, krypton, and xenon that occur today are much too small for an origin in a solar dust cloud. These elements will not combine with anything else and are too large to diffuse away into space, so that they could not have been lost gradually. Therefore, the only reasonable hypothesis for the origin of the present atmosphere is that it escaped from the interior of the earth when the primitive crust separated from the mantle.

What might this atmosphere have been like in its pristine state? There must have been argon in it, produced by the decay of radioactive potassium, as there still is. It should already have contained essentially all of its nitrogen, because there is no reasonable way in which really large quantities of this relatively inert gas could have been added later. Volcanoes might have contributed some sulfur and hydrogen, probably as the malodorous gas hydrogen sulfide. Finally, even the earliest Precambrian rocks indicate that weathering was already taking place on land. Because land plants, which are the main weathering agents today, had not yet appeared, this means that carbonic acid must have done the job. Carbonic acid is formed from carbon dioxide, and therefore we conclude that this gas was present in the atmosphere as early as 3.8 billion years ago. In fact, carbon dioxide must have been a good deal more abundant then than it is now, because the amounts of carbon laid down in the form of limestone, coal, and oil since then are so enormous. If we convert all of these carbon-containing sediments back into carbon dioxide and distribute it between ocean and atmosphere in the ratio 60:1 that prevails today, the carbon dioxide content of the atmosphere should have been more than 100 times greater than today. Even if we allow for some carbon dioxide to have been added by volcanoes during the following eons, this value of 2–3 percent is high. Because of the greenhouse effect (see Chapter 3), the earth should have been quite hot; yet we know it was not.

OXYGEN

Even though we really have no idea why the early atmosphere of the earth was so different from that of the other planets, we can make a reasonable guess regarding the source of its oxygen, that oddest of its components and, from our point of view, the most important. There are both geological and chemical arguments against the possibility that there was a significant amount of oxygen in the earliest atmosphere or ocean. There is, for example, no firm evidence that until about 2.5 or even 2 billion years ago, rocks were weathered in an oxygen-bearing atmosphere. Precambrian river, beach, or dune sands sometimes contain minerals that cannot exist in the presence of oxygen. Precambrian dolomites, sedimentary rocks consisting of magnesium-calcium carbonate, are rich in ferrous iron, incompatible with an origin in seawater containing dissolved oxygen. Locally there may have been a little oxygen at times, but that was surely all.

There are theoretical arguments against free oxygen as well. It is, for instance, difficult to imagine any process that would release oxygen during the primordial melting of the mantle and allow it to escape to the surface in the presence of so much iron, hydrogen, carbon, and sulfur, all ready to capture every passing oxygen atom. Similarly, the synthesis of chemical compounds indispensable for the origin of life would have been impossible in the presence of free oxygen in air or water. It may seem paradoxical that life, with few exceptions now dependent on oxygen for its proper functioning, would not have been able to arise if this element had been present in ocean or atmosphere in the beginning, but it is a fact.

How, then, do we proceed from the primitive oxygen-free atmosphere to the present one, which contains 21 percent of this gas? The process we seek had better be efficient, because it had to generate not only all of the oxygen now present in the ocean and atmosphere but also enough to satisfy all oxygen "sinks" in the world, the easily oxidized elements and minerals. Only after all the iron had been laid to rest as ferric oxide, all carbon oxidized to carbon dioxide, all hydrogen transformed into water, all sulfur into sulfate, and all ammonia, methane, and hydrogen sulfide converted to oxidized compounds could free oxygen begin to accumulate.

Two processes are serious candidates for this job. Under the influence of the ultraviolet (UV) rays of the sun, water breaks down into oxygen and hydrogen:

$$2H_2O + \text{UV radiation} = 2H_2 + O_2$$

But the end products will instantly recombine unless we remove one or the other. In the upper atmosphere, some of the hydrogen diffuses out to space, albeit sluggishly, and is lost, while an oxygen surplus gradually builds up. At present, this process generates about 2 million tons of oxygen per year, which sounds rather impressive. Had this been the only process, however, another 26 billion years would have to pass before all oxygen sinks would be filled and free oxygen would have risen to its present level. Moreover, this process of photodissociation is inherently self-limiting, because some of the oxygen combines to form ozone, O_3, which is an efficient shield against ultraviolet radiation. That is, in itself, a good thing, because ultraviolet light is very damaging to genetic material and to live tissues. Without an ozone shield, the earth's surface would be uninhabitable.

The other process is photosynthesis – one form of photosynthesis, that is, because there are two. Some bacteria and algae remove, with sunlight as an energy source, hydrogen from such substances as hydrogen sulfide. This form of photosynthesis does not produce free oxygen. The other, more familiar type of photosynthesis is used by most green plants. It splits water molecules, with the aid of energy from sunlight, and converts carbon dioxide into useful organic compounds, as follows:

$$CO_2 + H_2O + \text{sunlight} = O_2 + CH_2O$$

Provided there are enough green plants, the second type of photosynthesis generates copious oxygen, about 20 billion tons per year at the present time. Most of this oxygen is used up in weathering and soil formation. The remainder, about one-fifth of the total present in the atmosphere, is consumed for biological purposes or dissolved in the oceans, where it helps in the decomposition of organic matter.

With a bit of guesswork we can put together a global budget for the oxygen, carbon, and hydrogen involved in photosynthesis, stored in the atmosphere and ocean, buried in rocks, recycled, or lost to space (Figure 14.3). This budget accounts for all oxygen by photosynthesis alone. Budget making, however, involves numerous assumptions, and others may read the balance sheets differently. There are those who do not see photosynthesis as an early dominant process and stake their faith instead on ultraviolet rays to produce the initial oxygen content of air and sea. The localized and temporary oxygen for which there is sparse evidence very early might indeed have been produced in this manner. There are also some biological grounds to argue that the first photosynthetic path taken by organisms would more likely have been the one that does not produce oxygen. If that is so, early Precambrian fossil evidence for photosynthesis would not be an argument for biological production of oxygen, and we would be forced to call on ultraviolet light. With a little strain, the budget can be adjusted to this view, but it is not widely accepted, mainly because the slightly later Archean and early Proterozoic sediments provide evidence for such large scale production of oxygen (see Chapter 15) that the ultraviolet-light process seems hopelessly inadequate.

In summary, at this time photosynthesis holds center stage as the principal process that oxygenated our atmosphere. Hypothesis and rock evidence together have been interpreted to show a slow and gradual increase in the oxygen content of the atmosphere beginning around 2 billion years ago (Figure 14.4), reaching about 10 percent of the present level by the beginning of the Phanerozoic.

AMMONIA! AMMONIA!

Some years ago a cartoon circulated among appreciative scientists that showed an alien from outer space crawling on all fours through the lovely California landscape gasping pitifully, "Ammonia, ammonia!" Biochemists, perturbed by the rejection of a primitive atmosphere rich in the ammonia and methane so necessary to construct the building blocks of life, sympathize with this cry. Could we, by some means,

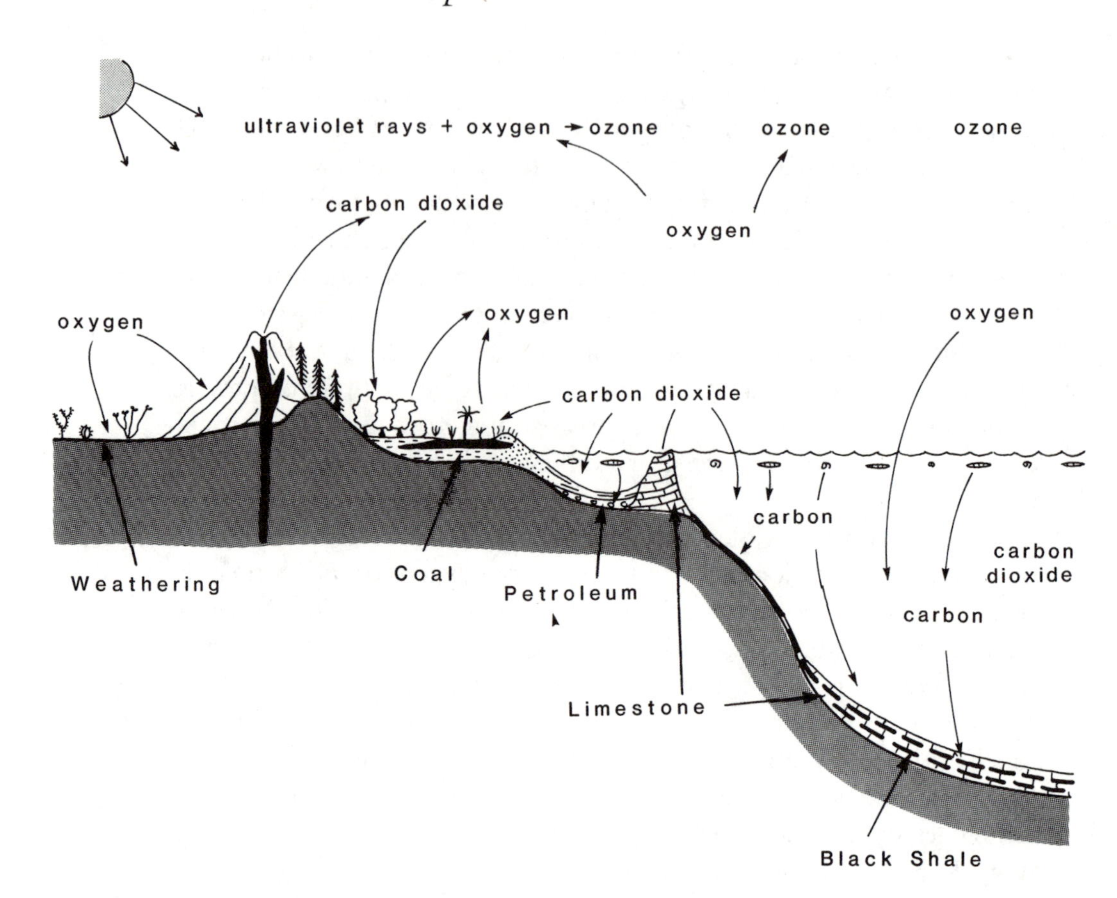

Figure 14.3. The carbon dioxide cycle in atmosphere, oceans, and life is complicated. The gas is consumed by plants in photosynthesis and converted into organic matter and oxygen. Much of the organic matter is permanently removed by burial in sediments. The oxygen enters the atmosphere, where some of it forms the ozone shield, much is used up in weathering, part is consumed in breathing, and a great deal is dissolved in the ocean. Carbon dioxide also enters the ocean, where a large part of it is converted by organisms into reefs, limestones, and calcareous oozes.

perhaps have just a little ammonia in the atmosphere or the sea, at least until the building blocks of life have been constructed? After all, not much is needed, and not for long, certainly much less than 1 billion years!

Ammonia may indeed have been released during the segregation of crust from mantle, because even today some comes out of submarine hotsprings. Its survival, however, would have been counted in at most millions of years, because it decays rapidly in sunlight. This is not a trivial point, because if we fail to find a source of ammonia, either we must forget about inorganic synthesis of basic substances of life (see

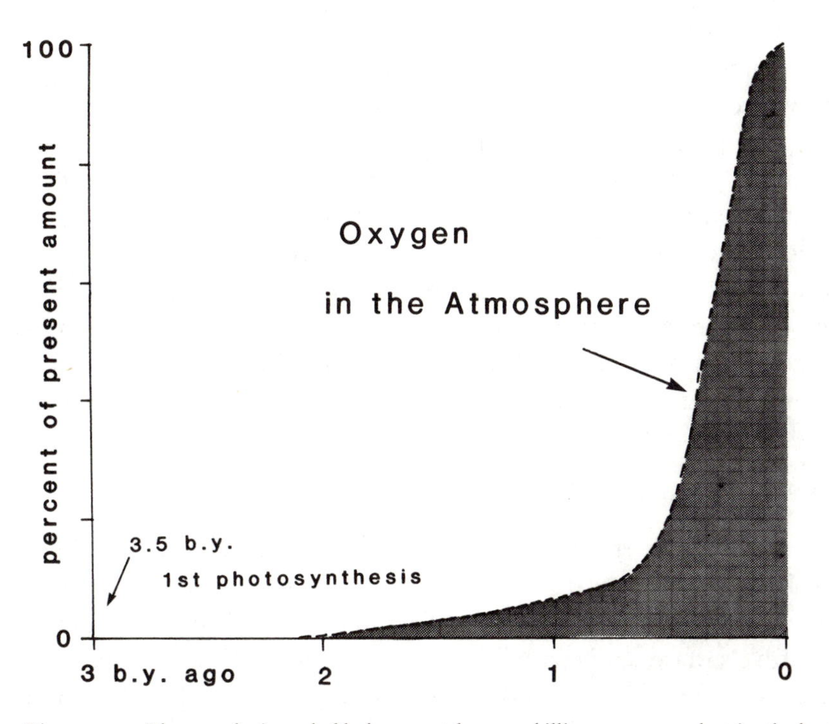

Figure 14.4. Photosynthesis probably began at least 3.5 billion years ago, but in the beginning the oxygen was used up in the oxidation of earth materials. Free oxygen did not enter the atmosphere until about 2 billion years ago, and at the start of the Phanerozoic the amount was still less than 20 percent of the present value.

Chapter 15), and import life from elsewhere in the universe, or we must manufacture them extremely early. Fairly complex biochemical compounds have been found in some kinds of meteorites, suggesting that this preliminary stage might have taken place soon after those bodies formed, but this does not necessarily mean that on earth it happened equally early and under the same circumstances.

There does appear to be a way out of the dilemma, however, even though it is hardly more than a theoretical possibility, and far-fetched at that. The element titanium, a component of several minerals found in sand, promotes or catalyzes the synthesis of ammonia from atmospheric nitrogen and water with the aid of sunlight. The sands of Imperial Valley in California are rich in titanium and, at times, the air above them contains a noticeable whiff of ammonia. The process works only on land, and after the minerals have been concentrated by weathering, and therefore ammonia from this source could not have appeared until the first continents had emerged above the sea.

Ammonia vented by the Precambrian equivalents of modern deepsea hotsprings is another and better possibility. The synthesis of organic building blocks would then have to take place in such springs, but they also furnish heat, hydrogen, and carbon dioxide, which are needed as well. In either case, ammonia would escape to the atmosphere in very small amounts, which is just as well, because the evidence against the presence of significant amounts of it is persuasive. In an ocean containing dissolved ammonia, one would, for example, expect calcareous sediments to be common and siliceous sediments rare, but the Archean rock record shows just the opposite. Small local ammonia concentrations are the best one can hope for, and the chemical evolution leading to early life must have made do with that somehow.

A WEAK AND PALE SUN

Except during the occasional ice age, the earth has never been very cold, nor has it become overheated as is the case on Venus, which most likely suffers from a runaway greenhouse effect. This remarkable fine tuning of the earth's surface temperature deserves comment, and not just because of the carbon dioxide problem we have mentioned earlier. There is good evidence that the sun's brightness, the solar constant, has increased by about 25 percent over the lifetime of the earth. Because practitioners of disparate disciplines communicate less than they should, this fact, well known to solar physicists, has until recently escaped the attention of geologists.

We have seen before that the temperature drop required to bring on an ice age is not large and that the oscillations of the Milankovich curve, resulting merely from changes in distance from the sun and in the orientation of the earth axis, were sufficient to bring on such a major climate change. Consequently, the impact on the earth's climate of a sun with only three-fourths of its present heat ought to have been quite spectacular. Our ignorance of such important factors as the area and distribution of Precambrian lands or the nature of their albedoes (probably high, because they must have been deserts), or of the ocean circulation of the time, makes any calculations of the consequences highly suspect, but it should have been a cold earth indeed. On the other hand, we are quite sure that if the earth would cool to the point that it is entirely frozen, the feedback effect of the high albedo would require a very large increase in heat, perhaps as much as one-third, to snap the planet out of it. That number is so large that we can disregard the possibility that the earth was ever entirely frozen, even for a short while.

What might have compensated for the lack of heat from the sun

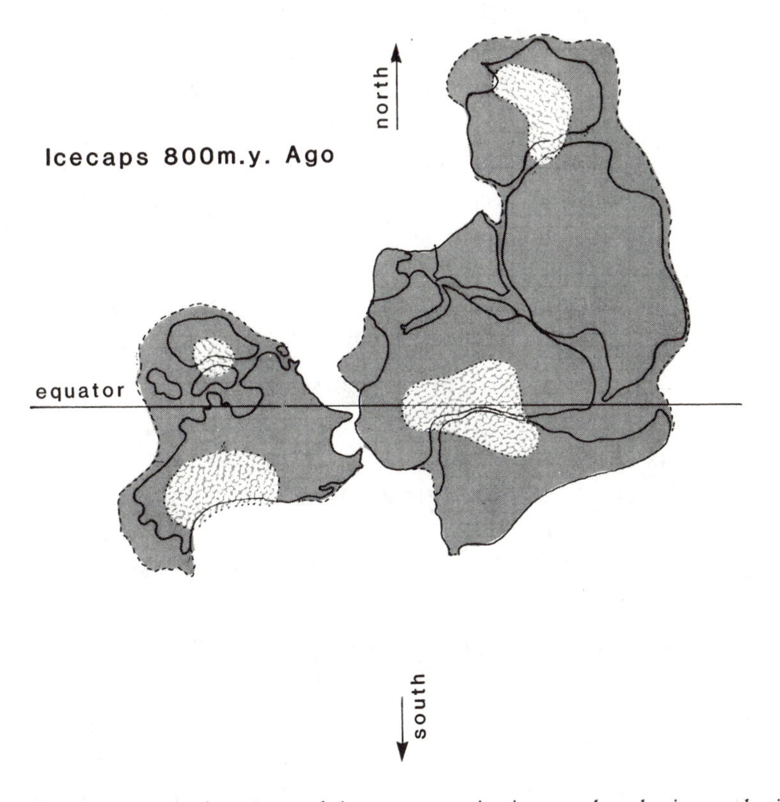

Figure 14.5. The late Precambrian saw a major ice age, long-lasting and with icecaps at astonishingly low latitudes. The planet may, in fact, have come close to freezing permanently. This ice age may, some think, have been caused by excessive consumption of carbon dioxide by planktonic life, at a time the sun was still weak and the greenhouse effect badly needed.

during the Precambrian? There would, of course, have been much more ocean and hence a lower albedo and a more even temperature distribution, but that would scarcely have sufficed. Clearly, rather than worrying about the carbon dioxide problem raised earlier, we should wonder if its greenhouse effect was beneficial and perhaps the principal factor that kept the temperature at the earth's surface so nicely within a narrow window around 25°C. The proposition is a simple one, and although difficult to prove, it has much to recommend it. It also has some disturbing aspects. One is that if we restore all available carbon to the atmosphere and the ocean by recycling coal, oil, and limestone, we almost certainly have too much of a greenhouse effect. That might be avoided by postulating that substantial amounts of carbon dioxide were added later by volcanoes or hotsprings, but that smacks unpleasantly of special pleading. Even more troublesome is that in order to keep the temperature within its rather narrow range, the removal of carbon dioxide from the atmosphere must have been matched to the increase in luminosity of the sun. How fine the required tuning is can be judged by our serious concern about possibly disastrous effects when our burning of coal and oil will have doubled the present carbon dioxide content of 0.03 percent in the not so remote future. Because the prime remover of carbon dioxide was certainly life, a finely adjusted control mechanism by feedback is needed. What that might be is not at all clear.

Understandably, the rock record is silent on the subject, except for the intriguing hint that at least once the feedback mechanism, whatever it may have been, partly failed. There were several Precambrian ice ages, but little is known about them. The last one, however, left abundant glacial deposits on all continents, and it was an odd one.

For one thing, it lasted an exceptionally long time, more than 300 million years. During that period there were three maxima, at 940, 770, and 620 million years ago, but even if each was actually a separate ice age, they were still long and unusually close together. Even more surprising is the geographic distribution of the glacial deposits (Figure 14.5). The data are, of course, not as profuse as we would like, but it is puzzling that, thus far, all deposits for which we have good paleomagnetic data are located below 55° latitude. The paleomagnetic data therefore clearly rule out the possibility that what we see is simply the result of various continents crossing the poles at different times. A global glaciation is therefore likely, because it is doubtful that one restricted to middle and low latitudes would be possible; most likely there was at the time little land in the polar regions, or perhaps we have not yet found or dated the glacial deposits that existed at high latitudes.

It is very tempting to suggest that this global glaciation during the latest Precambrian was caused by a large burst of biological activity around one billion years ago that consumed and buried far more carbon than was wise in view of the still somewhat feeble sun. The rather extensive limestone deposits of that time support this suggestion, but like others in this chapter, it remains tenuous.

15

The dawn of life

The first step in the evolution of life need not have involved life itself. It is probable that in the beginning, during a "chemical evolution," various nonbiological reactions produced the organic compounds from which the first living cells would be assembled. This event, if it happened, has left no trace in the geological record, and neither have the first steps of life itself. What happened during this critical time is thus speculative and controversial, but that is no reason to shun such an important topic. The arguments inevitably depend mainly on the rules of chemistry and on convictions held with regard to the environments of the early earth, their chemical condition, and the available sources of energy. Obviously, the boundary between chemical and biological evolution depends on what is and what is not life. I shall not attempt a thorough definition here, but shall assume that life presumes a systematic pattern of biochemical functions and a reliable means to reproduce and pass on to the offspring the processes that allowed the parents to exist.

CHEMICAL EVOLUTION

Even the simplest forms of life use a wide range of biochemical compounds, but for the manufacture of many of them a biological factory is not required, even though they are usually synthesized within an organism. The right amounts of carbon, nitrogen, hydrogen, oxygen, water, and some trace elements, with application of suitable energy, will produce many of them outside a living body. Besides in living tissue, such basic compounds have been observed in meteorites and in space. Most of them will not form in the presence of free oxygen, and few will survive if that element is there.

A key component of the living cell is the molecule deoxyribonucleic acid, or DNA. It contains genetic information as well as instructions for the biochemical functions of the cell, instructions that are conveyed to the place of execution by the companion molecule RNA (ribonucleic acid). DNA and RNA are constructed of long strands of simpler molecules, the nucleotides. Other important substances in the cell also consist of simpler molecules strung together in long chains. The proteins, for example, are chains of amino acids. Many of these basic substances have been synthesized in the laboratory under conditions that, we believe, reasonably resemble those of early Precambrian environments.

Amino acids, for example, form easily with the help of heat; the type produced depends on the temperature. Warm volcanic pools might be a good place, or deep-sea hotsprings. Besides heat, ultraviolet light or electric discharges provide suitable energy. The first experiment in prebiological synthesis was carried out years ago at the University of Chicago by Stanley L. Miller, who flashed artificial lightning through a flask filled with a mixture of gases thought to resemble the primordial atmosphere. Its success made him instantly famous. Since then, other experiments in a variety of simulated environments have yielded an astonishing array of biologically important substances.

On the primitive earth, energy was available in several forms, but the methane, ammonia, and hydrogen that were needed were, we now think, in short supply. Nevertheless, though the most probable atmosphere was quite unlike most experimental conditions, there is hope. Theory and a first few experiments indicate that nature is tolerant and that even our own atmosphere, provided it is shorn of its oxygen, would yield in small amounts many of the necessary substances during, for example, a thunderstorm. What we do not know is how much would be enough to set off the chain of reactions toward life.

Whatever the processes and the sites, at some instant in the remote past isolated pools of water contained enough basic compounds to permit the next step to occur. If most of the carbon now tied up in coal, oil, and limestone once existed as basic, simple organic compounds, leaving just enough carbon dioxide to keep the earth warm, the whole ocean would have had the strength of a very weak cup of broth. Of course, in smaller ponds and basins, the concentrations might have been somewhat higher. In such diluted fluids, how did the floating amino acids, peptides, and innumerable others combine to form more complex molecules like the sugars or the proteins? Mere chance encounters would never do, as many a statistically inclined scholar has pointed out, deeming the probability that life could have arisen under such conditions far too small to be taken seriously.

The argument, it seems to me, is specious. Not only did life demonstrably arise, but there is no need to call on mere chance. Catalysts favor and enhance certain reactions above others. Orderly mineral structures, such as those of the clays, may have functioned as templates on which simpler components could be patterned to form complex ones. It is a remarkable quality of large biological molecules that they easily combine in the proper sequence, seemingly possessing an internal drive toward the right structure. For example, Sidney Fox of the University of Miami has shown that if we evaporate a broth of amino acids under the right conditions, then wet them again, and so forth, perhaps in a small and admittedly rather special pond or tidepool in sun and rain, the amino acids will arrange themselves to form simple proteins. What is more, these proteins can form small, double-walled bodies, microspheres, that look like cells and possess the ability to absorb food from the broth, to grow, and to divide and multiply. Within the microsphere, even more complex molecules can form.

Large molecules also sometimes spontaneously form clusters called coacervate droplets, which are surrounded and protected by a tightly bound double layer of water molecules. Like the microspheres, coacervate droplets absorb substances from the liquid and can synthesize internally more complex molecules. Other processes that lead to similar results include, presumably, the still unknown ones that yield nucleotides, DNA, and RNA.

There is no reason why some or all of these processes should not be going on at present, but if they are, their products are consumed instantly by a teeming multitude of organisms. No such problem existed during the first half-billion years of the earth's history.

THE FIRST ORGANISMS

None of this is life yet. Life, as we said at the beginning of this chapter, presumes an orderly performance of biochemical functions and the means to pass on to the offspring the blueprints of the parents and a copy of their operations manual. This demands the step to DNA, which is a large one compared with the advance from primordial gas to amino acid. The nature of this great event is completely obscure at this time. Whether it happened on earth by processes not yet conceived by science or perhaps by import from outer space, some day, some long time ago single cells constituting the first true life were floating in the organic soup. Heterotrophs like us, they fed on organic matter that they had not themselves manufactured but that was, at least for the time being, reasonably plentiful.

We should not think of these first heterotrophs as simple or invariant. There may have been considerable variety from the start, and mutations, abrupt changes in the genetic code induced by ultraviolet light or radioactivity or simply by biochemical errors within the cell, may have increased it. If chance genetic alteration enabled an organism to subsist on a hitherto unused substance by means of a new enzyme, it would be able to survive better and longer than others who could not do so. Thus, step by small step, a complicated biochemical factory was constructed by many different organisms experimenting along many different paths.

A footnote seems in order at this point. Our view of the chemical evolution rests on plausible reactions between reasonable substances. Inevitably, however, much is beyond our understanding, because we know nothing about concentrations and little about reaction rates and must assume that many important reactions are still unknown to us. When did the organic soup appear? How concentrated was it, and how stable? How long did the various reactions take? With an atmosphere like our own, but with plenty of carbon dioxide and no oxygen, perhaps 3 million tons of organic substances could have been generated each year. Is that enough? How long might the first heterotrophs have been able to subsist on this soup, formed over a few hundred million years, before it was exhausted? Simple organisms are very good at adjusting rapidly to any available food supply, and they do deplete it in no time at all. Was there some mechanism, some combination of stable stratification and upwelling in the ocean perhaps, that limited the rate at which food was supplied to them? It would seem almost imperative, or famine would have snuffed out the prospect of lasting life on earth very soon.

Eventually the soup was gone, and to preserve life there was a need for organisms that could manufacture their own food. This is what autotrophs do with the help of a suitable energy source, using carbon dioxide, water, and various nutrients. Some bacteria obtain the energy from chemical reactions; others use sunlight and photosynthesis. Currently, photosynthesis, and thus ultimately the sun, provides virtually all of the food flowing through the pyramid of life from the smallest green plants to the largest predators.

PROKARYOTES AND EUKARYOTES

All organisms can be divided into two categories: the simple single-celled *prokaryotes* and the more complex single- or multi-celled *eukaryotes* (Figure 15.1). Prokaryotes carry a modest amount of genetic mate-

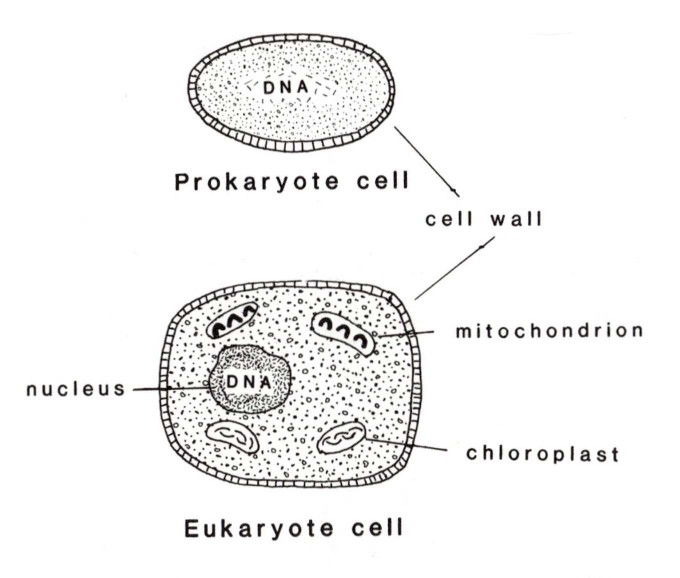

Figure 15.1. Prokaryote cells are small and simple, and their genetic material is diffused through part of the cell. Eukaryotes have a nucleus containing their DNA; they are usually much larger and carry a number of specialized bodies such as chloroplasts (plants only) and mitochondria.

rial disseminated throughout their cells, whereas eukaryotes keep much larger amounts arranged in packages called chromosomes and neatly wrapped in a nucleus. In addition, the eukaryote cell, but not that of the prokaryote, contains small bodies called organelles, each of which performs a special biochemical function. The best-known organelle is the chloroplast of green plants, which houses the chlorophyll responsible for photosynthesis. Chloroplasts resemble blue-green algae (nowadays called blue-green bacteria) and have some of their own separate genetic material that is passed on to descendants quite independent of the delivery of the main genetic package of the cell.

It was proposed long ago that chloroplasts were once independent prokaryotes that lived in symbiosis with eukaryote cells and gradually became so identified with their hosts that they now reproduce simultaneously and no longer have separate existences. Another organelle, the mitochondrion, which operates as the main energy plant of the cell, may well have begun in the same way: as an independent, symbiotic, respiring bacterium. It is even possible that the little tails and hairs that enable single-cell eukaryotes like the amoebae to swim in search of food may originally have been free-swimming bacteria captured through symbiosis (Figure 15.2).

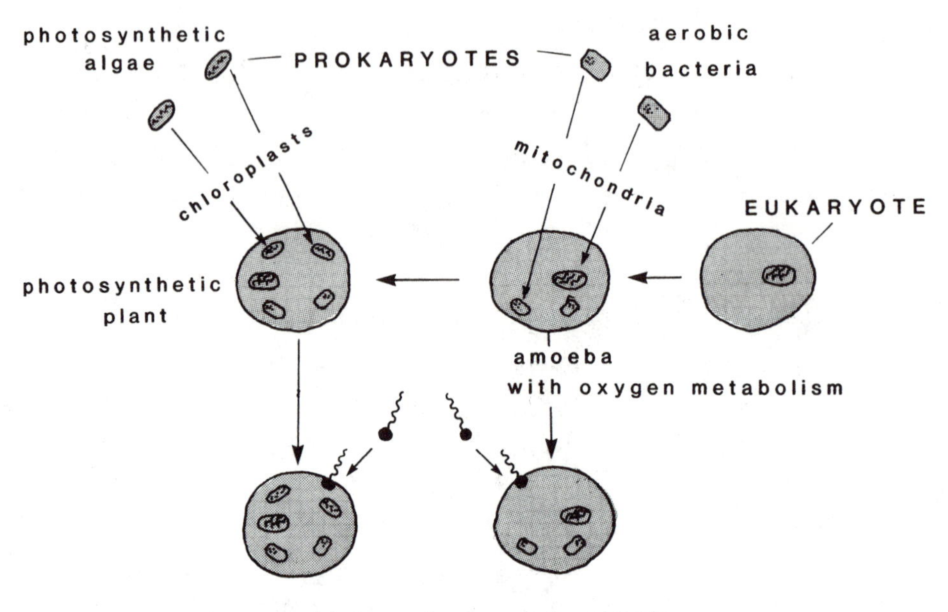

Figure 15.2. It has been proposed that many of the specialized features of the eukaryote cell began as prokaryotes living symbiotically within. Photosynthesizing algae may have become chloroplasts, aerobic bacteria took charge of oxygen-based energy housekeeping, and spirochetes with tails furnished a means of locomotion.

Eukaryotes also differ from prokaryotes in the way they reproduce. Prokaryotes usually multiply by cell division, each part receiving a full share of the parental genetic material. This form of reproduction is accurate, and variations come primarily from mutations. Eukaryote cells also divide, but by a complex and elegant process that splits the genetic material in such a way that a second step becomes possible, namely, sexual reproduction. In sexual reproduction, sperm and egg, products of cell division, each acquire half of the parental genetic material. The original amount is restored when fertilization takes place, but the parts come from different parents, giving each cell some qualities of one parent and some of the other. New genetic combinations are thus created, and change can be more gradual than with mutation. Lynn Margulis of Boston University has suggested that this important innovation was also the result of symbiosis with an organism that took charge of the reproductive system.

Must we then conclude that the eukaryotes, which may owe so much to prokaryote symbiotic guests, did themselves evolve from prokaryotes? This is indeed the prevalent view, but new discoveries in biochemistry (see Chapter 17) have cast doubt on such a simple heritage. The record is still being deciphered, but it is already clear that prokaryotes and eukaryotes have followed quite different paths for most of their histories (Figure 15.3). The biochemical differences between their DNA and RNA sequences are best accounted for by the assumption of a common ancestor.

In summary, it seems that life began with a variety of heterotrophs, ever better equipped through random mutations to take advantage of the varying food supply of their environment. Because the air was still without oxygen and there was no ozone shield to serve as protection against ultraviolet radiation damage, one of the early accomplishments must have been the ability to repair such damage, something some prokaryotes are still able to do today.

When the organic soup ran thin, the first autotrophs appeared, either on the chemical or on the photosynthetic pathway. We still have representatives of each, but the second mode uses a more widely available resource: sunlight. Photosynthesis soon came to dominate, first perhaps in its oxygen-free form (see Chapter 14), but when it became possible to split water, the other mode rapidly took over, because water is so much more abundant than other hydrogen donors such as hydrogen sulfide. The blue-green algae that acquired this capacity were very successful, but at the same time they became a major threat to all life because of the oxygen they produced. No doubt, some organisms perished, some developed biochemical protection against this poison, and still others went further and adopted an oxygen-based metabolism that had a considerable advantage in energy efficiency. Somewhere, sometime, in this setting, the eukaryotes appeared, rather unobtrusively at first as they built up their collection of symbiotic arrangements.

This is what theory, some experiments, and much thoughtful speculation suggest as a plausible but by no means proven early history of life. It has an elegant, contrapuntal logic to it. The first building blocks of life might not have formed without ultraviolet radiation as a source of energy and would not have survived in the presence of oxygen. Yet, as soon as life existed, ultraviolet light became a liability. Photosynthesis solved that problem with an ozone shield, but created a new one by generating free oxygen. Once biochemical means had been developed to deal with that menace, however, it turned out that a metabolism involving oxygen has many advantages over one that does not. Thus, the availability of oxygen in air and water became a major factor in the further evolution of life.

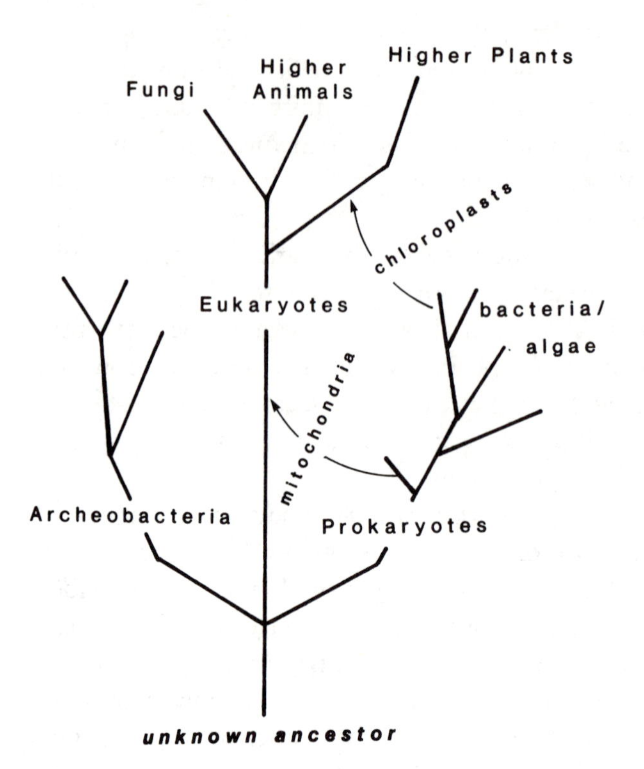

Figure 15.3. This evolutionary tree of Precambrian life, rather speculative still, of course, incorporates the idea that the eukaryotes benefited from many symbiotic arrangements.

THE RECORD OF PRECAMBRIAN LIFE

Let us now turn to the Precambrian rocks to find some commentary on this story. The oldest rocks, at least 3.8 billion years old, bear witness to the presence of land and sea. Were those seas empty, devoid of life? Organic matter has been found even in these oldest of all rocks, its chemistry suggesting but not proving that it might have been produced by photosynthesis. Unfortunately, the rocks were later metamorphosed at very high temperature, and we cannot be sure.

Better evidence appears a little later, in rocks between 3.5 and 3.3 billion years old. One should constantly remind oneself that the innocent decimal point precedes a number that indicates hundreds of millions of years; raise that number by one, and a time equivalent to the Cretaceous and Cenozoic combined passes. Much less severely altered rocks of this age are found in Australia and southern Africa, associated with true continental and shallow marine sediments. These sediments contain, rarely at first, but later in abundance, a peculiar kind of limestone called stromatolite. Stromatolites are bulbous masses of calcium carbonate, a few centimeters to a few meters across and finely laminated. The surrounding rocks are sometimes so beautifully preserved that we can reconstruct their environments in exquisite detail. Judging from the mud cracks and sand ripples in the sediments, the stromatolites formed in shallow or intertidal seas. They were abundant for a long time, but have been rare since the late Precambrian. Even today, though, algae and some bacteria deposit, as by-products of their photosynthesis, stromatolite-like bodies in the warm, shallow waters of the Gulf of California or western Australia.

We cannot strictly prove that the earliest stromatolites were of biological origin, but it seems very likely. Moreover, in the associated sediments, we often find the remains of what look like small, single-celled organisms, perhaps blue-green algae. The geochemical data, good because the rocks are so well preserved, support a biological origin and the occurrence of photosynthesis. Thus, if life at 3.8 billion years seems possible, at 3.5 to 3 billion years it is highly probable.

In this manner, life introduced limestone, a new rock type, to the record. In the Bulawayan rocks of Rhodesia, stromatolites are abundant from around 3 billion years ago. When the Proterozoic arrived a half-billion years later, stromatolites were everywhere, in Africa, Australia, North America, bearing evidence of the now ubiquitous practice of photosynthesis by what we believe were blue-green algae similar to those that still thrive in our ponds and puddles.

What happened to the oxygen generated by all this activity? Eventually, of course, it found its way into atmosphere and ocean, but not right away, because so many elements, minerals, and soils had to be oxidized first. The early history of oxygen has been laid down in a most characteristic rock, unique to the Archean and early Proterozoic: the Banded Iron Formation (BIF). This beautiful rock type consists of thin laminae of siliceous or more rarely calcareous sediment in many colors: blue, brown, red, black. The layers alternately contain abundant reduced (ferrous) and oxidized (ferric) iron. BIF deposits are voluminous on many continents and are a major source of iron ore, as in the Great Lakes region. They were laid down in large, shallow continental seas and occasionally in deeper troughs along the continental margins.

The BIF is a geochemical paradox. Ferrous iron, furnished by volcanoes and submarine hotsprings, must have been abundantly available in the early ocean, and so was silica. In this ocean, slightly acid because of the abundance of carbon dioxide, both remained in solution and

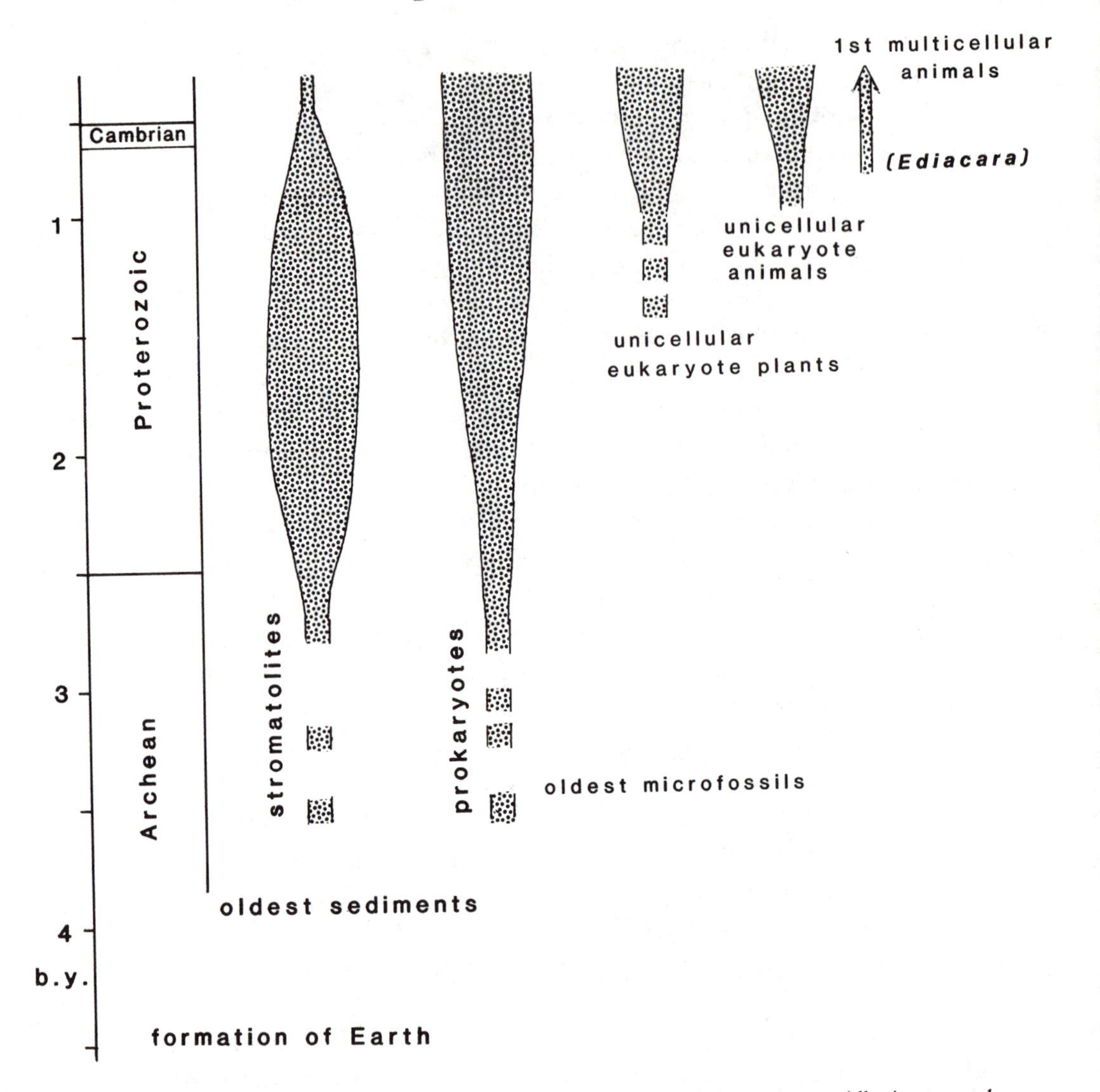

Figure 15.4. The fossil record of the Precambrian, although improving rapidly, is not nearly as satisfactory as that of the Phanerozoic. Nonetheless, this diagram, even with its uncertainty about the timing of most events, is much richer in information and much more likely to be approximately correct than we would have thought possible 20 years ago.

could migrate in that state everywhere. When, in shallow water, conditions began to approach saturation, silica precipitated, dragging some ferrous iron with it. In order to deposit iron in large quantities, on the other hand, oxygen is needed; it oxidizes ferrous iron instantly to the ferric state. In that state the iron is insoluble and settles out. How can we explain that ferrous and ferric iron alternately accompanied silica to the bottom in almost rhythmic fashion, if the two demand mutually incompatible conditions?

Now imagine this shallow sea, filled with small planktonic algae or bacteria busily photosynthesizing and generating oxygen. The oxygen bound the ferrous iron and sent to the seafloor a steady rain of silica-rich ferric iron mud. At other times, perhaps in other seasons, there was no plankton bloom, and hence no excess oxygen, and the silica settled mainly by itself, accompanied by just a little ferrous iron. The fine banding of the BIF certainly fits a seasonal mode of deposition by plankton blooms in open water, or algal mats on shoals.

In this way the BIF accommodated the oxygen supply for a long time. This may have been accidentally fortunate, because it provided time to evolve biochemical means of coping with the pesky element. On the other hand, the BIF might also have been a delaying factor. Until it ceased draining the oxygen, oxygen-based metabolism was not able to take off. In any case, the course of BIF history is well defined (Figure 15.5). The oldest BIF deposits appeared almost 3.8 billion years ago, and over time they became a major part of the shallow marine sediments of the Precambrian, peaking in abundance around 2.5 to 2 billion years ago, and abruptly disappearing shortly afterward. At about that same time, oxidized continental sediments such as redbeds, or gypsum in marine evaporites, suddenly became abundant and widespread. They show clearly that oxygen had invaded the atmosphere and ocean.

The disappearance of the BIF need not, of course, have coincided with the time that organisms learned to cope with the oxygen threat by themselves. That might have happened any time before the first free oxygen entered the atmosphere. It merely says that the ferrous iron demand had been satisfied, and the redbeds and gypsum tell us that was true as well for other major substances that required oxygen. There probably had been earlier instants when free oxygen had been available locally in air or water for a while, but all the evidence indicates that it was only around 2 billion years ago that this element became widespread, albeit in small amounts, stayed, and began to increase steadily.

For free oxygen to accumulate in air and sea, however, it was necessary that carbon be removed from circulation permanently. Otherwise, the release of oxygen by photosynthesis would simply have been matched by its uptake when the organic matter after death decayed once again to carbon dioxide. Thus, burial and transformation of organic matter to black shales, coal, and oil, as well as deposition of

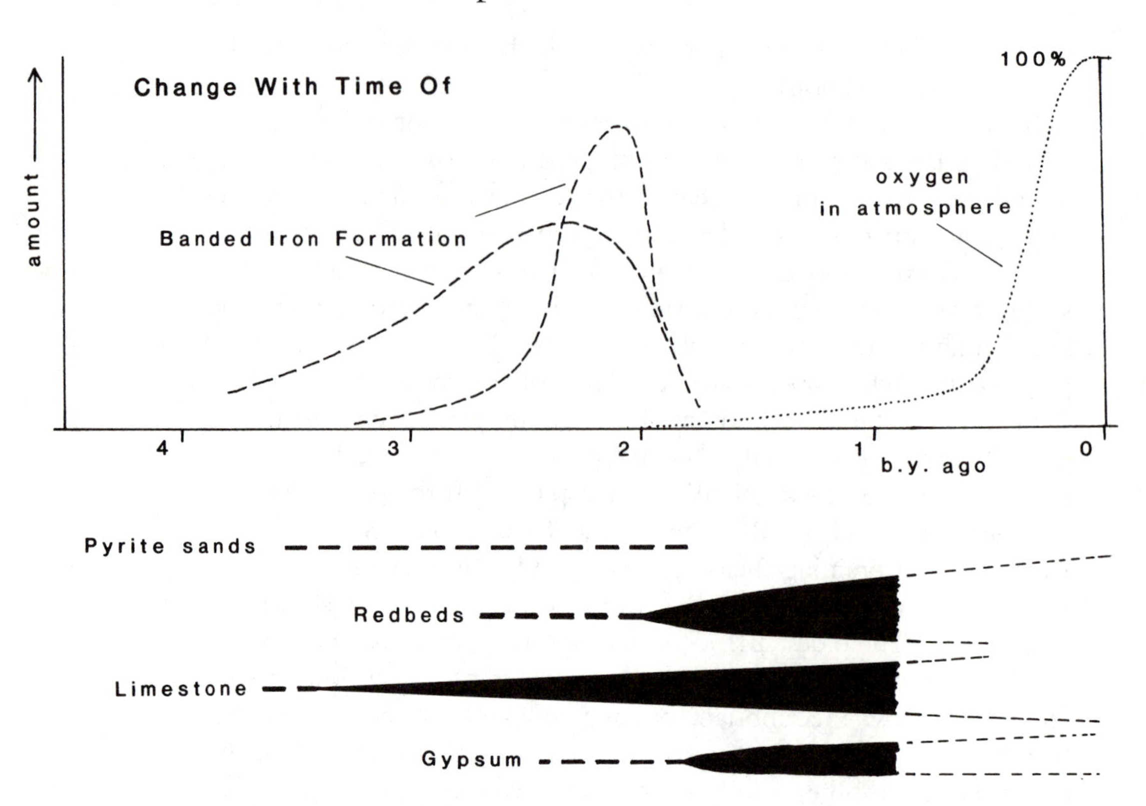

Figure 15.5. The history of oxygen is recorded in many sedimentary rocks. Limestones imply photosynthesis as early as 3.5 billion years ago, but pyrite sands, not compatible with free oxygen in air or water, do not disappear until almost 1.5 billion years later. At that time, redbeds appear as testimony to the presence of oxygen in the atmosphere, and gypsum deposits show that it was dissolved in the sea. The record of the Banded Iron Formation (BIF) remains under discussion. An earlier version with a pronounced peak interpreted as the result of a massive plankton bloom has been replaced by a more gradual sequence of events.

carbonates in the form of limestones and dolomites, were essential for the creation of a breathable atmosphere.

Oxygen, then, was in the ocean and atmosphere around 1.8 billion years ago, ringing in a new era, even though its abundance may not have been more than 10 percent of the present value as late as the beginning of the Phanerozoic (Figure 14.4). Somewhat before this small but momentous event, about 2.5 billion years ago, the great expansion of continents and shallow seas, accompanied by proliferation of life demonstrated by stromatolites and BIF deposits, had presented another challenge to evolution. Which, if either, of these two opportune moments was the one that introduced the eukaryotes to the world we cannot say, nor do we know when, much later, the first

Table 15.1. *Summary of the early history of life*

Time (billion years)	Events	Consequences
?	Chemical evolution	Organic soup forms, reduces carbon dioxide in the atmosphere
?	First life	Heterotrophs consume organic soup
?	Competition for basic organic substances	Ancestor of prokaryotes; genetic apparatus formed
>3.8	First autotrophs	Carbon dioxide reduced; first oxygen in sea
~3.5	Photosynthesis brings oxygen; neutralized by reaction with ferrous iron	Bacteria and algae lay down stromatolites and Banded Iron Formation
~2	1% oxygen in atmosphere; enzymes give protection against oxygen within cells	Ozone shield; redbeds; end of Banded Iron Formation; oxygen-based metabolism
2–1	Eukaryotes appear	??
~1	Sexual reproduction	Communities more complex and specialized; multi-celled organisms
0.7	Atmospheric oxygen 10%	Rapid diversification of higher plants and animals, though still soft-bodied
0.6	??	Cambrian explosion of hard skeletons and diversity

multi-celled organisms began the long road to Phanerozoic life and to ourselves.

The evidence is slim, and our present view of the later Precambrian history of life is hardly less speculative than that for the earlier part. There is, in California, a dolomite formed about 1.4 to 1.2 billion years ago on the western margin of the North American continent. This dolomite contains larger and more complex microfossils than had occurred heretofore; they may have been eukaryotes. Other finds of fossilized cells of larger size and greater complexity point in the same direction, suggesting that the first eukaryotes appeared on the scene around 1.4 billion years ago. That is, at the moment, all one can say, except that it must have happened before 700 million years ago. At that moment in time, an accident of fossilization preserved a rich fauna of soft-bodied, multi-celled organisms in the Ediacara beds in Australia. Similar faunas of the same age have since been found in several other places. Generally, these fossils have been regarded as the forerunners of what later became the worms, corals, and jellyfish of the Cambrian, but Adolf Seilacher, a paleontologist at the University of Tübingen in Germany, has suggested that they are not. Instead, they may have been one of nature's experiments in dealing with the transport of metabolic products and gases in large bodies. Modern organisms solve this problem with a vascular system of tubes, but the Ediacara fauna may have tried to accomplish the same thing by increasing the surface area of their bodies relative to their volume; they are all flat and leaf-like. The experiment must have failed, and the subsequent Cambrian fauna would therefore not be its direct successor but an independent line of evolution for which we have not yet found the earliest stages. Whatever the position of the Ediacara fauna may have been, about 100 million years later many animals suddenly acquired shells or hard skeletons, and an immensely diverse life achieved geological visibility.

Between 2 and 1 billion years ago, then, some major steps in the evolution of life were taken in obscurity. One wonders why they came so late, especially if the eukaryotes did indeed descend from the same ancestor that brought forth the much earlier bloom of the prokaryotes. What had they been waiting for? Oxygen? The right composition of seawater, favorable shallow marine environments, enough evolution to produce the correct types of prokaryotes for symbiosis? We shall probably not find out easily or soon. Fossilized billion-year-old eukaryotes look very much like any other small, shrivelled organic blobs. More and earlier equivalents of the Ediacara beds would be most welcome and may well be discovered some day. Until then, biochemists, biologists, and geneticists, rather than geologists, will have the field mostly to themselves.

When one contemplates the Phanerozoic, as we shall do in the next four chapters, it often seems that life has dominated the history of the earth. During the Precambrian, on the other hand, the histories of earth and life seem more fairly balanced, a minuet, a stately exchange of steps and bows.

PERSPECTIVE

The Precambrian lasted nearly 4 billion years. Though almost unimaginably long, it does not seem too long for the origin of the earth, the birth of the continents, the formation of ocean and atmosphere, and the first hesitant steps of life. What is astounding is that all this, except life, had been accomplished before the oldest known rocks were formed 3.8 billion years ago.

The Precambrian rock record, old, metamorphosed, mainly devoid of usable fossils, is difficult to interpret. Furthermore, the early earth was hot and its lithosphere thin. Thus, it stands to reason that its global tectonic processes were different from the present version, but what they might have been is so uncertain that this affords us little help. Physics and chemistry play a large part in Precambrian studies, but we have spent much time on what Einstein called "thought experiments," as much because they are such a wonderful and challenging way of thinking as for the insights into the Precambrian they have provided. Because we have drawn heavily on the better-known Phanerozoic, they have also been, in a sense, a test of our current understanding of the principles of earth history.

Life appeared on earth late in the first billion years. Its beginning is pure speculation, but speculation is preferable to the somewhat desperate view that it was imported from elsewhere in the universe. Its early history exhibits a marvelous dialectic: ultraviolet rays, useful in creating its biochemical precursors, were a deadly enemy once life had arrived. These rays were rendered harmless by oxygen, an accidental by-product of photosynthesis, but the oxygen itself then became a threat. Fortunately detoxified by ferrous iron in the sea, it was eventually put to use in an efficient energy metabolism that opened new opportunities.

Of all the elements that make up the history of the earth, the first few billion years of the evolution of life were the slowest. Then, less than a billion years ago, life suddenly blossomed and outpaced in its infinite capacity for change and variety every other phenomenon on earth. It is to that world of exploding diversity that we must now turn.

FOR FURTHER READING

The following books cover in more detail the subject matter of the last three chapters and still are quite easy to digest and are not mired in terminology:

Cloud, P., *Cosmos, Earth, and Man,* Yale University Press, 1978, 372 pp.; see Chapters 8–11.

Folsome, C. E., *Life, Origin and Evolution: Readings from the Scientific American*, Freeman, San Francisco, 1980, 148 pp.

Somewhat more thorough and hence more demanding are the next two, the first on the origin of planets, the second on the origin of life:

Murray, B., Malin, M. C., and Greeley, R., *Earthlike Planets*, Freeman, San Francisco, 1981, 387 pp.

Folsome, C. E., *The Origin of Life: A Warm Little Pond*, Freeman, San Francisco, 1979, 168 pp.

The following is a good summary of the current state of our knowledge of the nature of Precambrian flora and fauna:

Vidal, G., The oldest eukaryotic cells, *Scientific American*, 250(2):48–57, 1984.

Finally, much more technical, but rich and fresh, are the articles in

Schopf, J. W. (editor), *Earth's Earliest Biosphere*, Princeton University Press, 1983, 632 pp.

Life, time, and change

I will show you time in a handful of life.
(with my apology to T. S. Eliot)

Whence and whereto, stranger?
Homeric greeting

THE ENDLESS INTERACTION

Two things, more than anything else, are remarkable about Precambrian life: its enormous duration of more than 3 billion years and its apparent simplicity, certainly not entirely due to our ignorance. The history of life, from about 3.5 to less than 1 billion years ago, can be summed up quite succinctly. First there were prokaryotes, which were small. Then came the eukaryotes; they, too, were small. Sexual reproduction arrived, followed by the first jellyfish. Only 700 million years remained for the rest of evolution.

Very late in the Precambrian, the single-celled algae became quite diverse, diverse enough to be useful for stratigraphic purposes. From about the same time, the Ediacara beds in Australia allow us a glimpse of a rich and diversified fauna usually seen as the ancestors of jellyfish, worms, and other multi-celled creatures. A mere 100 million years later, all but a few of the major categories of life were present, and life in shallow seas was not so different from that of today. A remarkable progression it is, and a rapid one, even though we blame the incompleteness of the rock record for some of its apparent abruptness.

The evidence does support rapid diversification in the last few hundred million years of the Precambrian, just before the Phanerozoic begins. Not only are the earliest multi-celled animals, the Metazoa, already quite complex and built on a range of ground plans, but the first hard shells are also advanced. Moreover, the sparse imprints of Precambrian soft-bodied life are not their only remains. Metazoans crawl, walk about, burrow, and so leave tracks and holes that sometimes are more easily preserved than the animals themselves. In the Precambrian, tracks and burrows appear at about the same time as the earliest fossils of larger animals, and in the same order: crawlers first, diggers next. Thus, the explosion of higher organisms can hardly have begun much before the date of the Ediacara fauna, some 700 million years ago.

After this explosion, life evolved steadily toward greater complexity. By the latest Paleozoic it had managed to occupy every inhabitable corner of the earth and every ecological niche on land and in the sea. Then, spectacularly, it folded and narrowly escaped extinction, at least in the shallow sea. There followed another expansion, and another cluster of extinctions at the end of the Mesozoic, this one very well known because of the demise of the dinosaurs. Another recovery took place, and we find ourselves in the present time.

New views on an old planet

The gradual evolution of life toward ever greater diversity, its complete conquest of land, sea, and even air, and its sudden expansions and large extinctions raise many questions regarding the how and why of it. These questions range from the scientific to the philosophical and metaphysical. The full historical panorama of life far exceeds the scope of this book, but with our focus on the evolution of the earth as a whole, I shall stress the pull exerted by the environment, the molding of that environment by the evolving life, and the role of the currently ruling theory: continental drift.

16

Bones of our ancestors

We are interested here in the role of life in the history of the earth, not in the history of life itself. To explore this role, we must build on three foundation blocks: the organization of life forms in a hierarchy of categories, the history of life since the late Precambrian, and the theory of evolution, which organizes and attempts to explain the first two. The classification of organisms used nowadays was developed by Carolus Linnaeus in the middle of the eighteenth century. The history of life as we understand it today is the product of the labors of many, beginning in the sixteenth century, and it has been known in its essence since the first half of the nineteenth century. The theory of evolution was formulated by Charles Darwin more than 100 years ago; it continues to evolve as our understanding of biological mechanisms continues to deepen.

THE ORGANIZATION OF ORGANISMS

In the classification of organisms, the fundamental unit is the *species.* Species that are closely similar in form are grouped together in a *genus.* Each organism bears first the name of its genus, then that of its species: *Homo sapiens,* as contrasted with our extinct cousin *Homo erectus.* Morphologically related genera are grouped in *families* (Hominidae); families combine to *orders* (Primates) and orders to *classes* (Mammalia) (Figure 16.1). Living organisms are classified with the aid of many more characteristics than fossils, for which we usually have only the hard parts: a shell, bones, some teeth.

Morphological similarity, however, is a treacherous guide, no matter how comprehensive the description may be. Within some species there is great variability, whereas others, possessing distinct soft bodies, have indistinguishable shells. The formal biological criterion for a

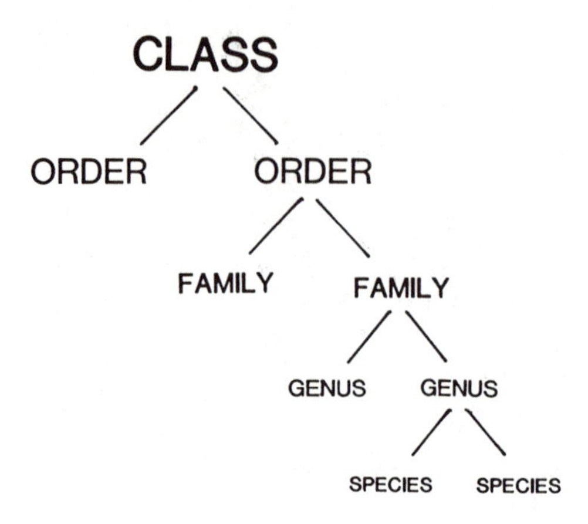

Figure 16.1. The taxonomic system used to bring order among the variety of forms of life is hierarchical, with many species at the bottom of the pyramid and a small number of classes at the top.

species is that its members are able to interbreed under natural conditions and to produce fertile offspring. That seems straightforward, but it is obviously of little use in paleontology, where the test cannot be applied; there are also exceptions, and asexually reproducing organisms present a problem. Therefore, imperfect as it is, the morphological classification still serves.

Morphological similarity can usually, although by no means always, be interpreted as an evolutionary, a *phylogenetic,* relationship. Species of the same genus can be thought of as having descended from the same predecessor. In that sense, the *taxonomic hierarchy* represents a phylogenetic tree connecting ancestors and progeny. The details of this tree at the level of species and genera contained in the geological record tend to be woefully incomplete, so that in practice we display and discuss evolution and the history of life mainly in terms of families and higher categories. A useful concept that we shall apply frequently later is that of *diversity,* the numbers of species, genera, families, as the case may be, for a given time, place, and condition – for the shallow Ordovician sea, for example, or for the whole of the Permian.

The most fundamental distinction we can make in the world of living things is that between prokaryotes and eukaryotes. Another, of nearly equal importance, is between single-celled and multi-celled eukaryotes. Among the latter, which we call, for no good reason, the "higher" plants and animals, we can distinguish those that manufacture organic matter by photosynthesis (the plants) from the fungi (which reduce organic matter to its constituents) and from those that eat plants, fungi, or each other (the animals). Five *kingdoms,* then, make up the world of life (Figure 16.2).

In principle, the number of ways in which a living organism might be designed is large, but in practice nature has adopted but few of the many possible plans. Instead of Kipling's "nine and sixty ways to construct tribal lays" we have only a dozen or so basic designs. Jellyfish and corals, dissimilar as they seem, are constructed on the same ground plan, which is quite different from that of the clams and snails, or that of the starfish and the sea-urchin. These fundamental architectural differences, often most clearly revealed in the embryo, divide the kingdoms of the fungi, higher plants, and animals into *phyla.* The phyla, in turn, divide into classes, and so forth.

Whereas species, or genera, families, and so on, more often than not correspond to phylogenetic descendant-ancestor relationships, this is not so for the phyla. There is reason to believe that the blueprint for the phylum of the higher, *vascular* plants developed four times along independent pathways, and those of the phyla of the higher animals, the Metazoa, at least three times.

AN OCEAN TO CONQUER

The sea was life's first environment, and its conquest was complete before plants and animals moved onto the land. In the sea, life occupies two distinct and in part separate realms: the upper few hundred sunlit meters of the ocean, where floating plankton and pelagic swimmers live, and the bottom, most densely inhabited in shallow water. Both zones were invaded early by life; we know of Precambrian planktonic algae and find encrusting ones in Precambrian shore deposits.

For much of earth history, however, we are poorly informed about life in the open ocean, and even more poorly about that of the deep-ocean floor. In the late Precambrian and throughout the Paleozoic, the planktonic flora of the open sea is represented by the acritarchs, precursors perhaps of the modern group of the dinoflagellates. There also were some grazing or predatory single-celled animals such as the radiolarians, but whether or not the pelagic realm included the diverse and complex community that lives there now we do not know. Only in the Mesozoic do we begin to see details, and we see them then for the same reason that life in shallow seas suddenly emerged with clarity in the Cambrian: Abruptly, many planktonic organisms, plants and animals alike, adopted hard shells and proliferated enormously. From that moment on, the record of pelagic life was kept in detail in the oozes of

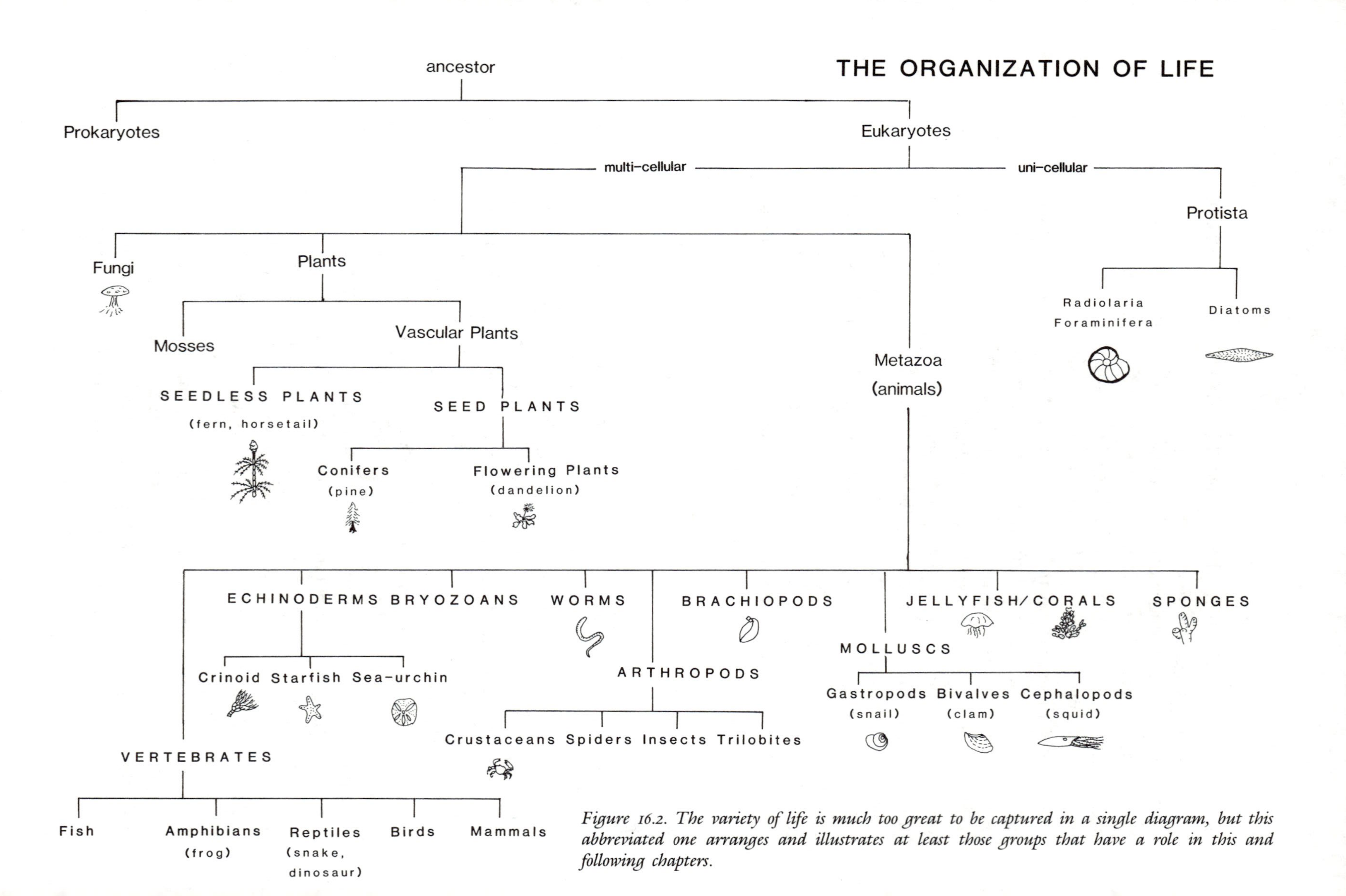

Figure 16.2. The variety of life is much too great to be captured in a single diagram, but this abbreviated one arranges and illustrates at least those groups that have a role in this and following chapters.

the deep-ocean floor, which are composed almost entirely of the shells and skeletons of microscopic plankton. We fail, as yet, to understand the cause of this sudden blossoming and wonder if it means that up to such a late date of less than 200 million years ago, the open ocean had been but sparsely inhabited. And if so, why?

Much better does the fossil record display the conquest of the shallow seas. This conquest began in earnest 700 million years ago with the Ediacaran fauna, followed by a veritable explosion of variety in the early Cambrian, when organisms suddenly appeared in virtually every shallow marine environment. We find dwellers on mud, sand, and rock, floaters, burrowers, crawlers, mud-eaters, filter-feeders, and scavengers. The only kind lacking were large predators; they did not turn up until early cephalopods, cousins of the squid and the octopus, assumed that role late in the Cambrian. Just before and during the Cambrian, all major categories of shallow marine organisms appeared, except one: the vertebrates. Still, notwithstanding the variety, the scene is dominated by two groups, both typical for the early Paleozoic: the now extinct trilobites and the superficially clam-like brachiopods, today a minor group. Both groups diversified in an astonishing number of directions; we find trilobites and brachiopods in just about every function and environment of the shallow sea, from filter-feeding to burrowing, from rock to sand.

We should not assume, however, that the Cambrian explosion of life was a success story throughout. Even a cursory look at Figure 16.3 shows that many lineages present at the beginning of the Cambrian perished before its end, so that it is fair to say that this was a time of experimentation, of trial and error. The following Ordovician brought the consolidation; the failures had vanished, and the successes took over the stage. After that, expansion was mainly at the level of the family and below; the fundamental patterns had been set and have served ever since.

One of the richest and most interesting of all shallow water environments is that of reefs, those hard limestone banks and ridges wholly constructed by organisms, by algae or corals growing upward from the seafloor or leaving piles of their skeletal debris. Reefs have been around since the Precambrian, when small calcareous algal masses were common, but they truly came into their own in the Cambrian and have, with few interruptions, been a major part of the geological record ever since. The Cambrian reefs were constructed by a sponge-like group, the Archaeocyathids, but the experiment failed, and the world had to do without reefs for 50 million years, until true corals arrived in

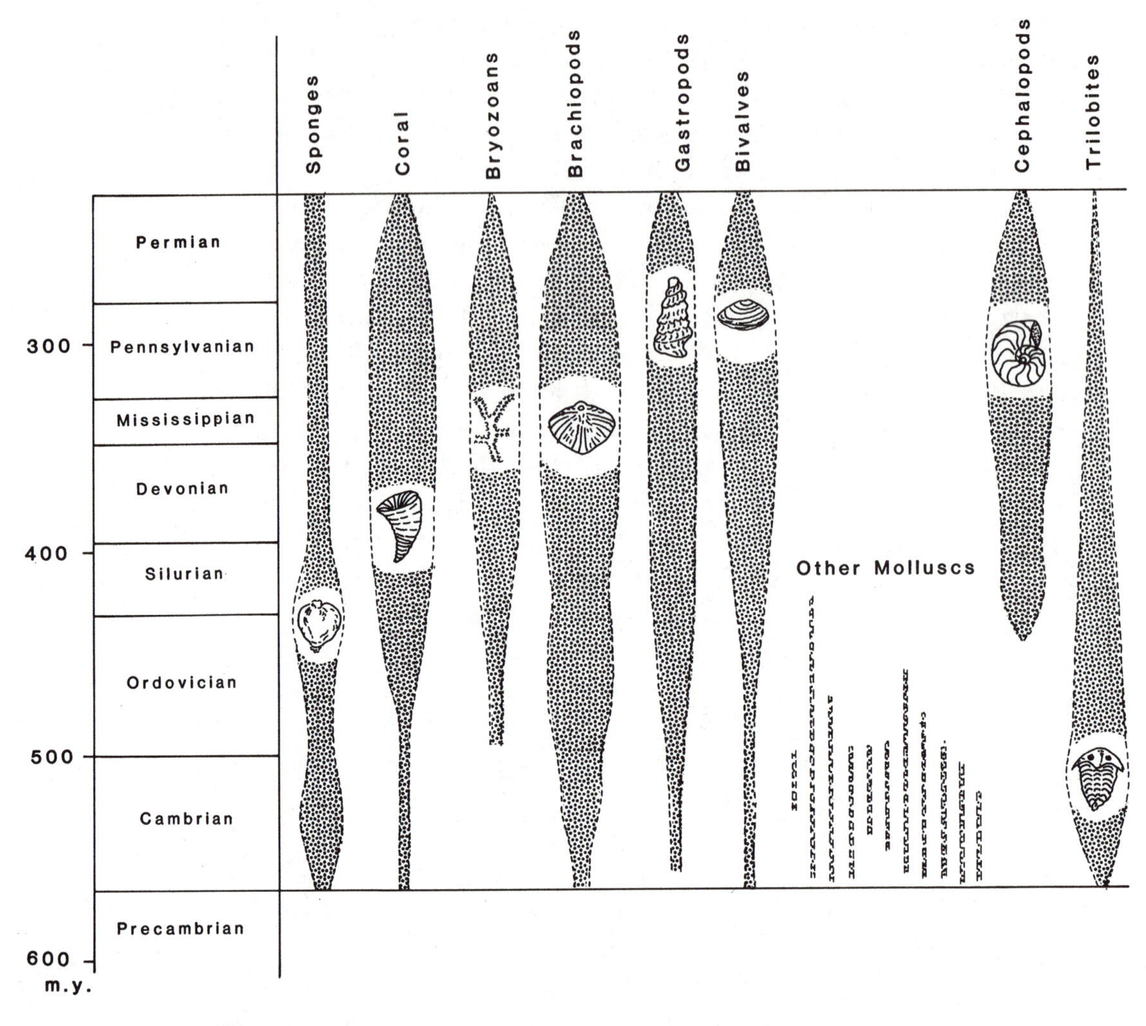

Figure 16.3. Early in the Paleozoic, all main groups of marine invertebrates made their appearance. There also was a good deal of experimentation, some of it unsuccessful, as the many now extinct molluscan groups show. The shaded columns in this and following similar diagrams depict the change with time of the abundance of each group.

the middle Ordovician. The Ordovician reefs were large and beautiful, with a fascinating fauna, and their limestones are some of the most durable sedimentary rocks, but they were not meant to continue any more than were those of the Cambrian. The Devonian saw the end of them in conjunction with a widespread but otherwise minor crisis of marine life. It took 15 million years for another start, only to have that one fail also in the largest marine catastrophe of all time: the great extinction of the late Permian (see Chapter 19). Obviously, however, the world is meant to have reefs, because new types of coral arose and flourished spectacularly in the Mesozoic seas, especially in the Tethys between Africa and Laurasia. The story is getting monotonous: another crisis at the end of the Cretaceous, another recovery, and we come to the reefs of today, once again some of the most beautiful scenery the sea has to offer.

This pattern of alternating expansion and contraction, of success and failure, is not unique to reefs. It shows us the vulnerability of life, but also its recuperative powers, the stubborn return of a life form well suited to its setting, in this case that of clear and fertile coastal waters.

THE FIRST TENDER GREEN

One of the most momentous events in the history of the earth occurred when life moved out of the sea and onto the land. This move was far from easy, because there are many advantages to life in the sea. The sea cradles: There is no need for supporting stems or skeletons, unless one moves about, digs in the bottom, or lives attached to rocks in rough water. The sea protects: against the drying of tissue, sperm, eggs, or embryos, against excessive radiation, even against enemies. The sea furnishes water and nutrients in a most convenient way, and sperm and egg can simply be set adrift to find each other. Lake and river are much like the sea, with one important exception: Their salt content, so much lower than that of body fluids, demands some form of protection against loss of salt or uptake of too much water. A tough skin and specialized organs like the kidney are needed. Migration from sea to land by way of freshwater is thus a useful, though not mandatory, route.

Our first intimation that life reached dry land comes from Russia and from the latest Precambrian. It may have been an abortive move, because we see no more for at least another 100 million years, until the first certain fossil land plants appear in the late Ordovician and earliest Silurian. They arrived, as they should, ahead of the first land animals, a rather unsavory-looking group of arthropods and a few species of snails, by more than 50 million years. Offhand, that seems unnecessarily long; given the availability of plant food on land, some freshwater animals, already protected by impervious skins or other means against loss of water, should have made it sooner. Blame the inadequacy of the record.

The innovations that made the step from sea to land possible for plants were major. Algae do live on land, but they are restricted to puddles or wet spots. Water is generally available within the soil, and so are nutrients, but for photosynthesis, leaves must be held up to the sun. Mosses and liverworts have roots to tap the soil and stems to support leaves, but they lack a transport system to carry water and nutrients up and products of photosynthesis down. Their size is therefore severely limited. Moreover, they have to place their sperm and eggs in water for reproduction, even if it is only a single drop. These limitations were decisively overcome by the vascular plants, which developed a transport system of tubular cells to distribute water, nutrients, and metabolic products throughout the plant, as well as a set of spores or seeds to permit reproduction away from open water. Today, land vegetation consists overwhelmingly of vascular plants, and the first fossil land plants we know about also belonged to this category. It does seem likely, however, that there was a preceding phase of algae and mosses.

Once plants had a foothold on land, their expansion was rapid. The Devonian was still a time of large seas, but land plants of many sizes and shapes proliferated on emerging shores, setting the stage for the spectacular forests that formed the Carboniferous (Pennsylvanian) coalbeds a little later (Figure 16.4). Giant horsetails, ferns, treeferns, and many other strange and wonderful species colonized the coastal lowlands where water was plentiful. Subsequently, plants moved out of such convenient settings, and by the end of the Mesozoic, virtually all climate zones, mountain tops, hills, semi-arid lands, and river plains, had acquired a cover of vegetation.

The history of land plants (Figure 16.5) is mainly a history of perfecting the means of reproduction in a hostile setting. The long road leads from the mosses, which must reproduce in water, to the seedless ferns and horsetails of the middle Paleozoic and, by way of the now extinct seedferns, to the Mesozoic conifers, cousins of the pines and firs of our own forests. The gingko tree, which shades many a suburban street but is extinct in the wild, the Cycads that decorate greenhouses, and the treeferns of tropical jungles all are survivors from the forests of 200 million years ago. Our present flora, however, consists overwhelmingly of a new group that arose only in the last 100 million years: the flowering plants. One of the many keys to the success of this group was a flower that, in addition to using gravity and the wind, which have served the fertilization role since time immemorial, enlisted the help of insects, birds, and other animals for fertilization. Another was a seed that by refined adaptations to dispersal and germination ensured the colonization of distant soils far beyond the reach of what

Figure 16.4. The Carboniferous was a time of swamp forests that left behind peat deposits. The peat later turned to coal. The forest here is not so lush and not likely to have become a major coal deposit, but it is typical of the exotic nature of the Paleozoic flora. The hills in the background are barren; plants capable of dealing with high, dry ground did not yet exist.

falls straight down from pine, spruce, or fern. In turn, the rise of the flowering plants with their dependence on co-evolving animals led to great diversification of animal life to take advantage of the opportunities so offered. The Cenozoic expansion of the insects, and to some extent that of the mammals and birds as well, can largely be attributed to the flowering plants.

The flowering plants developed a great variety of small herbs, shrublets, and grasses to settle environments not well filled before. Ferns require moist conditions, and conifers are mainly trees; so it seems that the drier Paleozoic and Mesozoic lands must have had a limited and open plant cover with little undergrowth. Only the flowering plants offer something green for all seasons, for all places, for grassy banks, for the floor of dry woods, cactuses for the desert, thorn shrubs for the land of winter rains, succulents for coastal cliffs. Rocks, saltflats, places without soil, mountain peaks whipped by vicious winds – they all turned green during the last 100 million years.

The rapid radiation of new species of flowering plants took place late in the Cretaceous and early in the Cenozoic. It was followed by a large expansion of the mammals, just as the coastal forests of the Devonian and the Carboniferous prepared the world for amphibians and reptiles. Especially important was the arrival of grasses, which

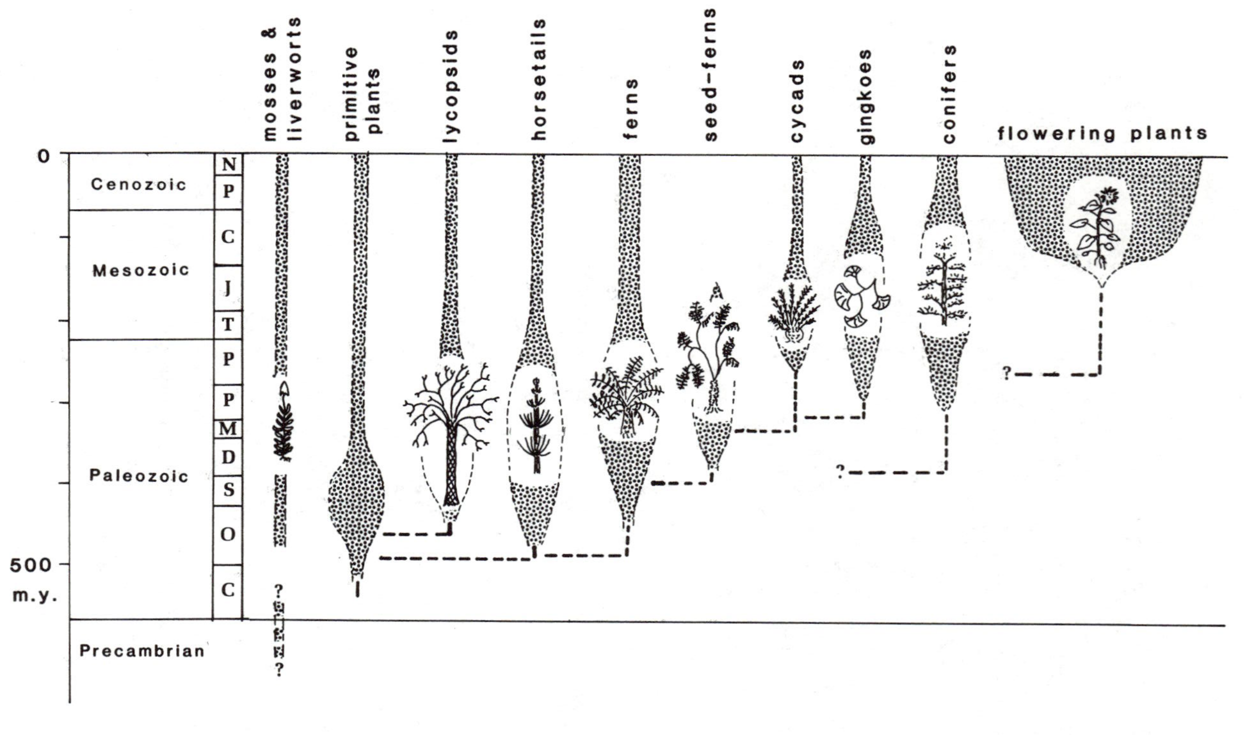

Figure 16.5. The evolution of land plants began with mosses, still needing water for reproduction, and ended with the proliferation of flowering plants that is today the delight of every nature lover. It is evident from this diagram that, more often than not, we are in the dark concerning the major branching points of evolution, even where the general trend is clear.

turned semi-arid lands into steppe and prairie, the environment of some of the world's richest faunas of large mammals. Incidentally, they also increased the resistance to erosion much above what it had been before.

ACROSS THE SHORE AND INTO THE HILLS

Animals attempting to move from sea to land encountered the same problems that plants had, with the need for locomotion, for seeing, for jaws to chew food thrown in for good measure. Many invertebrate groups, such as sponges, corals, bryozoans, and molluscs, although by no means all, successfully made the transition from sea to freshwater, but only a few, some worms, the gastropods (snails), and particularly the arthropods, primarily the insects, took the next step to dry land. The insects are by far the most successful adaptation to land life ever, accounting for about three-fourths of all animal species existing in the world today, and they are quite likely the last group of animals to become seriously threatened by man's indelicate attempts to rearrange the world to suit himself.

Some animals took advantage of adaptations that had served them well in the water, such as the shell of the snail, which can be closed tight against the risk of drying out at low tide, or the impervious chitinous armor of the crab or the lobster, cousins of the insects. Among the most innovative were the vertebrates, who managed to get an infinity of ingenious uses out of a skeleton. Originally intended only to anchor muscles or protect the body, it has been found equally useful to provide support, to walk and grab with, to chew food, to house a brain, or to play the piano.

When we first encounter the vertebrates, the last major category to show up on the scene, they had already fully separated from their still mysterious ancestors. We assume that the development of a skeleton began with something far simpler than a set of articulated bones, perhaps a rod analogous to the spine of those worm-like, fish-like, or simply gelatinous little organisms the tunicates or lancelets. Further back, the prime candidate for our progenitor is, somewhat surprisingly, the group that contains the starfish and the sea-urchin, the echinoderms.

Once established, the vertebrates, at first only fishes, did well. They began hesitantly in the Ordovician but diversified mightily in the Silurian and Devonian, the true "age of fishes" (Figure 16.6). The early groups had an experimental flavor; we also noted that for the inverte-

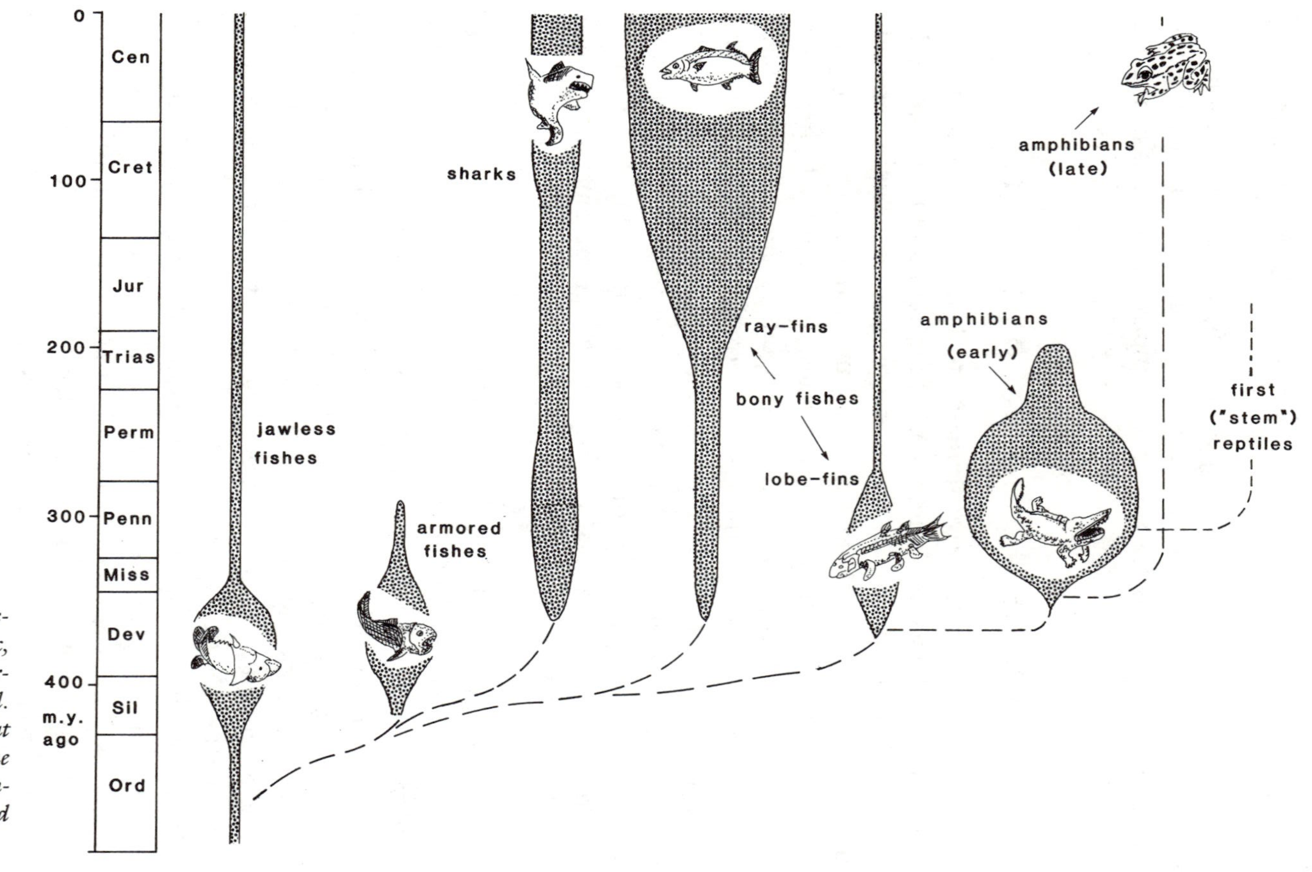

Figure 16.6. The fishes were latecomers in the early Paleozoic, compared with the marine invertebrates, but soon very successful. As before, there are many gaps at key points in the record, but one major step, that from fish to amphibian by way of the lobe-finned fishes, is clear.

brates. They were small, odd-looking, armored beasts that had no jaws and must have slobbered soft food, their formidable appearance notwithstanding. They were not an evolutionary success and soon gave way to others, still armored, but equipped with that most versatile tool ever invented before the hand: the jaw. Eventually the armor disappeared, and as the internal skeleton developed further, the sharks sprang from this stock, then the ray-finned fishes, which are the ancestors of all modern ones. There also arose a small, inconspicuous group, the lobe-finned fishes, surviving in that famous living fossil the coelacanth *Latimeria*. These fishes had short arrays of bones that supported their limb-like fins and eventually evolved smoothly into the legs of the amphibians, the reptiles, and finally the mammals. Some ray-finned fishes manage to get around on their fins, such as Florida's walking catfish, but this is not very efficient, and the tactic has not caught on.

Amphibians, the first vertebrates to live successfully on land, must return to the water to reproduce, because they failed to develop suitable protection against the drying out of their eggs. The reptiles solved that problem by means of a tough eggshell, and with that step the conquest of all parts of the land by vertebrates became a foregone conclusion (Figure 16.7). The evolutionary road, as usual, went via a few unsuccessful groups that are now extinct. One of these groups merits special attention, because it was from the rather tepidly named mammal-like reptiles that sprang the mammals.

Before the turn of the mammals came, however, the world was dominated for more than 200 million years by reptiles. They were completely adapted to life on land, and some even developed a primitive form of internal care for the embryo, as mammals do. Bizarre as many of them appear in the reconstructions everyone has seen, they evolved an enormously useful series of adaptations to a vast range of habitats and functions. Some flew, some swam in the sea, some ran across the land; some were carnivores, others herbivores. There were reptilian equivalents of whales, sharks, tigers, and cattle, and probably even a high-latitude fauna adapted rather better than one might expect from cold-blooded animals to what passed during the Mesozoic for a cold climate. Only after 200 million years, several times as long as the mammalian domain has thus far lasted, did the reptiles begin to decline. The survivors are but a shadow of their Mesozoic glory. However, the direct descendants of their most famous branch may still be with us in disguise: When we feed birds on our windowsill, we may be feeding the last of the dinosaurs.

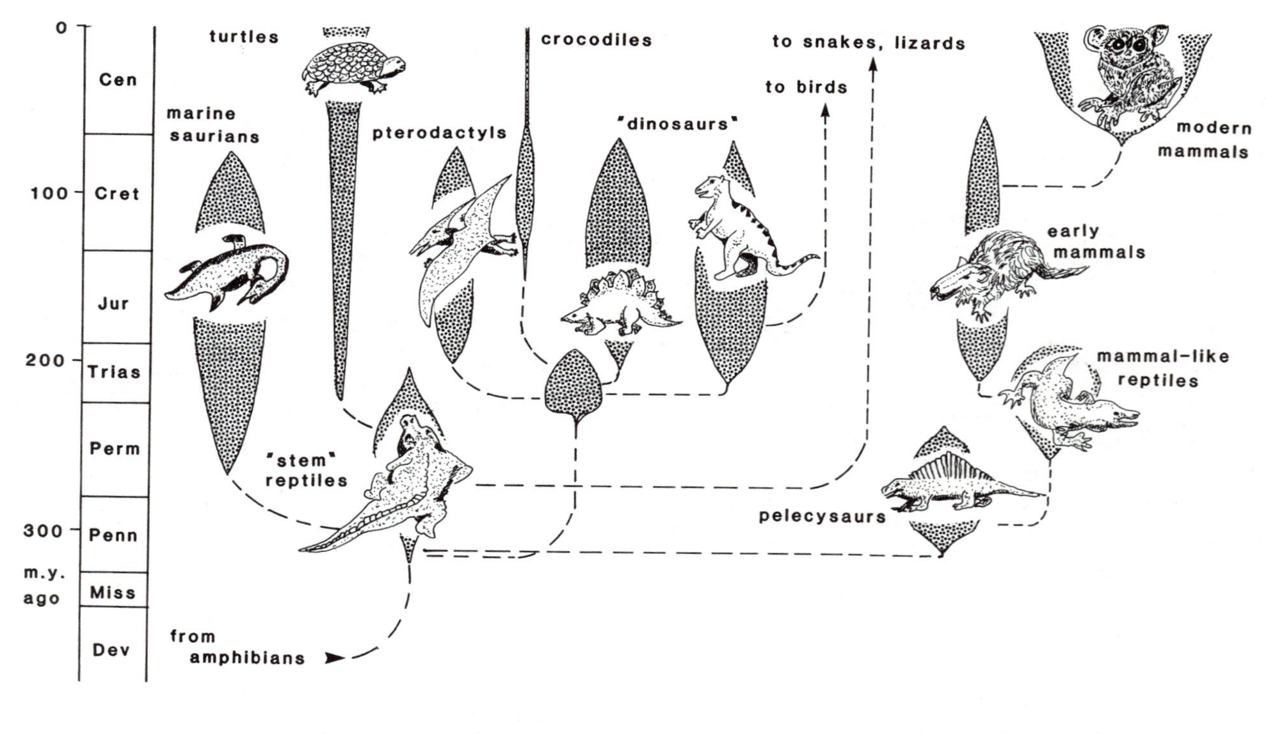

Figure 16.7. The wonderful variety of reptiles is only inadequately displayed here. There were many Paleozoic reptiles, but the Mesozoic was clearly their era, although they are not unimportant even today. The mammals came quite early, but as seems so often to have been the case, they had to wait in the wings for a long time until, at the beginning of the Cenozoic, they came into their own.

FROM MOUSE TO MAN

Mammals have worked out some problems of living better than the reptiles did. One is an improved system of maternal care. Another is control over body temperature that provides the means to put on a burst of energy, to get started quickly on a chilly morning, or to live through a winter season without being dormant, a dangerous state. The dinosaurs, living over as wide a geographic range as mammals do, had the advantage of a warmer Mesozoic climate and, for some, of the heat retention that is possible with a large body. However, they were not specialists in energy management as the mammals are. The mammals did not invent the technique of thermal control, because their ancestors in the group of mammal-like reptiles show by their bone structure that they were to some extent already able to regulate their body temperature, but they perfected it. It served them well when the climate began to turn cold later in the Cenozoic.

It is not easy to tell from fossil evidence just where the amphibians stopped and reptiles began; neither can we draw a clear dividing line between reptiles and mammals. Some mammals still living, such as the duck-billed platypus, are true members of the class, but do lay eggs, thus warning us not to be too rigid about what separates us from snakes and lizards. There are, of course, other differences – the placenta, maternal feeding, legs placed under rather than beside the body, specialized teeth or hair – but few are unique to the mammalian class.

The first step in the evolution of the mammals was a quiet one. After a great bloom during the Permian, the mammal-like reptiles left the Mesozoic to the real reptiles, among them the dinosaurs, but not before they had planted a time bomb in the form of some tiny, inconspicuous, shrew-like creatures that appeared early in the Triassic. After waiting quietly in the wings for more than 150 million years, these mammals, small in size and few in number, crossed the border into the Cenozoic to spawn the elephants and mice and pigs and men of today.

17

Beyond Darwin

In 1859, Charles Darwin published *The Origin of Species,* a work unique in science for its impact on the spirit of the time and of all times following. Darwin's thesis was startling, although not wholly novel, and it has retained public attention for almost a century and a half. Works of equal importance, and even more directly relevant to the human condition, have not always weathered as well.

The theory of evolution is simple, straightforward in its premises, and direct in its consequences. More than a century of biological and paleontological study has added depth and refinement but has not clouded its basic structure. Virtually all scientists today accept it as true. Lately, some of its aspects have been vigorously debated, and new supplementary ideas have sprouted, but this is a debate of renewal, not revolution.

Although the theory of evolution touches our lives in many ways, it has been less the scientific content that has been troubling than its philosophical consequences. Foremost among these is the realization that if evolution is really the product of a random selection of variants created by random processes, it is a deeply materialistic concept that leaves no room for loftier views. It is, of course, possible to think that a Creator created a world ruled by randomness, but for many that would be a disconsolate thought. To reject a vital force or a creational design is as difficult emotionally as it is to accept that the evolution from blue-green algae to man does not represent progress but merely an increase in complexity. The matter greatly bothered Darwin himself and may well explain in part why he waited 20 years before he published, and then did so only because A. R. Wallace had independently arrived at the same conclusion.

PRINCIPLES

Among scientific theories, certainly among those dealing with complex issues, Darwin's theory is a simple one, consisting, as Stephen Jay Gould of Harvard has said, of two undeniable facts and one inescapable conclusion. The two facts are that organisms vary and that some of the variation is inherited by their offspring, an offspring too numerous to survive in full. The conclusion is that those descendants who vary toward greater compatibility with their environment will be less likely to die (Figure 17.1). Natural selection causes better-adapted variants to accumulate and thus will steer life in that direction.

However, this says no more than that natural selection weeds out the less fit. To see it as a molding, a shaping force, as Darwin did, it is also necessary that the variation arise without pattern, just as likely to be of negative as of positive adaptive value. If that were not so, a process internal to the organism would be driving it to improved adaptation, and natural selection would merely eliminate failures.

It follows that evolution has no purpose, and no consequence other than to ensure survival of those individuals best suited to their environment or, when the environment is changing, best able to respond suitably to the change. There is no long-term plan, and any appearance of a trend beyond what is imposed by environmental change is an illusion.

What lies behind the variability of organisms? Mutations, for one thing: small, sudden changes at specific points in the genetic sequence. Genetic recombination also produces new variants, when half of one parent's genes join half of those of the other. Finally, a small population can lose genes from its pool by accident in a process called genetic drift. These are but the most common means by which a menu of variations is put together for nature to choose from.

Most changes so produced are small; they may thicken the pelt when the climate turns cold, or cause blue eyes to become extinct. Is that all there is to it? Must we accept that by infinitely small steps the light-sensing cell on the rim of a starfish evolved into the human eye?

There are two separate issues here. First, there is the question whether or not changes exist that are not adaptive until they have been fully developed. If the evolution of some complex organ requires many small steps of no immediate use before it can function at all, how can we accept that each of those steps was judged by nature on its adaptive value? Apart from the fact that it has been difficult to point to actual examples of nonadaptive variation, we have two possible answers to

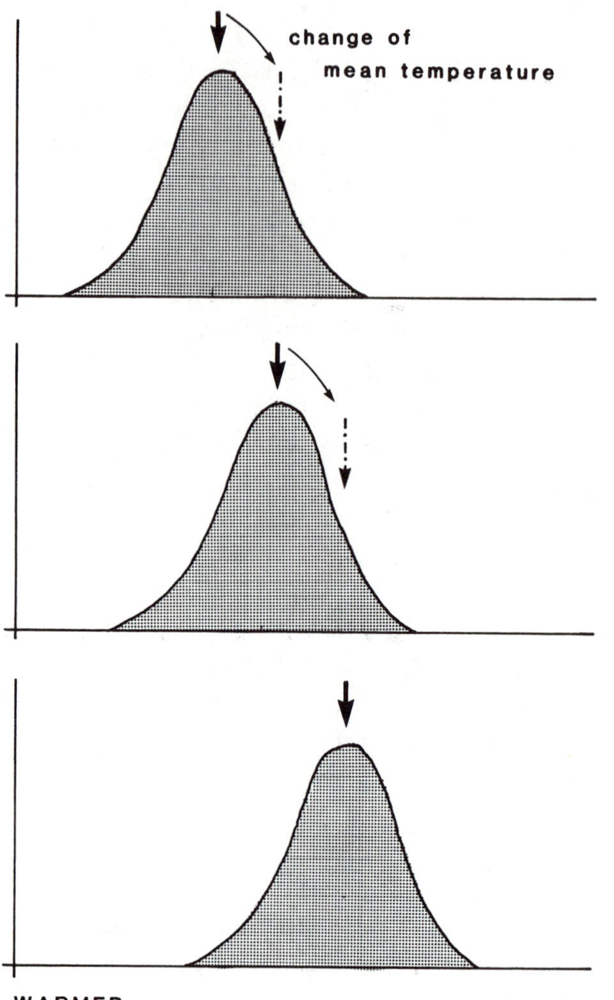

Figure 17.1. The concept of environmental selection from a set of individuals of varied characteristics is illustrated here. A range of individuals is able to exist at a given mean temperature, although some would prefer it a little warmer, whereas others can tolerate colder conditions. When the temperature drops, this latter group finds itself well accommodated and soon begins to dominate, whereas those at the warm end of the range do not survive.

this question. Commonly, it is not a single gene that causes change, but an entire array. This implies that a nonadaptive change, or a not-yet-adaptive one, might ride piggyback on an adaptive change: As it gets colder, the pelt thickens, and turns a little lighter in color as well. Alternatively, an organ may evolve toward one purpose, then be re-adapted to quite a different one: The jaw of the early fishes was fashioned quickly and conveniently from the bone arch designed to support the gills.

The second issue is that of speciation. How does one species evolve into another? Will the selective breeding of dogs some day produce a new species in the genus *Canis*? If that is so, when and how does crossbreeding stop, the flow of genes become interrupted, to satisfy the condition that only members of a single species may interbreed? The simplest way to accomplish this is to place two populations of the same species in a situation in which they are no longer physically able to mate. The imposed genetic isolation will inevitably cause the two populations to drift apart genetically, until their differences become such that crossbreeding is not longer possible. Behind the divergence may lie the effects of environmental differences, of a diverging chain of random mutations, or of genetic drift.

Isolation of populations can be achieved in many ways, but most easily by geography, by migration to an island, by being cut off by a desert or a mountain range, or more spectacularly by continental breakup and drift. If the isolated population is small, genetic change will be rapid and will speedily lead to a large divergence from the parent population. The classic example is that of Darwin's finches on the Galapagos Islands: Many species quickly evolved from very few as a result of geographic isolation.

Are there means to bring about major changes other than by isolating groups of organisms? The major work of biochemical evolution was completed long ago, and today the possibility is remote that a chance variation might produce a new protein or enzyme of value. Now that the days when the basic blocks were created are behind us, evolution at the primary level mainly reshuffles the deck, alters gene patterns, or changes the timing of events in the development of an organism. Here lies the opportunity for occasional major change. Many genes specify activities, such as the manufacture of an enzyme needed to help turn the fertilized egg into a mature individual. Others, regulators, see to it that these actions happen in the right order and at the proper time. A change in a regulatory gene will not introduce a new chemical or a new process, but it may insert an activity at a new stage, delay one, prolong another. A good part of the evolution from Miocene primates to modern man has to do with the gradual delay of the maturation process. Man's larger skull permits a bigger brain, but it passes through the birth canal only with difficulty. Retard the growth of the skull and this problem is solved. Intuitively, one senses that most large random changes, most "genetic jumps," are unlikely to be beneficial; there is little chance that these "hopeful monsters" will survive, although once in a while one just might.

ELABORATIONS

Historically, the search for mechanisms that could accomplish change in big leaps has failed, and it might be wise to abandon the idea. The compelling reason that the issue refuses to die, however, is that, as the phylogenetic charts of the previous chapter so clearly showed, we find ourselves quite able to document minor gradual evolution, but embarrassingly short of links between major categories. We are forced to interpolate at virtually every major branching of the tree. Darwin was well aware of this lack of paleontological support for his Lyellian view that evolution was slow and gradual, but he felt inclined to blame the incompleteness of the geological record. We do so today also, or add that paleontologists, bent on defining species, tend to assign intermediate forms to one kind or the other, and thus reduce our opportunity to recognize transitional states. There is no doubt that the main strength of the theory of evolution is in biology and genetics, not in paleontology and geology.

Scientists, like everyone else, learn to live within the confines of their opportunities and the limits of their theories. The twentieth century has clarified and confirmed the mechanisms of evolution so strikingly and in so many ways that concerns such as the one just expressed have remained in the shadows. However, discomfort, even if only intuitive, cannot be repressed forever, and a wave of questioning regarding the validity of gradualism in evolution has lately arisen.

An unbiased look at the fossil record shows that it is quite compatible with a model of evolution that consists of long times of little or no change interrupted by major divergences so sudden and brief that they might almost be called instantaneous. To a small group of challengers of a somewhat reluctant establishment, the branches of the phylogenetic tree are horizontal and vertical rather than slanting (Figure 17.2). Their view, called "punctuated equilibrium," has evoked strong resistance, in part because of the problem of finding suitable genetic mechanisms and in part because large "instantaneous" jumps have been very difficult to demonstrate.

The debate, to be sure, is about the origin of species, not about natural selection, random variation, or evolution itself. It does not involve the theory of evolution but only the rate of its processes. What might produce near-horizontal branching of the evolutionary tree? The

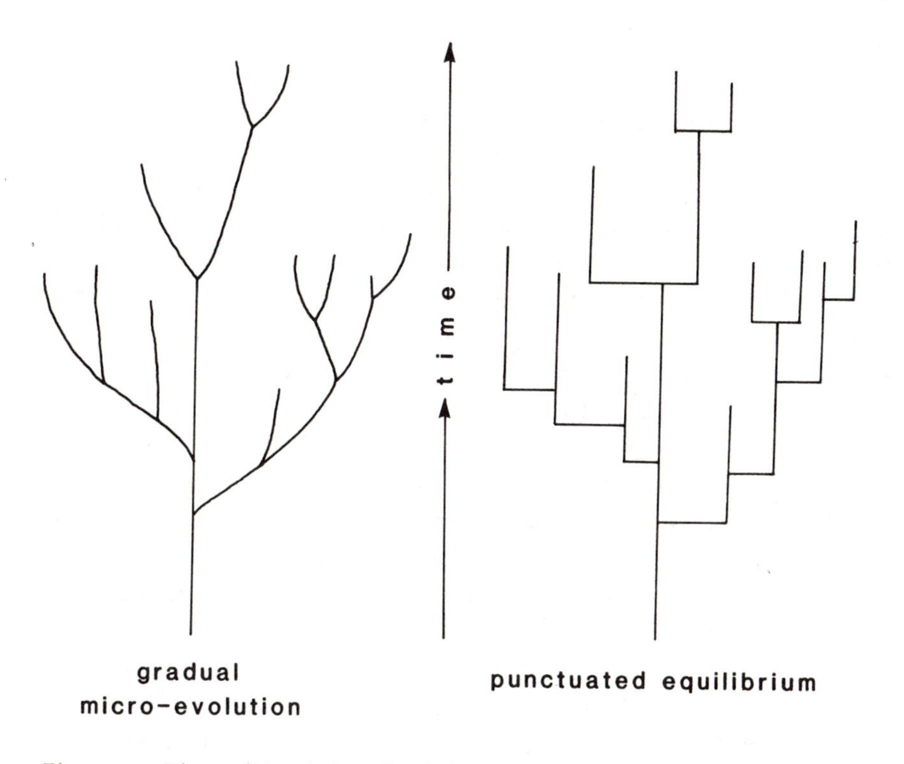

Figure 17.2. The traditional view of evolution sees a gradual divergence of one species from another, and continued change as time passes. This yields an evolutionary tree with sloping branches (left). Proponents of punctuated equilibrium, on the other hand, regard the formation of a species as a large and instantaneous step followed by stagnation (right).

most obvious candidate is geographic isolation; I have already alluded to its power. By virtue of the speed of this process, and the small size of the area in which it happens, such an event might be undetectable in the geological record (see Chapter 1), and the resulting change would thus appear to be instantaneous. Another way would be a genetic jump, but we are far from sure that those are possible.

If we cannot actually point the finger at examples of sudden genetic change, perhaps we might at least demonstrate that stagnation occurs between major steps. Only limited success has been achieved here also. Some Paleozoic invertebrates indeed remained unchanged between times of major diversification. In other cases, gradual small-scale evolution is quite evident in the record. The question remains open.

A problem of definition bedevils the discussion. What is instantaneous to the geologist, say 10,000 or even 100,000 years, is unlimited time to the geneticist. Speciation can be very rapid; six new species of a small fish living in a lake in Uganda have appeared in the last 4,000 years. Perhaps we need not postulate alternating jumps and stagnation, but only variations in the rate of evolution, a far less profound revision of the conventional model.

Because the issue centers on the origin of species, genera, or families, it is not surprising that some proponents have suggested that nature selects for fitness not only the individual but also entire species. If a new species, suddenly evolved, represents a viable adaptation to its environment, it may remain unchanged until the environment itself changes. At that point, the species, unable to adapt further, becomes extinct, and its place is taken by another species that is more flexible. As a result, lineages that have high rates of speciation will have an adaptive advantage over those that do not.

This, of course, implies that a jump, large as it is, must have equal opportunity to be adaptive or counter-adaptive; otherwise, it will not be nature that selects, but once again some internal mechanism that steers change in a specific direction. By proposing, or at least implying unfavorable as well as favorable jumps, punctuational evolution differs fundamentally from an acceleration of evolutionary rates.

The debate is not soon to be resolved. My personal preference, and probably that of most geologists, tends toward varying rates of evolution rather than jumps and natural selection by species. Undoubtedly, however, the discussion, lively and relatively free of constraints until observation and experimentation can catch up, is enormously fertile, though at times a little contentious. Thus far, little of real substance seems to have been changed in our thinking, but from the sea of red herrings and floating straw men might eventually emerge a fresh and revitalized form of the theory of evolution, though the paleontological record is likely to remain as ambiguous as ever.

NEW APPROACHES

Describing the fossil record has always been a somewhat intuitive process, with strong emphasis on the selection of type specimens for each species, chosen on the basis of criteria judged valid by the investigator. Once a set of type specimens has been selected, all fossils must be assigned to one or another of these. It is obvious that, even if great care is taken, there might be a natural bias against intermediate forms. The evident alternative is to begin with a careful, unprejudiced description of as many specimens by as many criteria as practical. Such descriptions, if quantified, can be sorted by statistical techniques into a hierarchy of categories based on diminishing degrees of similarity. This more quantitative form of the Linnean way of defining species

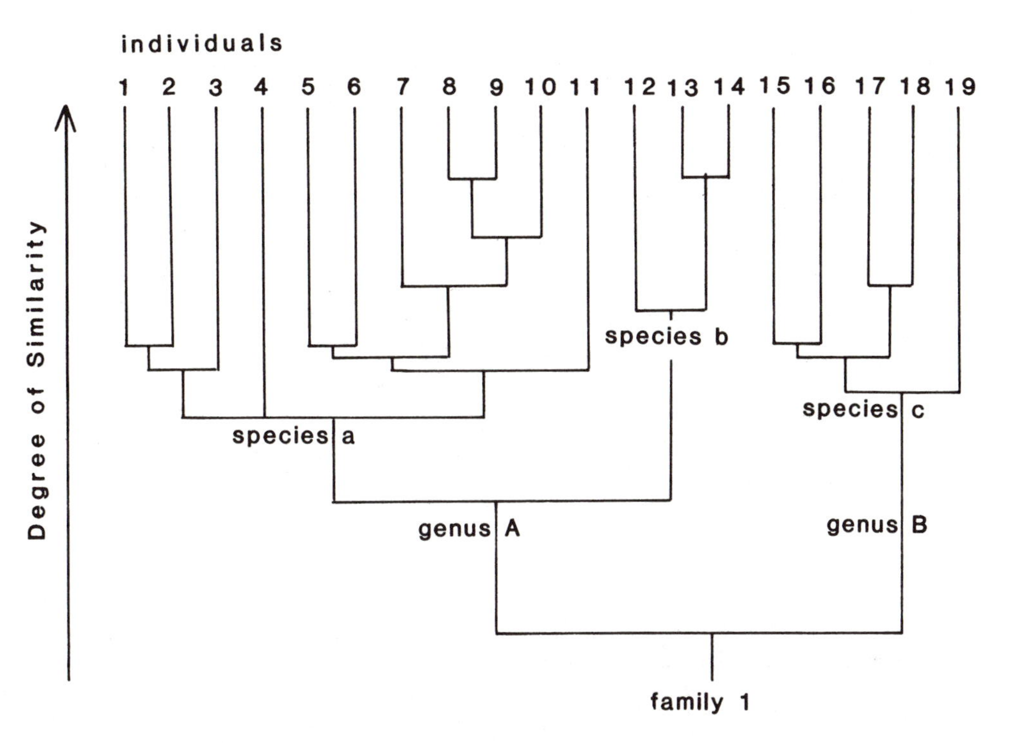

Figure 17.3. If we describe, preferably in numerical form, all characteristics of a large set of individuals, we can use statistical methods to determine how similar or dissimilar each possible pair is. In the resulting similarity hierarchy, clusters of very similar individuals may be labeled species, and at the next lower level of similarity, species can be combined to genera. The levels of similarity that separate individuals from species and species from genera are sometimes self-evident and sometimes have to be chosen arbitrarily.

derives strength from its lack of preconceived notions of what species should be like, or even what the relevant distinguishing criteria might be. On the other hand, it is not possible to describe anything, no matter what, in its entirety, and the unavoidable choices may still introduce bias.

Another problem with fossils is that many essential criteria are missing, because only hard parts are usually available, and hard parts can be quite alike for organisms with very different internal structures. In practice, the results of the quantitative approach have not deviated all that much from the more traditional way of dealing with classification. Of course, the structuring of quantitatively defined species into higher units (Figure 17.3) has been much admired by some because so often what requires computation appears to be more worthwhile than what does not. The approach has merit, but it remains a matter of interpre-

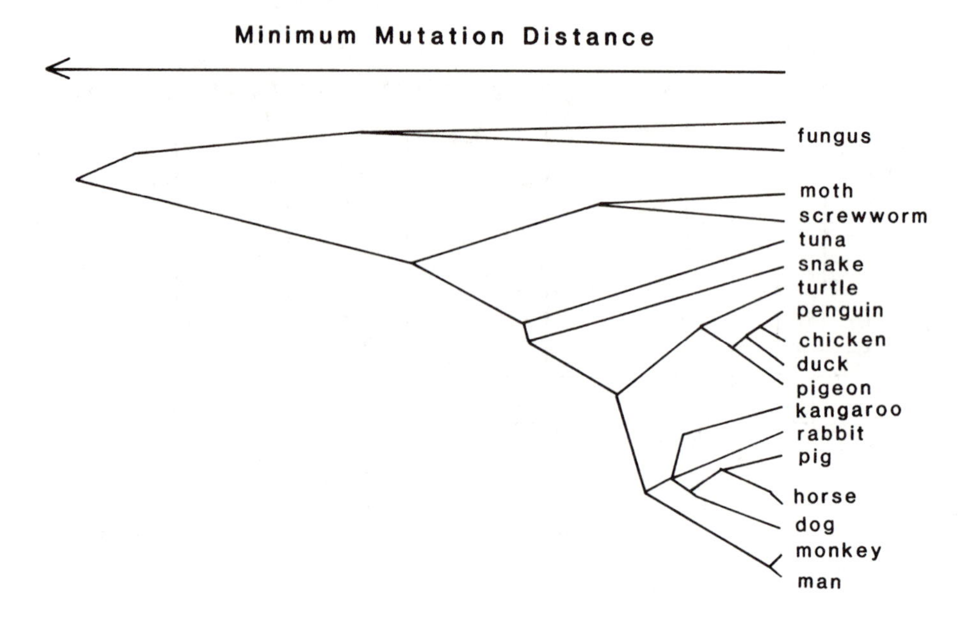

Figure 17.4. Organisms belonging to different species or genera are biochemically distinguishable by substitutions in the amino acid chains that make up their proteins. From the number of such substitutions the degree of relationship can be determined. Horse and pig are more closely related to each other than to the rabbit, and man and monkey branch off from a main stem at the same level as the kangaroo.

tation whether the similarity-dissimilarity tree so constructed is genetically more valid than one obtained in the old-fashioned way. A common ancestor might explain the similarity, and so would a simple ancestor-descendant relation, but it might also stem from analogous solutions to identical environmental problems by totally unrelated organisms.

Quite a different approach holds promise for clarifying the genetic and evolutionary record. It is also based on careful quantitative description, but here the properties described are biochemical. The method, briefly alluded to in Chapter 15 in connection with the ancestry of prokaryotes and eukaryotes, is known as molecular genetics. Over time, small changes accumulate in the amino acid chains of proteins. When species diverge, so do the amino acid sequences of their proteins, and the degree of divergence is a measure of their affinity to each other and to the ancestral species.

The sequencing of amino acids is controlled by components of DNA, and the number of DNA substitutions necessary to produce the observed changes can be determined. The more substitutions implied, the greater the genetic distance between two proteins (Figure 17.4).

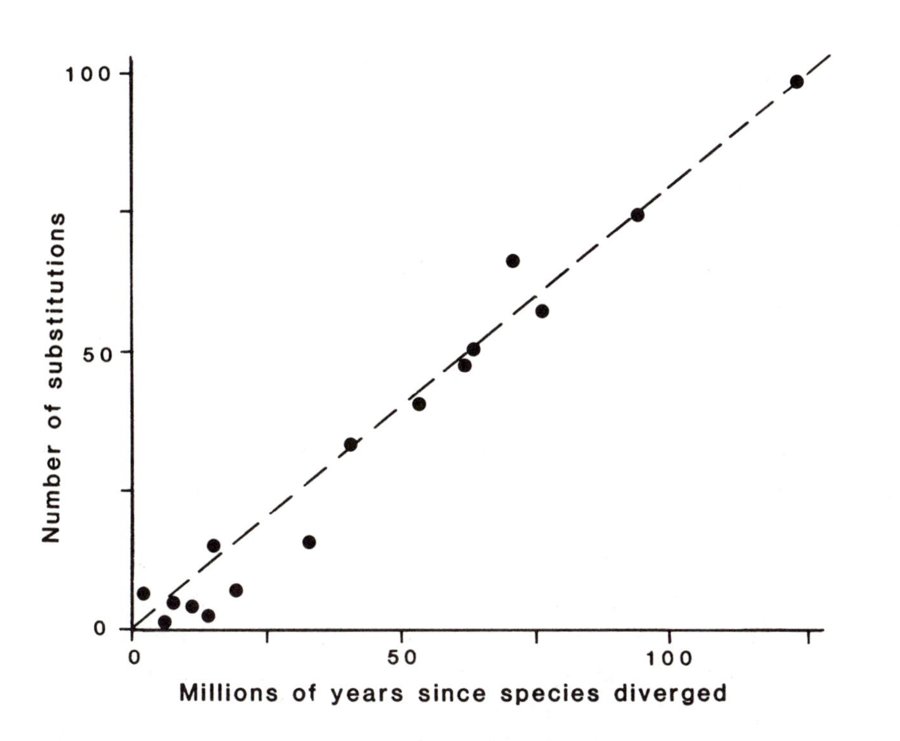

Figure 17.5. The number of amino acid substitutions, determined, of course, on modern organisms, appears to depend on the time elapsed since two species diverged. The lapsed time is furnished by the fossil record; notwithstanding the defects of this kind of information, a remarkably close correlation with the degree of biochemical affinity emerges. The diagram clearly supports the existence of a biochemical clock.

We can base an estimate of phylogenetic distance on more than one protein; the correspondence is usually good and enhances our confidence in the method. Of course, the method works only for living organisms, but we can compare the results with the fossil record and connect them in this way with extinct groups.

Having established phylogenetic trees by this means, what can we say about the time dimension of the evolution? Obviously, we might call once more on the fossil record, but in view of its much-discussed shortcomings, it would be most convenient to have an independent way to date the branches of the tree.

Such a way seems to exist, but it is rather startling and has, not surprisingly, met with a good deal of skepticism. It is based on the claim that molecular evolution, the accumulation with time of small changes in amino acid sequences, proceeds at a steady rate and can be used as a clock. The clock does not tick at fixed intervals but functions rather like a radioactive clock, in that statistically a certain number of substitutions can be expected to occur within a given time. Thus, counting the number of amino acid substitutions would be analogous to counting the number of decaying carbon-14 atoms.

It is not self-evident that amino acid substitution should behave in such a controlled manner. There is no reason why evolution should proceed at an even pace, and there is plenty of evidence that it does not, or at least not always. Some brachiopods did not change in hundreds of millions of years, but their rise and extinction took just a few million. This objection can be countered by noting that minor substitutions in amino acid chains are not the same as evolutionary steps and need not have evolutionary consequences. With a few exceptions, one hemoglobin functions as well as another. There need not be a close relationship between rates of amino acid substitutions and evolutionary rates.

The ticking of the protein clock has not been demonstrated to everyone's satisfaction, but new data keep coming in and appear to be in reasonable accord with the paleontological evidence. Geologically determined branching times for several groups (Figure 17.5) show a decent correlation with their molecular "distances," particularly if we remember that the chronology of fossil branching is also beset by quite large uncertainties. There seems to be good reason to believe that the clock runs, but it is entirely possible that it is not highly accurate.

Quantitative classification of fossils, to some extent, and much more the study of molecular evolution should clarify the structure of the phylogenetic tree considerably over the next decade or so. Whether or not this will impose limits on the swirling debate over the evolutionary records of specific groups of organisms, as well as the mechanisms by which evolution proceeds, is an altogether different matter.

18

Evolution and environment

Darwinian evolution is a process of the moment: It remembers nothing before yesterday, and it does not see beyond today. Natural selection evaluates currently available variation against present conditions, but because it is a weak force, what is here today was mostly also here yesterday. If we see a trend, it can only be because there was a trend in the change of the environment. The major role is for the environment, and important aspects of the history of life can be written in terms of an environmental drive.

We may consider the opportunities presented by the environment in three categories. First, there was once, but is not anymore, the potential of entirely unoccupied territories. Second, there always has been and still remains the opportunity to oust, by greater specialization, the occupant of an environmental niche. Finally, there is the challenge of environmental change, a change either within the physical and chemical environment or within the community of life. Mountains may rise to increase the aridity downwind, or predators may evolve to alter the nature of the food chain. Life and environment are mutually dependent, and the living partner is no less important than the lifeless one. Nevertheless, forces greater than life, a changing sealevel or climate, for example, impose a pattern that, within the context of our inquiry, is the proper perspective for this chapter.

CONQUEST OF THE SEA

Environments change on many different scales of space and time. On land, stability is usually brief and homogeneity very local. Beyond the coast, changes tend to be more gradual, so that many adaptations to the planktonic and benthic modes of living, although not the species themselves, have been with us since early Paleozoic time. For instance, photosynthesis in the sea, once performed exclusively by prokaryote blue-green algae, is now the function of eukaryote coccoliths and diatoms; also the top predators have sometimes been sharks, sometimes reptiles, sometimes mammals. Notwithstanding this, life in the Mesozoic ocean, for example, was in most ways very much like life in the present one.

We know so little about the first colonization of the sea in the Precambrian that we shall attempt a thought experiment instead. We have seen that the early earth had little land and much ocean and that there was life in those vast waters even before 3.5 billion years ago. How shall we imagine this ocean to have been? Little interrupted by land barriers, it should have had a simple current pattern, gentle surface temperature gradients, and a sunlit, warm upper layer mixed by the wind. The deep water, circulating slowly except during an occasional ice age, probably contained abundant ferrous iron, silica, and such poisonous substances as copper, arsenic, and mercury, furnished by a plethora of submarine hotsprings. Above all, the deep waters remained free of oxygen long after those at the surface had already been oxygenated. Except for bacteria living in the hotsprings, the abyss was probably devoid of life.

Upwelling in this early ocean of few shores must have been limited, and therefore so was the supply of nutrients to the surface. Inevitably, the Precambrian ocean was both stable and infertile, much more so than the present or even the Mesozoic ocean. For organisms, small size and large surface area would ensure that they would not sink to the poisonous deep, and populations must have been small and dispersed to match the stable but sparse nutrient supply. Consequently, there would have been little incentive for the development of grazers, let alone predators.

Around 2.5 billion years ago the continents had achieved reasonable size, and large shallow seas had become widespread. Coasts and shoals impeded the global circulation, but did enhance upwelling on appropriately oriented shores; rivers, at last draining large areas, brought increased quantities of nutrients to the coastal waters. Thus, one might expect a major expansion of primary producers to take advantage of the greater nutrient supply and also diversification to adapt to the many environments of shallow seas. This, indeed, is what the abundance of BIF and stromatolite limestone tells us. Because upwelling, on the global scale, was patchy, the ocean environment itself would have been patchy also, and geographic isolation might have begun to work toward a diversity of species. Denser populations invite grazers, and grazers attract predators; together they provide the potential for the evolution of larger, multi-celled organisms.

It has been customary to attach great importance to the arrival of freely available oxygen in sea and atmosphere around 2 billion years ago as an evolutionary drive. It is just as reasonable, however, to assign the major role to the large new land masses and the upwelling and greater nutrient supply they brought, relegating the development of an oxygen-based metabolism to second place. Life in surface waters might thus have faced a clear road toward diversification as early as 2 billion years ago, but it seems to have taken its own good time about it, because we have no evidence of important developments for another billion years. Why, we do not know.

But what about the abyss? Given so little recycling of organic matter, combined with the absence of oxygen in deep water, the typical early Precambrian oceanic sediments must have been black shales, a sink for nutrients and carbon. No Precambrian deep-sea sediments have been preserved, but even in the Paleozoic we still find a lot of black shales, deposited, as far as we can tell, at depths ranging from a few hundred meters to the deep-ocean floor. These shales suggest that even in the Paleozoic the ocean contained, at least part of the time, little dissolved oxygen at greater depths. As we have seen, black shales are uncommon in the Mesozoic and exceptional in the modern ocean. It is tempting to interpret this record as an indication that the abyss over time has become progressively more oxygenated and may not have been fit for continuous habitation until the Mesozoic. Indeed, the evidence for benthic life below a depth of a few hundred meters is weak until then.

THE LURE OF MUDDY BOTTOMS

The explosion of life at the beginning of the Phanerozoic and the accompanying rapid adoption of hard shells and skeletons have always seemed to demand a special explanation, some drastic biological or environmental event. Actually, two events may have been involved: one biological, to produce the sudden diversity of higher animals, the other chemical, and responsible for the opportunity to manufacture hard body parts. The first, as the Ediacara fauna shows, preceded the second by more than 100 million years.

J. J. Sepkoski of the University of Chicago has argued that the sudden expansion of life was not itself the key event, but a mere consequence of something that took place a good deal earlier. Diversification proceeds in much the same way as population growth: by a curve like that in Figure 18.1. Once set in motion, population and diversification will grow at an ever-increasing rate until the environment puts a limit to it. This is the standard growth curve that also characterizes a laboratory culture of bacteria, as well as the bloom of plankton in the spring. From the perspective of 600 million years, the steep part of such a growth curve looks like an explosion.

This suggestion has much merit, but what started the expansion in the first place? A few years ago an interesting idea flourished briefly before it died from a faulty timescale. It may nevertheless help us here. At some time during the Precambrian, when the area of continental crust had grown large enough, the world finally knew wide shallow seas, a whole new environment, fertile because of the river input of nutrients, and densely populated with plankton and perhaps a bottom-dwelling community. What would be more plausible than that this abundance of food should produce more and larger consumers? Moreover, much of the organic matter was likely to end up in the mud on the seafloor, where crawlers and diggers would be well suited to make use of the excellent supply. Crawling and digging are best performed by multi-celled organisms, and the dependable abundance of food would encourage them to grow to large size.

The continents, as we know now, arrived too soon for this scenario, about 2 billion years too soon. Perhaps the opportunity could not be grasped then because there was not enough oxygen in the water; perhaps it simply took a very long time to evolve the necessary types of eukaryotes. In any case, muddy bottoms were not really exploited until very late in the Precambrian.

When life finally did occupy the shallow seafloor, did it do so because of the evolutionary pull of this new and promising territory? Some say no and suggest that the big step to the Metazoa was purely biological, crediting as one possible mechanism the arrival on the scene of herbivores. It is one of those paradoxes of nature that a community sensibly exploited by grazers and predators (and under most conditions such exploitation is sensible), will flourish and diversify. This is because cropping restrains the dominant and most prolific species, clears room for others, and puts a premium on adaptations that minimize the damage. In a simple world inhabited only by primary producers, the incentive for change is limited, because the only major factors are geography and climate. The suggested mechanism is attractive, because it accounts for the observations, it rests on known biological principles, and it satisfies those who prefer a dominant role for

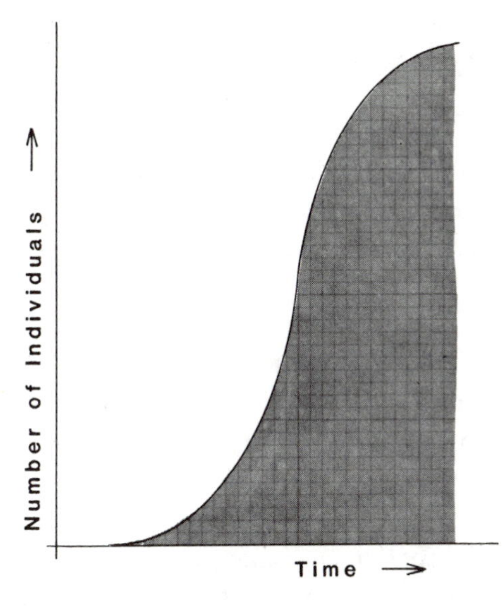

Figure 18.1. Population growth is a matter of continuous doubling of numbers, and doubling again and again until further growth is limited by the resources of the environment. Initially the growth is almost imperceptible, but an explosion inevitably follows, as human history so painfully bears out.

biological factors. There is not, however, even a scrap of evidence that this is what actually happened, and the question remains: Why did it take so long to get the job done?

To a physical scientist the preference for a biological mechanism seems less compelling, but it is not easy to think of an environmental factor that might have brought forth the Metazoa. Oxygen, then present in amounts so small and so slowly rising, is not likely to have been the lost factor. Let us approach the question from a different direction.

Clusters of cells, as some Metazoa are, can float, eat, and even swim, but they are poorly suited to dig in mud, attach themselves to rocks, or resist the tearing force of the surf. Most main groups of the Metazoa, however, are better designed for strength and leverage, because they possess a body cavity contained within a double wall, a *coelom,* which provides strength in the same way that a firehose or a bicycle tire has strength.

This might be a design useful for living on or in the sea bottom, but why should one actually do so? James W. Valentine has pointed out that upwelling provides a rich source of food, but it is a variable source, and adaptation to seasonal feast and famine is difficult. The seafloor beneath an upwelling zone, on the other hand, stores the organic rain and supplies an enduring, dependable reservoir of food, provided there is a means to get at it. Digging and crawling are the answer, and because this stirs up a considerable murk, it also permits filter-feeders sitting on the bottom to share in the feast.

The sequence (Figure 18.2) may thus have begun with small planktonic primary producers floating in the surface waters. Once upwelling became strong and widespread, and eukaryotes had appeared, grazers would have blossomed and evolved into floating, filter-feeding Metazoa like the jellyfish. The next step would have been a coelom, perhaps because it would allow better swimming. This, in turn, would have permitted feeding on and in the bottom, where a more dependable supply of food could be found throughout the year. Because the seafloor has a greater variety of environments than the surface water, rapid diversification to take full advantage of the special characteristics of each environmental niche may have followed. This scenario assigns a strong evolutionary drive to the shallow seafloor, but not until it became accessible as a result of the evolution to the multi-celled Metazoa. A billion years for such a history would not seem too long.

Whether or not this plausible piece of reasoning bears any resemblance to reality we do not know. Subtle changes took place in the early Paleozoic that warn us to keep an open mind. During the Cambrian, a distinct shift took place from mud-eaters to bottom-dwelling filter-feeders, such as brachiopods and sponges. Somewhat later this peacefully settled community was disturbed and largely replaced by biological bulldozers, which plowed the sediments, as sea cucumbers, crawling bivalves, and worms do today. The ensuing disturbance of the mud, a fact of benthic life ever since, led to a major reorganization. The filter-feeders moved mainly to the rocks, and the bulldozers monopolized sand and mud. We have no clue regarding the reason, but the observation tells us that a simple and elegant logic is not always sufficient.

Likewise, we know little about that other major event during the Precambrian-Cambrian explosion of life: the sudden widespread adoption of mineralized body parts. The utility of this move, for protection or as an anchor for muscles, is easy to see, but the direct incentives and pathways escape us. There is evidence of much initial experimentation; with time, the kinds of minerals that were used changed a great deal. In the beginning, about two-thirds of all organisms involved used phosphates, but in less than 20 million years calcium carbonate became the dominant substance as it remains today. In doing this, organisms

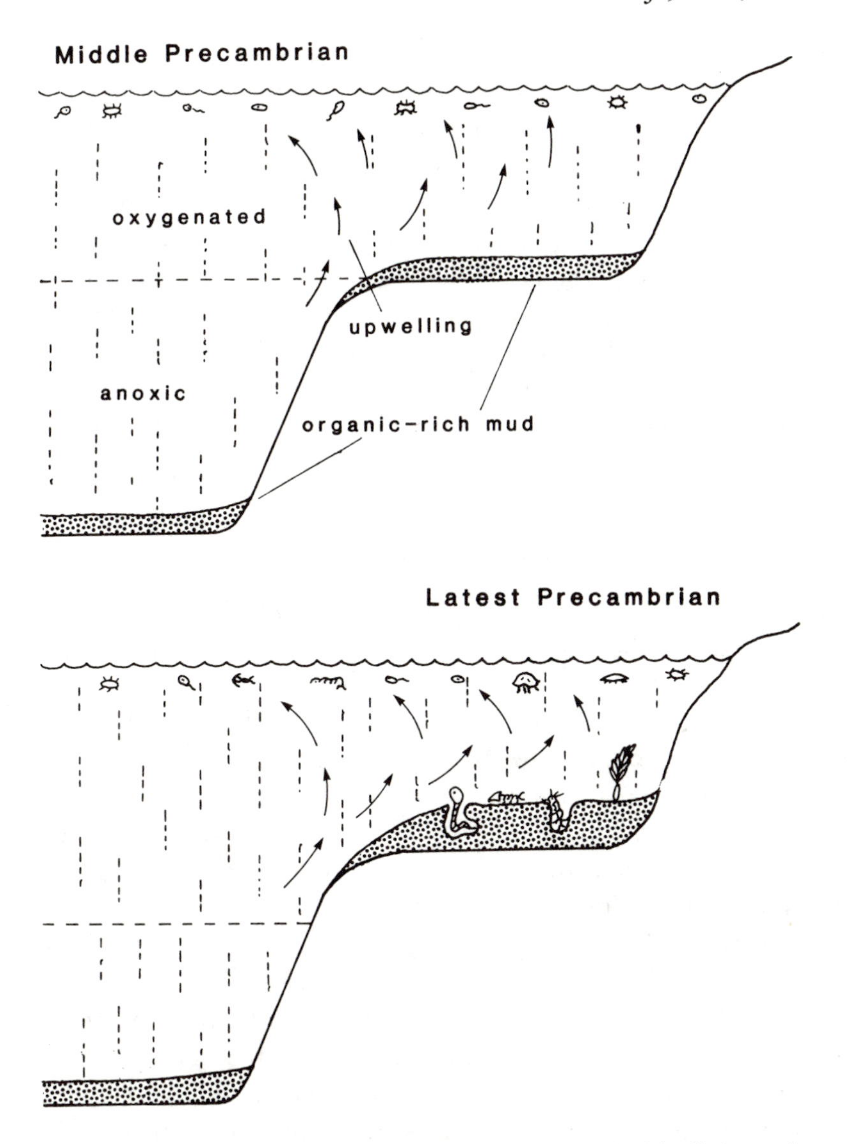

Figure 18.2. During the middle Precambrian, the continents had finally become large enough to exert a major influence on the circulation of the oceans. This included a great enhancement of upwelling and hence of the fertility of the surface waters. This event, doubtlessly welcomed by a rapidly expanding population of plankton, led to the deposition of abundant organic-rich sediments. Their exploitation, however, had to wait until late in the Precambrian, when multicelled organisms had developed body structures suited to the arduous job of crawling and burrowing in mud.

fashioned the beginnings of a biological control over the calcium and carbon dioxide cycles in the ocean that became virtually absolute with the expansion of calcareous plankton in the Mesozoic. Phosphate remained in use for bones, and silica was adopted by a few sponges and later by some major plankton groups, but most of the 30-odd minerals with which life had experimented were abandoned or survived only in rare instances. The importance of the questions why and how these hard parts were adopted is illustrated by the explosion of siliceous and calcareous oceanic plankton during the Mesozoic that so profoundly altered the chemistry of the ocean and even affected the atmospheric carbon dioxide content. Why? Like so many questions, this one remains to be answered.

EVOLUTION AND CONTINENTAL DRIFT

Many are the environmental changes that have pushed evolution one way or the other. There was the climatic deterioration of the late Cenozoic that greatly altered the flora and fauna of the northern continents. Then there was the formation of the warm, sheltered Tethys sea, temporarily located between the two halves of Pangaea, which housed a reef fauna so spectacular that by comparison Australia's Great Barrier Reef is but a faint shadow. Another example is the late Cretaceous transgression that covered the heart of North America and created two land masses for long enough to produce distinctly different faunas. And numerous are the extinctions and diversifications of shallow marine faunas that can be blamed on sealevel changes.

Some of the clearest examples of the influence of environment on evolution can be drawn from continental drift. They illustrate two separate phenomena: progressive divergence, as faunas became isolated when continents drifted apart, and the consequences of immigration, when lands bearing different faunas and floras collided.

In a science in which nature often seems bent on frustrating our efforts to understand her, the effect of increasing distance between drifting continents is pleasingly clear. Statistical analysis of the faunas of separate parts of the modern world shows that about half of their differences can be attributed simply to the distance between them. Not surprisingly, then, we find that as we plot the increase with time of the differences between faunas on opposite sides of the widening Atlantic Ocean, the relationship is very simple (Figure 18.3). In the North Atlantic, a connection persisted between North America and Europe until the Miocene, and the similarities between faunas on opposite sides

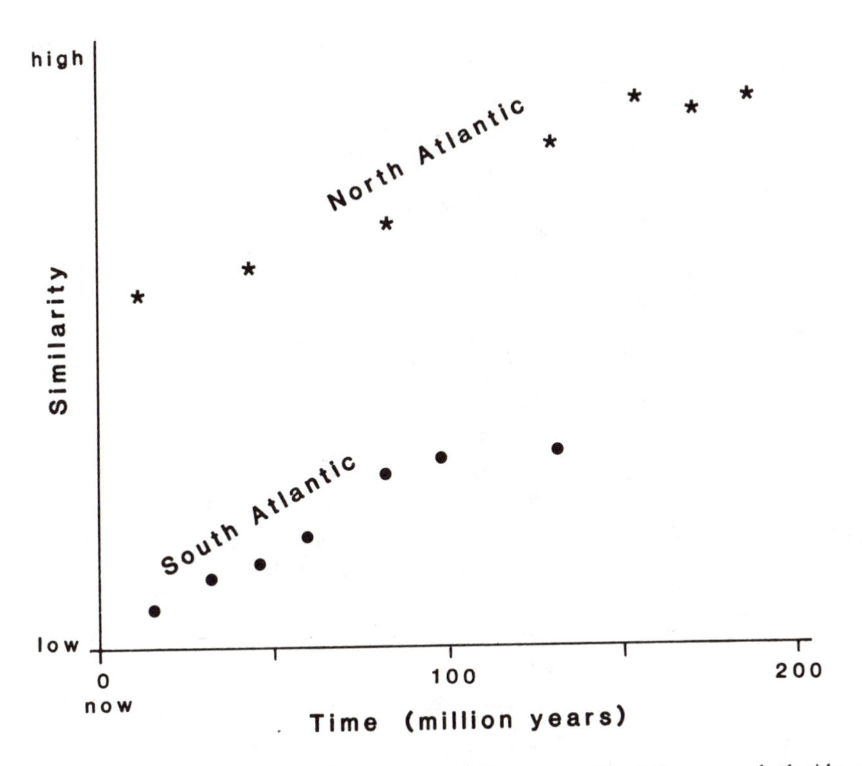

Figure 18.3. The Mesozoic and Cenozoic development of land faunas on both sides of the Atlantic furnishes a clear example of the impact of continental drift on evolution. Identical early in the Mesozoic, the American and European-African faunas gradually diverged in direct proportion to the width of the Atlantic Ocean. A landbridge in the far north kept the level of similarity in the North Atlantic higher until the bridge vanished in the Miocene.

remain higher than in the South Atlantic, where complete separation became a fact much sooner.

When we trace the fate of individual animal groups, the consequences of the Mesozoic disruption of the continents become more vivid. Many of the main reptile groups evolved early, before intercontinental distances became too great, and so managed to spread quite widely. When the Mesozoic began, the Tethys sea lay between Laurasia in the north and Gondwana in the south, and the crocodiles and dinosaurs evolved in Gondwana, whereas the ancestors of turtles, lizards, and snakes arose in Laurasia (Figure 18.4). Crossing the Tethys must have been easy, however, because in the middle Mesozoic we find descendants of all these groups just about everywhere.

Mammals, on the other hand, though originating in what was then an undivided Pangaea, did not truly begin to diversify until the continents had fully separated. Thus, their radiation started from many

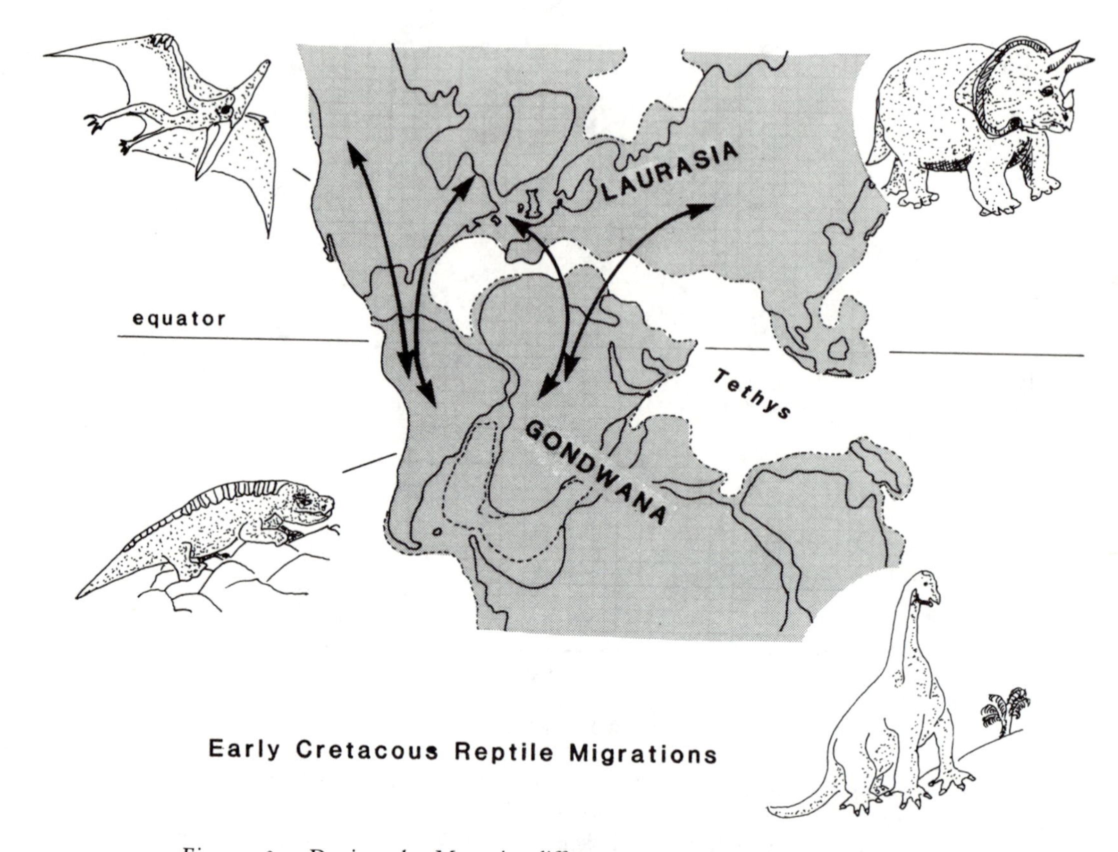

Figure 18.4. During the Mesozoic, different groups of reptiles arose in different parts of Pangaea. Their even distribution over all continents shows that the Tethys, the first of the seas that divided Pangaea, was not an effective barrier even in the Cretaceous.

isolated centers, including some that were kept apart by the Cretaceous transgressions. Shallow seas divided North America into two parts, Africa into three, and separated Asia from Europe. Between North America and South America there was an ocean, but South America remained joined to Australia by way of Antarctica until 45–50 million years ago. That presents us with a puzzle, because only primitive marsupial mammals reached Australia, even though other, more highly evolved mammal groups were present in South America long before the connection was broken. Otherwise, however, our current understanding of the progress of continental drift tallies well with the vertebrate fossil record.

Continents collide as well as separate, and when they do they sometimes bring together quite disparate faunas. The classic example is what happened when North America and South America were joined

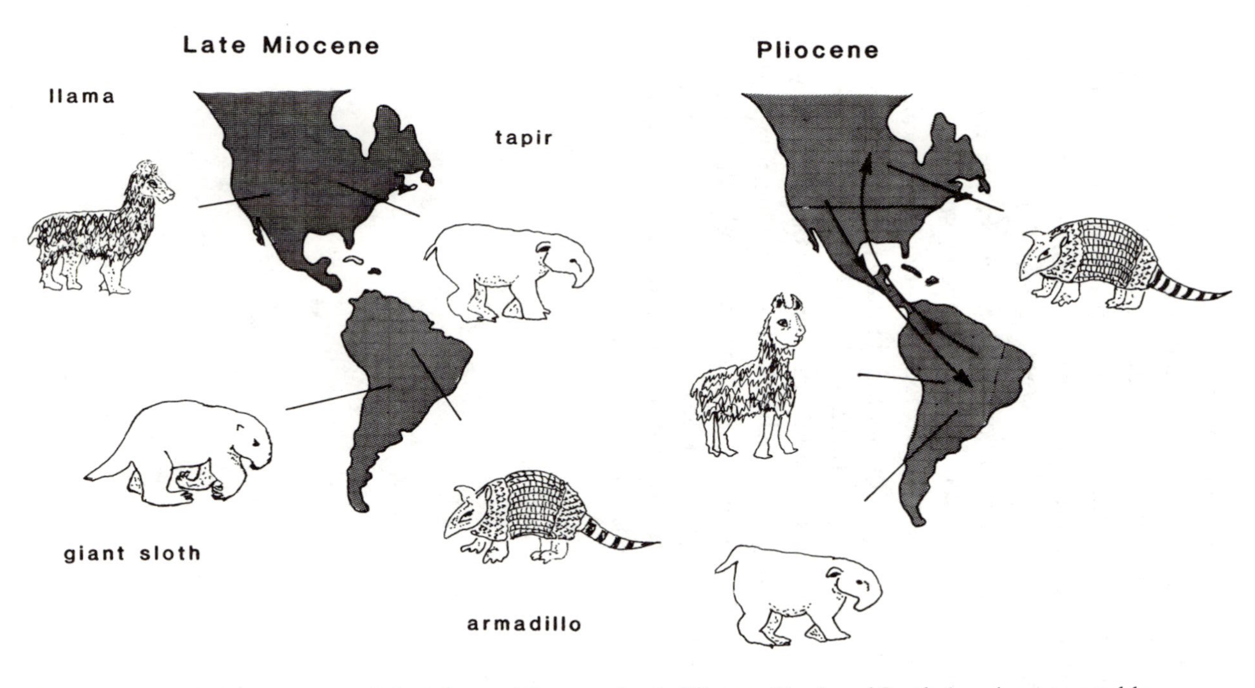

Figure 18.5. Until the emergence of the Isthmus of Panama in the Pliocene, North and South America, separated by a seaway too wide to be crossed by large land animals, had very different faunas. When the bridge was completed, the immigration of competitive northern animals made short work of the exotic marsupial fauna in the south. Such typical South American species as the tapir and the llama originated in the north; the only southern animal to make the reverse trip and succeed was the lowly armadillo.

3.5 million years ago. A rather bizarre fauna rich in large marsupials, but with few voracious predators, had evolved in South America during the long isolation of that continent (Figure 18.5). When the inhabitants of the north finally made their way south across the Isthmus of Panama, a host of apparently quite efficient herbivores, as well as some aggressive carnivores, made short work of their South American competition. Four entire orders of the local fauna became extinct in no time at all. Just a few southerners managed the trip north, and only the armadillo acquired a permanent foothold there. Ironically, South America is now a refuge for several initially northern species that subsequently became extinct there, such as the llama and the tapir.

The consequences of continental drift are thus obvious, but we must guard against attributing too much to it. For our education, nature has provided five anteaters on four continents, strikingly alike, but derived from wholly different ancestral stock. They present an example of convergence of form resulting from similar ways of living, and a fine example of a common paleontological trap.

19

Crises and catastrophes

Among the most striking events in the history of life are the brief but great contractions, sometimes near-collapses, that drastically reduced the diversity and, one suspects, the abundance of life for a while. Expansions, except for the earliest one, can be regarded as recoveries from the disasters that preceded them.

We can find comfort in these disasters. Lately it has become customary, perhaps for the first time in human history, to feel guilty about the damage we are inflicting on nature, guilty to such an extent that we suspect that the extinctions we have induced have had no counterpart in earth history and may well become the final ones. We readily believe that the paleo-Indians were to blame for the vast post-glacial extinctions of large mammals in North America, and were a skeleton of *Homo* to turn up in a late Cretaceous shale, we might well accuse him of the demise of the dinosaurs.

Nature, capable of great havoc without our assistance, has caused several memorable extinctions, but each time the recovery has been complete, even more than complete. We should not regard this as encouragement for our profligate ways, but may find some solace in it if we are willing to adopt a very long perspective.

It is easy to see the great dyings as catastrophes. Catastrophes stir the imagination and demand extraordinary causes. It is well to resist this temptation until we have examined just how catastrophic these events really were and how unusual their causes need to have been.

DEATH AND RENEWAL

The diversity of shallow marine life increased greatly during the early Paleozoic as specialization and new adaptations permitted access to an ever-widening range of environments (Figure 19.1). Stability, inter-

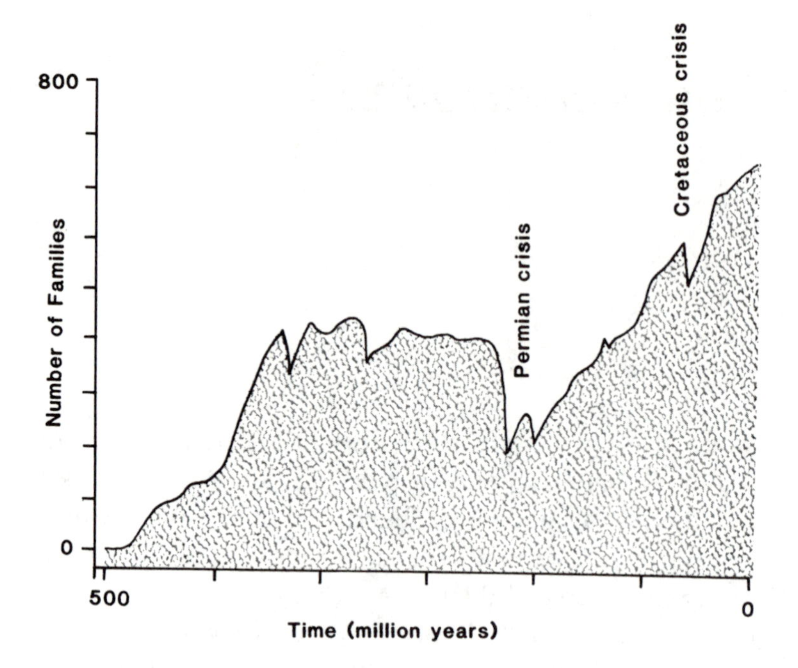

Figure 19.1. The diversity of life, expressed here as the numbers of vertebrate and invertebrate families living in the shallow sea, has greatly increased with time. There have been setbacks, but all were temporary, even the large ones of the later Permian and the Cretaceous.

rupted only by two moderate extinctions, lasted for about 200 million years, from the late Ordovician to nearly the end of the Permian, but then diversity dropped, slowly at first, then calamitously, until perhaps only one-tenth of all species crossed over into the Triassic. The subsequent recovery was slow but steady, and except for another, but quite different, contraction at the end of the Cretaceous, the diversity of marine life has increased ever since.

Another example is useful. Shortly after the last glacial period, extinctions of large mammals took place on all continents as the climate changed toward its post-glacial optimum and the environment changed with it. In North America alone, a dozen species vanished in about 3,000 years, including three kinds of mammoth, two kinds of bison, the camel, and the horse. The blame for this has been laid on the early Indians, but the final verdict is not in, because, among other things, a third bison species, also good hunting and later known as the American buffalo, survived and flourished. If man's influence was minor, as seems probable, we have a fine example here of the rapidity and thoroughness with which natural disaster can overtake a diverse and seemingly well-established group of organisms. This kind of crisis is common, although we are seldom in a position to determine how abrupt it actually was. There were, for example, four crises in the history of coral reefs and three major calamities that struck the trilobites. Many such extinctions are due to the failure of a limited group and to its replacement by more successful competitors. The nautiloids, shelled, swimming predators related to the squid and now represented by just a few species, lost out in the late Paleozoic to their cousins, the ammonites. When the ammonites in turn became extinct at the end of the Cretaceous, their place as top predators in the marine food chain went to the bony fishes. In each case the cause was most likely improved efficiency.

Behind other extinctions lie climatic shifts or regional changes in the environment. Minor spatial or temporal instabilities can have major consequences if the environment is a complex mosaic and its inhabitants are highly specialized. A rapid sealevel rise, faster than the growth rate of coral, will greatly impoverish reef faunas. Extinctions of this kind may seem random, their minor, brief causes lost in the shadow of time. This led the late Thomas J. M. Schopf of the University of Chicago to suggest that for most extinctions it may be futile to seek a specific cause; species become extinct when their time is up, just as people die when theirs has come. It is not important in the demographic sense whether one's death is due to heart failure, an automobile accident, or "natural causes," but only that people become "extinct" past a certain average age.

If extinction, like accidental death, often results from a brief but fatal mismatch between a species and some minor quirk of its environment, we may regard it as a random event. Such extinctions ought to be distributed randomly over time, and there is plenty of statistical information to suggest that this indeed is the case. But what about the major catastrophes, the incidents of concentrated large-scale dying that fueled the ideas of the eighteenth- and nineteenth-century catastrophists? How, in particular, do we account for the great faunal collapses of the late Permian and late Cretaceous? What causes lie behind these and other smaller but still major crises, all such fine stratigraphic markers for the working geologist?

The great extinctions were not all the same. Some were sudden, some more gradual; some affected selected groups, and some devastated all of the inhabitants of certain environments. Some seem almost arbitrary, wiping out most pelagic plankton but not the life in shallow seas, eliminating the dinosaurs but doing little damage to the vegetation. Some can be seen to coincide with large changes in the environment, but other major environmental remodelings of the earth seem to have had little impact.

Neither random extinction nor the usual kind of environmental change appears to be sufficient to explain the truly major crises. Something extra seems needed: At the end of the last glacial, man the hunter comes along and assists a cold-adapted fauna to its doom in the warm sun. A truly great sealevel lowering puts the shore at the top of the continental slope, and no shallow seas remain. The earth might pass through the poisonous tail of a comet, killing everything that breathes the air.

Before we embrace such magnificent notions, however, what is it, really, that demands an explanation – the extinction itself or some earlier, less conspicuous event that made the eventual crisis inevitable? We have used a similar approach to explain the Cambrian explosion of life. Can it be applied to extinctions as well?

Although many groups of organisms exist for a long time without any real change, once such a group begins to diversify, it usually rises rapidly to a peak of diversity. The decline is often equally precipitous. This stands to reason, because even though increasing specialization may be successful at first, eventually a price must be paid in the form reduced tolerance for natural variation in the environment. The stage must then be yielded to a more flexible competitor. When we plot the curves of the rise and fall of many families, we find them to be remarkably similar and also, more surprisingly, clustered in time (Figure 19.2). It seems as if specialization proceeds at much the same pace in many different groups and that there are times that especially favor expansion. Evidently, extinctions will then also cluster, and the real question may be why so many groups expand in unison, rather than why we have crises of extinction. What makes certain moments in time, spaced crudely 30 or 60 million years apart, so much more favorable than others?

This notion has not yet been greeted with much applause. It is easy to quibble with the data, because they are sparse and the statistics poor. Moreover, we are much more accustomed to viewing the past as a series of extinctions rather than as intermittent blooms of opportunity. Consequently, the record has not been scanned nearly as carefully for the good times as it has been searched for disasters. It should be a satisfying task, but we must leave it now and concern ourselves with catastrophes.

THE PERMIAN MARINE COLLAPSE

In terms of the numbers of species, genera, families, orders, and classes of animals lost, the Permian crisis was the largest of all. It began about

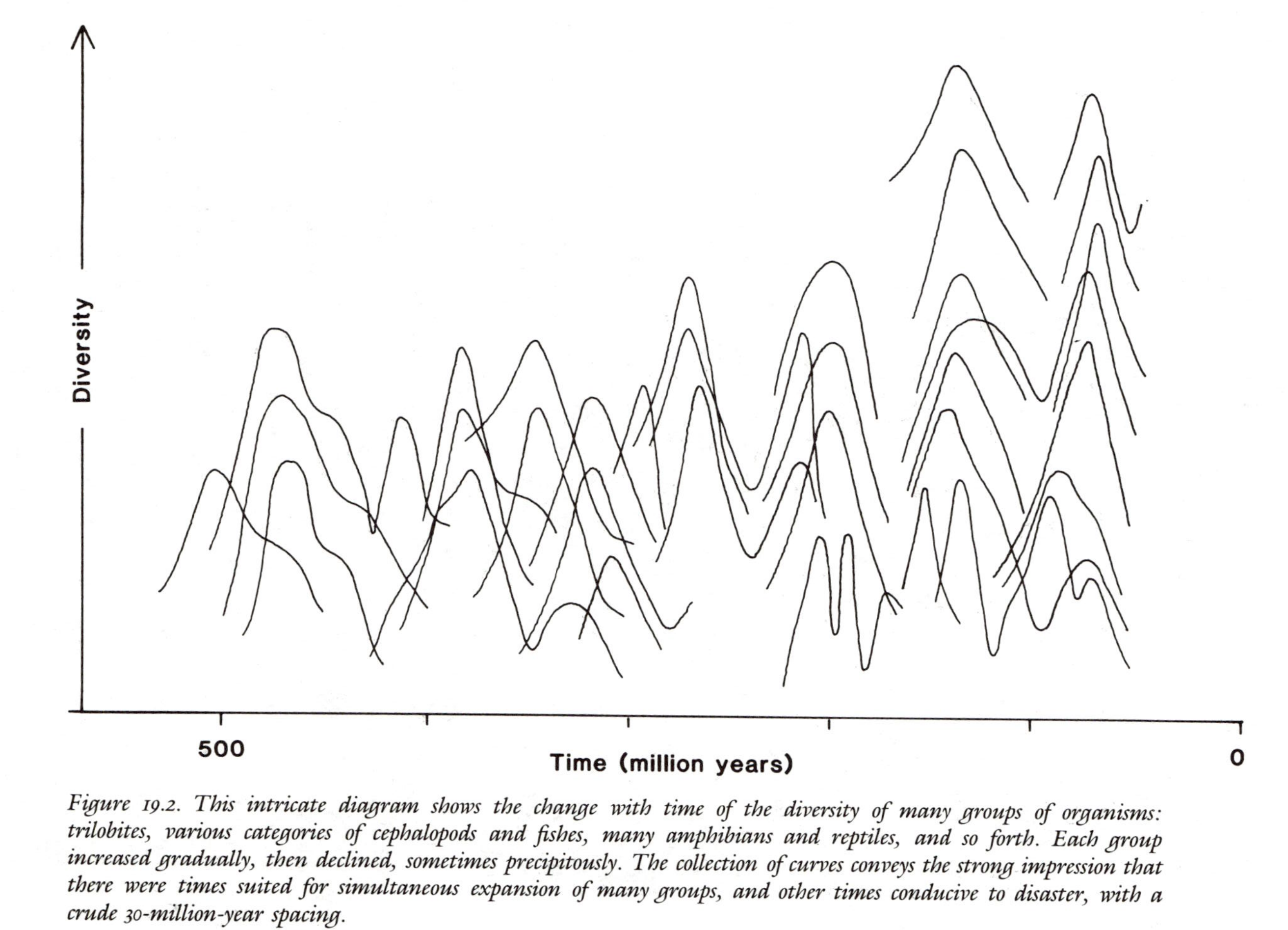

Figure 19.2. *This intricate diagram shows the change with time of the diversity of many groups of organisms: trilobites, various categories of cephalopods and fishes, many amphibians and reptiles, and so forth. Each group increased gradually, then declined, sometimes precipitously. The collection of curves conveys the strong impression that there were times suited for simultaneous expansion of many groups, and other times conducive to disaster, with a crude 30-million-year spacing.*

250 million years ago and left the world so much poorer in marine life when the Triassic arrived 20 million years later that total extinction in shallow seas was only narrowly averted. No real crisis occurred on land, but in the sea the 109 orders and 377 families of the Middle Permian shrank to 89 orders and 181 families. The disappearance of so many higher categories implies an even more drastic loss of genera and species. We cannot count those losses, because the record is too fragmentary, but we estimate with some confidence that about 65 percent of all genera and an almost incredible 85–95 percent of all species in shallow seas may have vanished. Out of an initial 50,000–250,000 species, only a few thousand may have made it into the Triassic, like scattered shipwrecked immigrants straggling to the shores of an uninhabited island.

Unquestionably it was a disaster, but it was a slow one and was limited to the shallow sea. This severely restricts our choice of explanations, of course, but it has not greatly limited the inventiveness of baffled geologists. Changes in environmental stability, reductions in the area of shallow seas, major sealevel changes, a drop in atmospheric oxygen, freshwater spilled from a large enclosed basin to reduce salinity, and even asteroids and meteorites have been proposed. Of these suggestions, the least exotic and perhaps most sensible ones have a plate-tectonics flavor.

Let us consider what might happen to marine life when conditions alternate between stable and unstable food supplies. Seasonal upwelling is a good example of a rich but unstable food supply: feast when it occurs, famine in the off-season. Large swings from abundance to scarcity and back again favor large populations with short lifespans and with a high reproduction rate to allow them to recover from mass mortality. Such a situation places a premium on tolerance for all sorts of conditions and on adaptation to all kinds of foods. Flexibility, low diversity, and fluctuating numbers are characteristic of this situation. In a stable environment, on the other hand, such as a reef, food may not be plentiful, but it is dependably available. Selection will be for a range of specialists, each precisely suited to take advantage of one out of the many narrow environmental niches. The population is diverse, but the numbers of individuals are small, and their adaptability is low. Go from here to an unstable condition, and extinctions will be widespread as the narrow specialists find themselves unable to adapt. Convert an unstable environment to a stable one, and the genetically flexible population will explode in a large variety of specialized forms.

James W. Valentine has married this concept to continental drift to explain the Permian marine extinction. A supercontinent, with its vast interior, is likely to be dominated by a monsoon climate. Monsoons produce large seasonal wind contrasts, with strong offshore winds during the cold season. The normal upwelling produced by ocean currents aided by the Coriolis force is enhanced by these winds, which seasonally blow surface water away from the shore. Thus, a supercontinent should have more widespread conditions of unstable food supply than the dispersed lands of the earlier Paleozoic or of the present. One would expect to see a high diversity of life when continents are scattered, a lower one when they congregate, and to some degree that appears to be true (Figure 19.3), although less so than the preliminary paleogeographic reconstructions available to Valentine several years ago suggested.

The proposition is plausible, but it is difficult to believe that the effect would be rapid enough to account for so much damage in a mere 20 million years. After all, continents join and separate quite slowly. Moreover, upwelling is common even in our present world of dispersed lands, and an increase of truly major proportions seems unlikely, even if all continents were joined together. Perhaps we need to consider other aspects of the difference between dispersed and aggregated continents.

Late Permian time was notable not only for its simple continental configuration but also for its rapid emergence. Between 250 and 230 million years ago the area covered with shallow seas diminished from 43 percent of the total land to about 13 percent, much as it is today. The reduction paralleled a decline in diversity (Figure 19.4), expectedly so, because fewer species can be accommodated in less space. The number of bird species nesting in one garden is less than that found in the entire city. On close inspection, however, it turns out that the diversity drop followed the reduction in area by a fair amount of time. At best, then, we are looking at a minor contributory factor.

Thomas Schopf, recognizing that this explanation was inadequate, proposed another, also related to continental drift. Given a supercontinent surrounded by an ocean, the minimum number of shallow marine biogeographic provinces, each distinct and bounded by climatic and oceanographic limits, would be eight (Figure 19.5). In a world of our present geographic complexity, on the other hand, the variety of shallow marine environments cannot be accommodated with anything less than 18 or 20 provinces, even if we ignore such special cases as oceanic islands. A large diversity decline would follow the joining of our conti-

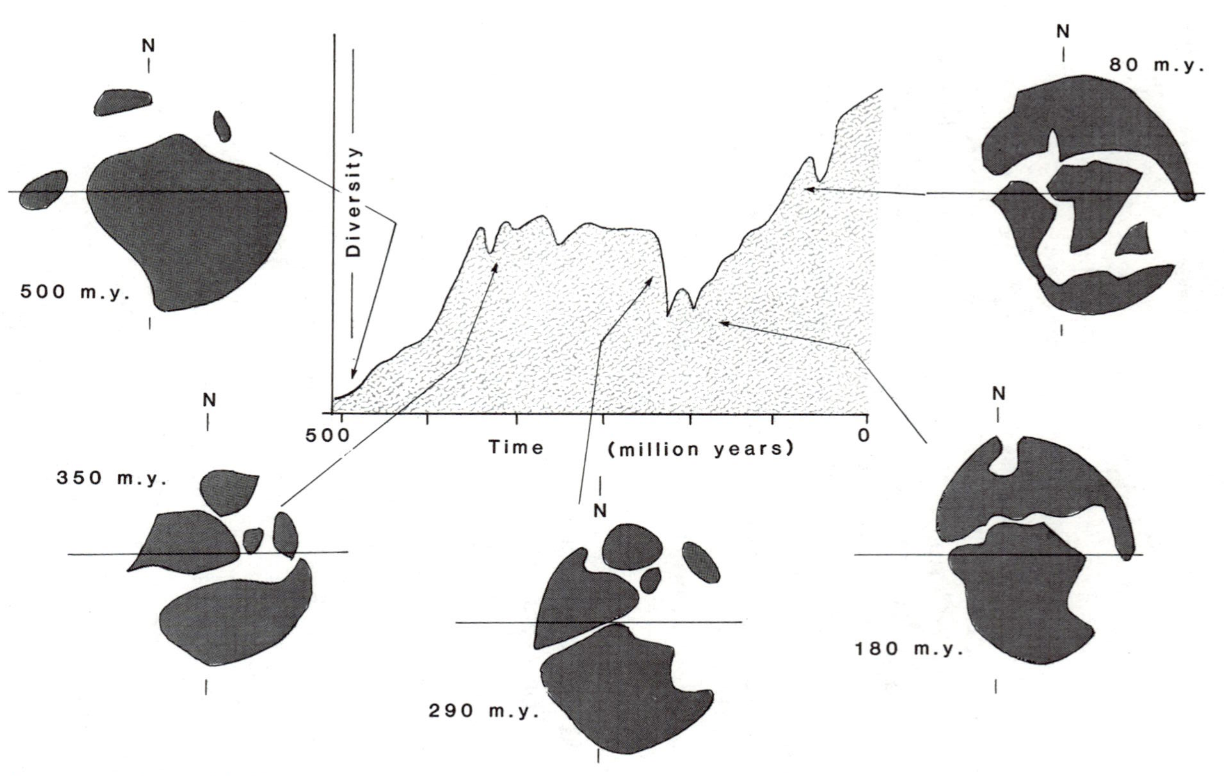

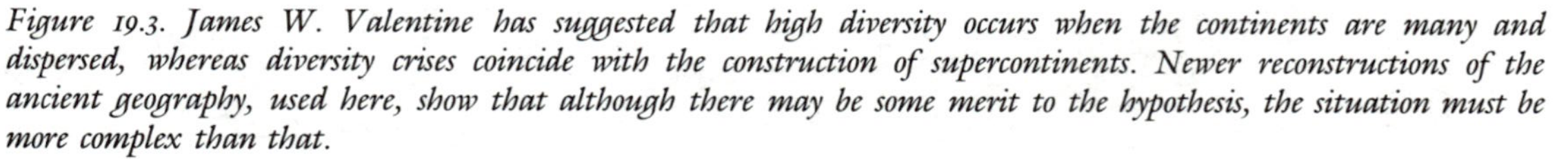

Figure 19.3. James W. Valentine has suggested that high diversity occurs when the continents are many and dispersed, whereas diversity crises coincide with the construction of supercontinents. Newer reconstructions of the ancient geography, used here, show that although there may be some merit to the hypothesis, the situation must be more complex than that.

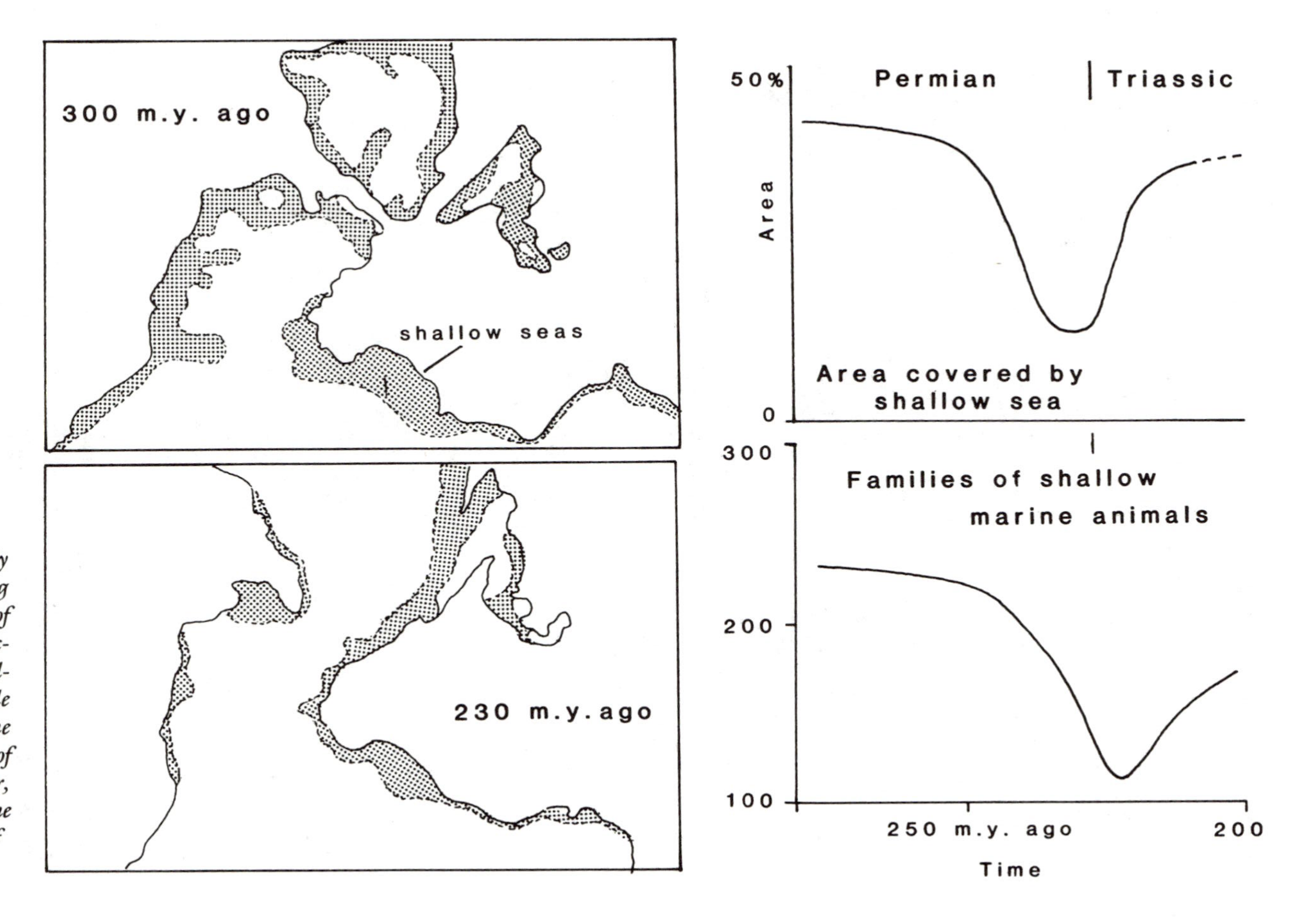

Figure 19.4. Pangaea finally came together during the closing years of the Permian at a time of low sealevel. The resulting reduction in the area covered by shallow seas has been held responsible for the crisis that afflicted marine life at that time. Comparison of the graphs on the right, however, shows that the extinctions came later than the decrease in shelf area.

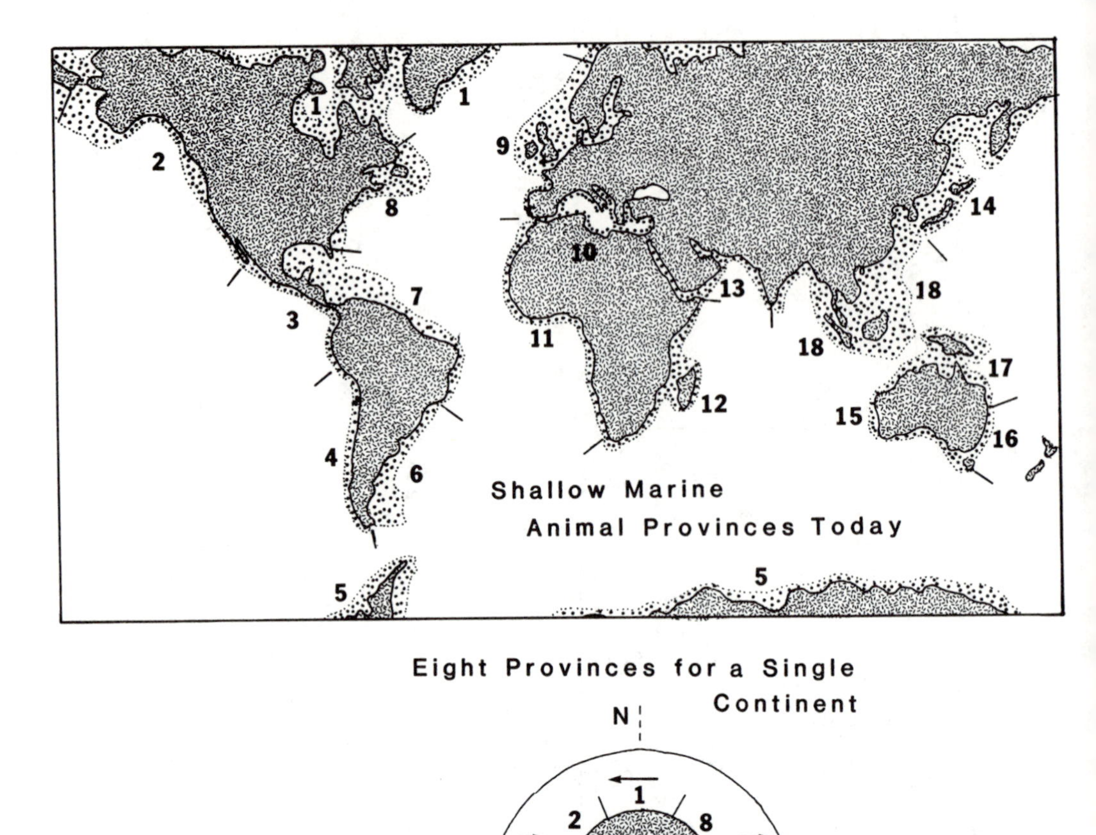

Figure 19.5. Another cause of the Permian marine catastrophe might have been the diminished range of environments that resulted from the simplified coastal geography of a supercontinent (bottom). A mere 8 provinces would account for the environmental range of Pangaea's shelves, whereas a minimum of 18 is needed for the modern world (top).

nents, simply because the number of provinces would be so much smaller. Statistics derived from modern faunas show that the effect of this change would indeed be large, larger by a factor of three than that of a reduction in area. I am thus inclined to accept Schopf's proposition as a simple and straightforward explanation of the late Permian marine diversity crisis, with food stability and territorial reduction thrown in as supplementary factors. Admittedly, this explanation lacks the dramatic flair appropriate for a crisis of such monumental proportions, but it has the merit of simplicity.

THE GREAT CRETACEOUS DYING

A wholly different catastrophe occurred at the end of the Cretaceous. It was less complete than the Permian marine crisis, but it is more intriguing to us because it affected so many different groups in so many different ways. Except for some considerable extinctions of marine faunas in the Tethys sea, then in the process of major tectonic remodeling, most shallow marine invertebrates crossed over into the Cenozoic without much damage. In the pelagic realm, on the other hand, the ammonites and belemnites, large cephalopod predators, disappeared completely over a period of about 10 million years. In contrast, the severe extinctions suffered by the oceanic plankton right at the Cretaceous-Cenozoic boundary seem nearly instantaneous.

The best known of the Cretaceous extinctions is that of the dinosaurs (and some other reptile groups of less popularity). Their demise has often been depicted as sudden and dramatic, but in reality the data are so sparse that all we can say with certainty is that the heyday of the great beasts was over long before the Cretaceous came to an end. In Africa, Asia, and South America, what we know of the disappearance of the dinosaurs is entirely compatible with a slow decline, and no survivors into even the later Cretaceous are known. Only in North America do we find dinosaur bone beds of late Cretaceous date in the stream deposits of the plains left behind by the vanishing Cretaceous inland sea. Much has been made of the "mass mortality" said to be implied by these torrential concentrations of bones, but in each case the immediate cause may have been no more than a devastating but local flash flood, a phenomenon not unknown to present inhabitants of the southwestern deserts. With good cause one might also suggest that the final demise of a score of dinosaur species is a trivial event that requires no recourse to an extraordinary catastrophe. The plankton extinction, on the other hand, may have taken less than 10,000 years, but our chronological resolution is poor so far back in the past; it might have taken 10 times as long. Nothing like this kind of rapidity can be demonstrated for the other victims of extinction. The key to the late Cretaceous extinctions is really how instantaneous they were, and how close in time to each other.

Still, the plankton crisis is indeed remarkable. It involved all major fossilized groups, calcareous and siliceous, primary producers and grazers, and it put an abrupt, albeit temporary, stop to the deposition of calcareous and siliceous pelagic sediments. In most places the interval of extinction is marked by a clear unconformity, elsewhere by a slowly deposited layer of brown clay much like the sediment forming today in the deepest, most infertile, most remote parts of the oceans. Above the unconformity or the clay the oozes return, but with a very different fossil assemblage. Only relatively few species survived (Figure 19.6), and there is a hint that the survivors were more broadly tolerant of salinity and temperature variations.

It is difficult to say how close in time to each other the various extinctions were. Dinosaur bone beds in river plains are not found in juxtaposition with pelagic oozes, and the principal correlations rest on magnetic reversals. Much work is being done on the subject, but at the present time it appears that the bone beds are at least one entire magnetic reversal away from the plankton extinction, a separation equivalent to 10,000–100,000 years and perhaps more.

Let us summarize. The shallow marine extinctions in the Tethys can easily be attributed to local oceanographic conditions as the continents converged across this seaway and reorganized the entire basin. Land plants declined, but mainly in higher latitudes, and a moderately cooler climate or an increase in seasonality would seem to be a perfectly good explanation. The dinosaurs went, but it was for them the thin and local end of a long decline. Though the plankton extinctions were sudden and drastic, we cannot demonstrate that they were synchronous with any of the other events. The sea withdrew gradually from the continents during the latest Cretaceous, and we know of a moderate climate change, but there were no major tectonic upheavals.

A sober and pragmatic judgment would thus be that simple and gradual causes, a cooler climate, a sealevel change, an altered vegetation cover, can explain everything except the plankton crisis. The evident desire of the public, and of some geologists as well, to adopt a more dramatic scenario is not compelled so much by the evidence as by reluctance to accept that a group of animals as impressive as the dinosaurs would just quietly vanish. Few scholars and no journalists at all have been content with simple causes for this romantic event.

Some geologists have proposed that a buildup of carbon dioxide in the atmosphere was the culprit and have held this up as a warning against our burning of fossil fuels. On land, the heat might have done the job, although the evidence suggests that the climate turned cooler. At sea, the increased carbonic acid might have dissolved calcareous

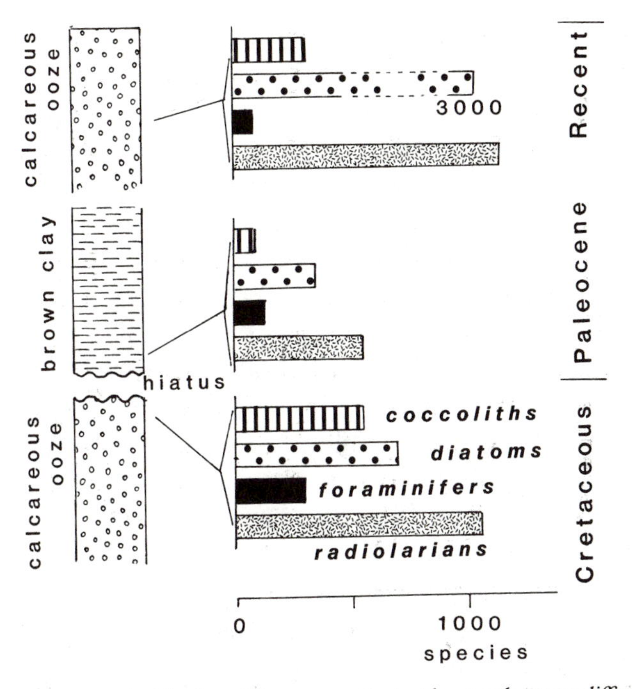

Figure 19.6. The late Cretaceous catastrophe struck many different groups in different ways, and some not at all, and involved extinctions at different rates and different times. Most dramatic was the reduction in numbers of species of oceanic plankton, both animals (foraminifers and radiolarians) and plants (coccoliths and diatoms), with either calcareous or siliceous shells. The diverse late Cretaceous flora and fauna (bottom) were replaced almost instantly with a much impoverished one (center), and the return to a diversity comparable with that of today (top) was much delayed. A hiatus commonly marks the Cretaceous-Cenozoic boundary, and the first Cenozoic sediments in the oceans were commonly brown clays rather than the calcareous and siliceous oozes of the preceding period.

shells, but this does not agree with the fact that siliceous organisms suffered just as severe an extinction.

Others have blamed magnetic reversals. This already rather venerable notion rests on the assumption that when the magnetic shield temporarily fails during a reversal, cosmic rays might do great genetic damage. Intuitively this makes sense, but the number of extinctions throughout history that coincide with reversals is no greater than what we would expect from chance alone.

More extravagant suggestions are supernova explosions, the poisonous tails of comets, and even the fascinating proposition that overcrowding made the dinosaurs so irritable that reproduction dropped below the minimum replacement value. For none of these hypotheses is there any evidence; in fact, for some, one wonders what evidence there could possibly be.

A rather less dramatic mechanism proposed a few years ago illustrates how difficult the problem really is. Suppose that there was an enclosed Arctic basin during the Cretaceous. Its waters should have been practically fresh, because so many rivers drained the surrounding temperate lands. Suppose that the Atlantic rift, steadily growing northward, finally connected this basin with the world ocean. Dense saline water would have rushed north and expelled the fresh water, which, being light even though fairly cold, would have spread over all oceans as a 50-m-thick layer. Much plankton, intolerant of the reduced temperature and salinity, would have died. This cool, brackish surface layer might have persisted for several years before it was fully mixed, and it would have reduced evaporation. A global drought would thus have occurred, and the world might have been a bit colder, as the observations indeed suggest. This mechanism obviously would work well to explain the plankton extinction; on land, the impact would be through climate. One does retain, however, the uneasy feeling that the opening of a passage into the Arctic basin could hardly have been instantaneous. There also remains the matter of proof.

Quite a different proposition has recently been much in the news. In its simplest form it postulates the impact, probably somewhere in the ocean, and at a speed of about 100,000 km/hr, of an asteroid some 10 km in diameter. There must have been a large blast and a shock wave, of course, but the major damage is attributed to a dust cloud thrown up so high that it might have obscured the sun for months on end. Photosynthesis on land and in the sea obviously would temporarily have come to a halt; there would have been little to eat for the vegetarians among the dinosaurs, and most of the plankton, with their short lifespans, would have become extinct. In fact, it is not quite clear how any primary producers in the sea, living generally only days or weeks, would have survived at all, unless the duration of the darkness has been grossly overestimated. The return of the sun in a few months would have been too late even for those dinosaurs that survived on the carcasses of their plant-eating cousins.

There is some evidence to support the contention that an extraterrestrial object hit the earth about 65 million years ago. The brown clay that occasionally occurs in deep-sea sediments at the boundary between the Cretaceous and the Cenozoic has been found to contain, in a small but growing number of cases, an anomalously high concentration of the element iridium, as well as a ratio of isotopes of the noble metal osmium, inferred to be alien. Some still unknown earthbound mechanism might account for this, but an extraterrestrial source for the iridium and osmium has appeared to many to be more likely. The exotic nature and wide distribution of the anomalies have been taken to mean that it must have been a large object that fell, although several smaller ones spread out over a somewhat longer time cannot be ruled out.

The main difficulties with this hypothesis are two: We cannot and may never be able to prove that the impact, if it happened at all, coincided with any of the various extinctions, nor can we show that those extinctions were synchronous, as the hypothesis demands. Moreover, even if an asteroid hit at approximately the right time, it would be impossible to demonstrate that it killed either the plankton or the dinosaurs; it might merely have made their inevitable demise more miserable. As time goes on, geologists have divided neatly into camps on this subject, some showing that iridium anomalies are not as uncommon as once thought, attempting to demonstrate that terrestrial mechanisms can account for them, or suggesting that other evidence for an extraterrestrial origin is lacking. Others have dug in and assembled supporting evidence for the proposition. The issue rests there and probably will continue to do so for some time.

What lessons can we draw from these tales of death and destruction, apart from the one that scientists, too, have their freewheeling ways of thinking? First, it is clear that nature is capable of causing large and sometimes rapid reductions in the richness and abundance of life. Second, the renewal after such devastation, although usually slow, invariably leads to new heights of diversity and complexity. Finally, when viewed with the proper perspective of time, most or perhaps all catastrophes are gradual and call only for simple terrestrial causes. Extraordinary, dramatic, or exotic events cannot always be ruled out, but they do not appear to be badly needed.

PERSPECTIVE

The history of life is the history of evolution, and it rests on two uncomfortable truths: The great biological panorama is the fruit of random variety, selected by the environment and its changes, and the road from bacterium to man is not progress, but merely improved adaptation.

The general course of life, from the Precambrian prokaryotes to the flowering plants, from the thriving fauna of shallow Paleozoic seas to the human condition, is well documented. There is no doubt about the trend, and there are no inexplicable precursors, no Cambrian turtles or Cretaceous primates. Diversity, except for the occasional setback, has continuously increased, and with it the demands on organisms living in an ever more complicated setting.

The reversals have been dramatic, however: Catastrophes that, intermittently afflicting life, have challenged the imagination of scholars and public alike, sometimes beyond prudence or common sense. These catastrophes illustrate not only that nature is vulnerable and fallible but also that science is not always as dispassionate as it ought to be.

The major transitions between life forms are often obscure, but that need not trouble us unduly, because the geological archives are full of gaps, and the fossil record is no exception. We need not be surprised that the theory of evolution rests more firmly on the biology of living organisms than on the paleontological record of the past.

The first evolutionary challenge was the colonization of unoccupied territory: the surface waters of the open sea, the muddy bottom near shore, the land. As each came to be colonized, the challenge shifted to competition, with the better equipped taking over from the less endowed. Throughout, there were continual changes in the environment, the propensity of the earth being never to remain the same for very long. The Paleozoic vividly illustrates the occupation of new territory, and continental drift and climate provide examples of the power of environmental change. We should like to know more about each, but that knowledge will come in time.

There remains for me now the task of drawing some conclusions, and seeking value beyond mere acquisition of facts, no matter how interesting. Inevitably, such a search for meaning is rather personal, but scholarship, after all, is a human enterprise.

FOR FURTHER READING

There is, of course, a vast literature in the field of paleontology and the history and evolution of life, much too vast to cite or even to know. That is also true for the nontechnical side, and the following list is to some extent arbitrary. It does, however, provide an introduction and further references.

Folsome, C. E., *Life, Origin and Evolution: Readings From the Scientific American,* Freeman, San Francisco, 1979, 148 pp.

Gould, S. J., *Ever Since Darwin,* Penguin Books, London, 1980, 284 pp.

LaPorte, L. F., *Ancient Environments,* Prentice-Hall, Englewood Cliffs, N.J., 1978, 176 pp.

LaPorte, L. F., *Evolution and the Fossil Record: Readings from the Scientific American,* Freeman, San Francisco, 1979, 222 pp.

McAlester, A. L., *The History of Life,* Prentice-Hall, Englewood Cliffs, N.J., 1977, 167 pp.

Rudwick, M. J., *The Meaning of Fossils,* Elsevier, New York, 1973, 258 pp.

Stanley, S. M., *The Evolutionary Timetable,* Basic Books, New York, 1982, 196 pp.

Epilogue

Many and strange are the universes that drift
like bubbles in the foam upon the river of time.
Arthur C. Clarke, *The Wall of Darkness*

When, in 1830, Charles Lyell set out to slay catastrophism, what he laid in the grave with a stake through its heart was already a ghost. His act of exorcism, however, was so persuasive that our memory of catastrophism has remained forever distorted.

Catastrophism, although a distant cousin of an older, theological view of the history of the earth, had in Lyell's time long evolved into a serious science, even by today's standards. Its major practitioners, Louis Agassiz, Georges Cuvier, Roderick Murchison, and Adam Sedgwick, were scientists, not holdovers from medieval metaphysics. Not for them were the supernatural cataclysms that extinguish life wholesale, nor repeated acts of creation, nor even an earth a mere 6,000 years old. They were keen observers, and their theories were based on quite detailed knowledge of the many crises of life we have discussed. Viewed without prejudice, the evidence was, and is, mostly on their side. The gradualists for whom Lyell spoke must invoke extreme incompleteness of the record to explain away the evident staccato mode of earth history.

Lyell was, of course, well aware of this, but his target was the public rather than the scientific perception of catastrophism, and he deemed a strong and unambiguous statement to be necessary. He did not foresee that from this statement would spring a seemingly indestructible caricature, one that cast its spell even on Lyell himself, who to the end of his life refused to accept the Ice Age because it smacked of catastrophism.

Uniformitarianism has been the subject of much exegesis since then, but catastrophism was permanently relegated to the history of science as an experiment that failed. Lyell's uniformitarianism consisted, as Stephen Jay Gould has pointed out, of several elements, some still accepted, others no longer regarded as reasonable. There is first a statement of method agreed to by every scientist, the statement that the laws of nature are invariant in space and time. If rocks might have fallen upward in the Precambrian, geology ceases to be possible. Science, any science, can exist only if we assume that nature does not at will violate its own laws. From this statement of method follows one of procedure: The processes we observe in action today are likely to have operated in the past as well, and must be the first ones to be invoked in explanation. Early nineteenth-century catastrophists accepted all this as much as modern geologists do, but Lyell went one step further by claiming that the present is not just the first and most likely key to the past, but the only one. Few of us would today define uniformitarianism quite so narrowly, even though we all prefer to be parsimonious with deviations from this principle.

With processes come rates, and here the difference between Lyell and the catastrophists becomes a matter of principle rather than degree. Lyell unequivocally stated that throughout geological history, change had been slow and gradual, never cataclysmic. From this he drew a second important conclusion: The configuration of the world has remained essentially the same since the beginning. As far as he could see, the course of history consisted of small, slow variations on a single theme, a view not even fully compatible with the evidence available in his time.

The catastrophists rejected both the last two points. Theirs was a world of major change, effected mainly in brief dramatic instants, revolutions, explosions of life, great extinctions. They saw the long patient work of erosion swiftly undone by the rise of new mountains, the slowly matured landscape swept away by a sudden rise of the sea. Their history of the earth resembled the punctuated equilibrium debated now by the paleontologists (see Chapter 17), and the record of the rocks is compatible with it.

Part of the disagreement is, of course, a matter of definition, of semantics. Catastrophe, crisis, and cataclysm are words with strong emotional content; they convey suddenness, totality, inevitability, random destruction. Yet an earthquake or a flash flood, both instantaneous, are neither more nor less catastrophic than a ten-year drought. The impact of an asteroid in England would be no more catastrophic than the Black Plague of the fourteenth century, which lasted nearly a hundred years. Catastrophes must be intermittent to deserve the name, but totality is not required, and the perception of their instantaneousness must be tuned to the perspective of time. The catastrophic post-glacial extinction of large mammals must have been seen by contemporary man as a slow, almost imperceptible decline.

Even with such a tolerant definition, however, catastrophism would have been anathema to Lyell and subsequent gradualists. On the other hand, modern geologists, while affirming the death of catastrophism, often invoke quite similar concepts under other names. Shorn then already of its supernatural causes and theological affinities, early nineteenth-century catastrophism greatly resembled what has been set forth in this book. How close we stand individually to a strictly gradualistic or to a more punctualist view is a matter of degree and personal preference, and to draw boundaries serves little purpose.

In the preceding chapters we surveyed the historical evidence, noting time and again the episodicities, sometimes even the intimation of periodicity. In the 1930s, the great German geologist Hans Stille thought that he saw rhythm in mountain building, and a few years later his Dutch colleague J. H. F. Umbgrove, felt the "pulse of the earth" in the subsidence of basins, sealevel changes, and climate as well. Alfred G. Fischer of Princeton University believes that the ocean, normally warm and only weakly stratified, changes about every 35 million years to a colder and more stable state, each change accompanied by a corresponding shift in the diversity of life. Although Fischer's oscillations appear dubious to the critical eye, there can be little doubt that the pendulum does swing, that episodicity is a feature of many of the earth's processes. On the other hand, one might suspect that we desire to see a pulse of the earth in part because Lyell's unvarying history is so unbearably monotonous.

Let us grant that ice ages come and go, that brief periods of mountain building break the stately procession of drifting plates, that expansions of life alternate with contractions. Are these merely random events, or do some or all repeat themselves periodically, rhythmically? Even more important, is the earth quiet for long times, whereas at other, briefer moments all hell breaks loose simultaneously in seafloor spreading, climate change, and ecology? Is there an underlying harmony, or are we listening to the tuning of an orchestra?

It is impossible at this time to answer this last question. It is statistical in nature and needs a statistical answer, because if we deal with several parallel series of random and mutually independent events, coincidences in time will occur by chance alone. Strange as it may seem, a record 4.5 billion years long is too short for many kinds of events; it contains too few ice ages, supercontinents, and catastrophes of life to determine whether they appear and disappear periodically or merely randomly. It will help if we can show some day that certain events are truly synchronous, but even then, simultaneous does not necessarily mean causally related. Our best chance to find out whether the processes of the earth march to one drummer or to many would be to conceive of a unifying cause, and then test that hypothesis.

There is a much-neglected aspect to Lyell's uniformitarianism.

Lord Kelvin perceived it as a fatal flaw when he pointed out that the earth cannot be a perpetual-motion machine. There must be a beginning, there will be an end, and the middle in which we live now should be different from either, or the second law of thermodynamics will be violated. Whether one accepts a gradualist or a catastrophist view of earth history, it must have direction; the engine must eventually run down. Geologists have given much less time to recognizing and describing this inexorable march of time toward the end than to debating the possible deviations from the path.

Our ground is not secure when we attempt to discern direction in the evolution of the earth, direction in a nonteleological sense, of course. The diminishing production of radioactive heat firmly says that there is one, but apart from the unresolved debate about the nature of Precambrian plate tectonics and the evidence for a gradual and diminishing growth of the continents, we have little to go by and have thought about it less. The Precambrian past is too dark, the Phanerozoic present too short to reveal a clear trend in the dynamic behavior of the earth. Perhaps we shall grasp the problem better when our knowledge of the terrestrial planets deepens, planets frozen in earlier states of development than the earth.

There is, on the other hand, no doubt about the change of life toward ever-greater specialization and variety. The theory of evolution says that this course was inevitable once life had been set in motion and that it must continue as long as conditions on earth do not put an end to it.

This cool phrase conceals what is, to me at least, the greatest question of all. The earth differs fundamentally from all other planets of the solar system because of the presence of liquid water and a gaseous but not poisonous atmosphere and because conditions at its surface are so well tuned to life's requirements. The earth has maintained this narrow environmental window within which life is possible for more than 3.5 billion years without failing once. We are not free to argue that this is irrelevant and that had our planet been differently conditioned, life would have adapted in different forms, because the conditions that permit the necessary biochemical reactions to take place are narrowly defined and coincide closely with those the earth maintains. Nevertheless, and though life can be sustained because the earth is the way it is, life's origin on the primordial planet is a highly improbable event, and the persistence of favorable conditions over nearly 4 billion years is scarcely less so. Yet life came and stayed; to import it from elsewhere merely defers the first problem and does not solve the second.

This question seems so momentous because in 1543 Nicolaus Copernicus taught the world that there is nothing unique about the position of the earth in the universe and that hence man, its observer, does not have a privileged status. This Copernican principle was later generalized to mean that, apart from merely local irregularities such as galaxies, the universe appears the same to its observer no matter from which point it is viewed. The fact that scientific experiments are reproducible even though the earth, between one attempt and the next, moves through time and space is a powerful illustration of that point.

Of late it has occurred to some astronomers that we may have been too zealous in our application of the Copernican principle. No one suggests, of course, that earth and man really do, after all, occupy the center of the universe, but one can argue with some force that life's existence imposes severe limits on the origin, history, and nature of the universe that contains it. If that is so, we might be able to explain why the universe is the way it is by reasoning back from the presence of life and of ourselves, the observers. This is the anthropic principle, introduced by Brandon Carter of Cambridge University. The anthropic principle can be used to advantage to clarify a number of conditions of the universe that otherwise seem arbitrary, and scientists, from long experience, are wary of apparent arbitrariness in nature. Take, for example, the age of the universe. It should not have been 10 times less, or elements like carbon, needed to make physicists, would not yet have become available. Make it an order of magnitude greater, and the suns required to warm planets and sustain life would long since have been extinguished. Reduce gravity 10 times, and no planet would hold together for observers to stand on; increase it 10-fold, and the only suns in the universe would be blue giants, too ephemeral to allow life to evolve.

Is it plausible that the universe should be so special, so precisely fit to accommodate life, to house its own observers? There must be an infinity of possible universes in quite different states that would be equally probable. We would much more comfortably accept our special universe if other, different ones did also exist. As astronomers have explored the anthropic principle, they have begun to consider this possibility, finding that the concept of multiple universes also helps in elucidating many an oddity of our world.

Multiple universes, for instance, can be used to illuminate a puzzling aspect of quantum theory. This theory says that the position of an elementary particle, a photon for example, can be described by an equation that gives us for each point on the path of the particle the probability that we shall find it there if we attempt to detect it. In a light wave, this yields many positions of equal probability; yet we find the photon at only one of those, not at the others. This clash between a probabilistic world and a deterministic one is perturbing. If many positions are equally probable, why should the particle exist in only one of them? Perhaps, one reasons, the equation describes simultaneously an infinity of universes. Because we live in only one of them, we find the photon only in one spot.

If an infinity of universes is possible, some with and some without observers, are we to assume that they all exist? The answer should, in principle at least, be affirmative, but that would be difficult to confirm. In fact, one might argue that the only possible confirmation of the existence of a universe is that one observes it to exist. It follows, then, that only universes that contain observers, that hold "life," need be assumed to be real.

Here we are drifting across the border into metaphysics and must retreat. We might, however, carry this kind of reasoning over to the history of the earth. If the presence of life sets limits on the conditions of the universe at its birth, its presence on earth must equally limit the origin and history of our planet and, inevitably, of the solar system. Thus, the issue is not how, in the face of incredible odds, life arose here. Rather, it is the inverse: What can we say about the birth of the solar system and the earth knowing that life demands certain conditions for its origin and its persistence? What does that tell us about the early and later atmosphere, about the weak and pale sun, and about the many other things that might so easily have gone wrong and wiped out life, but did not do so?

Instead of reasoning forward from the beginning, we should perhaps try to work backward from the end result, in satisfying parallel with the structure of this book, if indeed it is true, as T. S. Eliot said in *Four Quartets:*

> The dance along the artery,
> The circulation of the lymph
> Are figures in the drift of stars.

Glossary

This listing contains the more common terms used throughout the text. For others, reference should be made to the index.

Albedo The proportion of the incoming solar radiation reflected back by earth surfaces such as desert, snow, or water.

Amino acids Organic molecules containing nitrogen; the basic building blocks of proteins.

Andesite An extrusive volcanic rock intermediate in composition between basalt and granite; major component in subduction zone volcanism.

Arc A string of volcanic or sedimentary islands lying behind a trench and above a subducting plate.

Asteroid A body of rock larger than a meteorite, orbiting within the solar system and presumed to hit earth occasionally, causing disaster.

Asthenosphere A hot, soft layer of the mantle, on which the lithosphere floats.

Azimuth Here used for the direction from a point on earth to the magnetic pole.

Backarc basin A depression lying behind the volcanic arc above the subducted slab, and apparently spreading. Not clearly developed in subduction under continents, but common behind Pacific trenches.

Basalt Fine-grained, dark extrusive igneous rock rich in iron and magnesium-bearing minerals and relatively poor in silica; characteristic of ocean crust.

Black shale Dark, fine-grained sedimentary rock rich in organic matter; often source bed for oil.

Catalysis Acceleration of a chemical reaction by a compound, a catalyst, that does not itself participate in the reaction.

Chloroplast Small body or organelle within a plant cell; the site of photosynthesis.

Chondrite, carbonaceous A type of meteorite rich in water and light elements such as carbon; assumed to resemble the original material of the mantle.

Chromosome Thread-like elements in the eukaryote nucleus that carry the genes.

Coelom A cavity within the body of the higher animals; it lends strength, among other functions.

Convection current A process of heat transfer by rising hot fluid; once cooled, the fluid moves away laterally and sinks again.

Coriolis force A force resulting from the rotation of the earth; it deflects a moving mass to the right in the Northern Hemisphere when viewed in the direction of motion, and to the left in the Southern Hemisphere.

Correlation The procedure of determining that two outcrops of rock, or two formations or fossil zones, are equivalent, either in age or by virtue of once having been connected.

Craton A stable portion of a continent, commonly of Precambrian age and not deformed for a long time.

Cross-cutting An intrusive rock or a fault that cuts across the normal bedding and thereby demonstrates that it is of later date.

Crust The upper solid shell of the earth, separated from the underlying mantle by the Mohorovicic discontinuity, and of lower density than the mantle.

Cyclothem A sequence of shallow marine and coastal sediments including coals, that repeats itself vertically; usually used for Pennsylvanian coal beds.

Deformation The folding and faulting of a rock sequence.

Diversity A measure of the variety of life forms in a region, environment, the world, or during a given interval of time; usually expressed as the number of species, genera, or families.

Echosounding A depth-measuring technique whereby sound is transmitted to the seafloor, and the arrival time of the returning echo is measured.

Endmoraine A wall of glacial debris formed at the margin of a glacier or icecap by melting.

Episodicity Designates here an event or process that repeats itself over time without a predictable pattern.

Equinoxes The dates in spring and autumn when night and day are of equal length.

Eukaryote A cell containing a distinct nucleus bound by a membrane, and organelles; the cell type for all higher plants and animals.

Eustatic A global change in sealevel.

Evaporite A sedimentary rock produced by evaporation of seawater and precipitation of its salts; contains a wide range of components, among which rock salt and gypsum are the most common.

Facies The characteristics of a rock, generally sedimentary, that reflect the environment of deposition in which it was formed.

Fault A fracture in a rock body along which one side has been displaced relative to the other.

Feedback Occurs when a process is self-reinforcing (positive feedback) or self-limiting (negative feedback).

Forearc A zone at the boundary of colliding plates just behind the trench, where sediments and to some extent oceanic crust are being deformed.

Formation A suite of rocks with common characteristics distinguishing it from other suites nearby; traceable or mappable over a reasonably large area.

Fracture zone A dislocation at right angles to a mid-ocean ridge, usually marked by a set of transverse ridges and troughs.

Gateway, or seaway, a connection between two ocean basins.

Genetic drift Random fluctuation in the genetic makeup of a small population that, in time, can cause that population to acquire different characteristics.

Geosyncline A large, elongated depression in the earth's crust in which thick sediments accumulate, followed by deformation into a mountain range.

Glacial An interval of cold climate leading to great expansion of continental icecaps.

Gneiss A coarsely crystalline, quartz-rich metamorphic rock usually formed from granite.

Granite A coarsely crystalline intrusive igneous rock rich in lighter elements and containing mainly quartz and feldspar; typical for continental crust.

Graywacke A poorly sorted sandstone rich in rock fragments (often of volcanic origin) or in clay.

Greenhouse effect A warming of the climate because of an increase in atmospheric carbon dioxide, which traps heat reflected from the earth's surface. Expected to become serious as the burning of oil and coal continues.

Greenstone belt Elongated, folded complex of slightly altered sediments and volcanic rocks of oceanic and intermediate character; usually of Precambrian age.

Hiatus A time interval not represented in the rock record.

Hotspot An anomalously hot area in the mantle below the lithosphere; source of volcanic activity in a plate as it passes over the hotspot.

Hydrothermal The activity of hot waters in the crust; in the present case, used for springs resulting from ocean waters circulating through hot, new oceanic crust.

Igneous Rock formed by cooling from a molten state (magma or lava).

Inclination The angle between a suspended magnet and the horizontal plane; measures latitude.

Interglacial A warm interval between two glacial periods of extensive ice cover.

Interstadial A warmer interval within a glacial period, but not warm enough to produce full retreat of icecaps.

Isostasy The rise and fall of the earth's crust as a result of changes in its buoyancy.

Isostatic compensation Subsidence or rise of an area in response to a change in its mass, in the product of its density and thickness; isostatic compensation occurs, for example, when an icecap depresses a continent, or when the removal of ocean water causes the ocean floor to rise.

Isotopes Elements with the same number of protons but a different number of neutrons in the nucleus; they have very nearly the same chemical properties, but different atomic weights. Isotopes can be either stable or unstable (radioactive).

Jetstream A high-altitude wind blowing from west to east at mid-latitude, producing major effects on seasonal weather as it changes its path from year to year.

Lava Molten rock, usually so called when it flows from a surface fissure or a volcano; when congealed, it forms an extrusive rock.

Magma Molten rock existing deep in the earth; when congealed, it forms intrusive rocks.

Magnetic anomalies Areas of the oceanic or continental crust where the magnetic field is either stronger or weaker than that predicted on the basis of the earth's field alone; caused by magnetic properties of the crust.

Magnetic polarity The direction of the earth's magnetic field; it reverses from time to time, causing magnetic north to become magnetic south.

Mantle The part of the earth between the crust and the core.

Maria Dark areas of congealed lava on the moon, probably formed by bombardment with space debris, asteroids, and meteorites.

Marsupial A mammal not equipped with a uterus suited for prolonged care of the embryo; the embryo is carried in an external pouch.

Metamorphism A change in composition, mineralogy, or structure of a rock resulting from changes in temperature and pressure; common during deformation or very deep burial, as well as near intrusions of hot magma.

Mitochondrion An organelle, a body within a eukaryote cell in which processes involving oxygen take place.

Mohorovicic discontinuity (Moho) The boundary between crust and mantle, marked by a sharp downward increase in the velocity of earthquake waves.

Obliquity The angle between the rotational axis of the earth and the plane of its orbit.

Organelle A distinct body within the eukaryote cell, with a special function, such as photosynthesis or energy housekeeping.

Orogeny Mountain building, involving deformation, metamorphism, uplift, and the intrusion of bodies of igneous rocks; takes place during plate collisions, especially those involving two continents.

Orogen A mountain range formed by deformation.

Periodicity Designates events repeating themselves predictably.

Precession The gradual shift with time of the points on the earth's orbit where the equinoxes occur.

Primary producers The photosynthesizing autotrophs that form the base of the food chain that leads up through grazers and several levels of predators.

Prokaryote A cell lacking a membrane-bound nucleus and organelles; bacteria and certain algae.

Radioactivity Decay of unstable isotopes converting a parent isotope into a daughter while radiation and particles are emitted.

Reflector A boundary between strata that reflects the energy pulse transmitted by an earthquake, or in artificial seismic studies such as seismic profiling.

Regression A seaward shift of the shore, due to a fall in sealevel, a rise of the land, or sedimentation.

Rift A trough between two fault zones, with a down-dropped central block; it often is the beginning of continental breakup and also occurs on mid-ocean ridge axes.

Salinity A measure of the total amount of dissolved solids in seawater.

Shale A fine-grained sedimentary rock consisting of silt and clay; tends to separate in thin sheets along depositional bedding planes.

Slate A hard, fine-grained metamorphic rock formed from clay; its fine lamination is due to pressure, not to depositional bedding.

Solar constant The amount of solar radiation arriving at the earth as measured just outside the atmosphere; presumed to have increased significantly since the earth was formed.

Solstice The date of the longest or shortest day; the days on which the sun is farthest from the equator.

Stadial A colder phase within a glacial interval.

Strike-slip fault A fault along which the displacement of rock bodies is horizontal.

Stromatolite A finely layered mound of limestone produced by an algal mat; particularly common in the Precambrian, and one of the earliest signs of the presence of life on earth.

Subduction The process by which a plate, always one consisting of oceanic crust, is dragged under an adjoining plate; subduction compensates for the continuous creation of new oceanic crust on mid-ocean ridges.

Sunspots Dark blotches on the sun indicating unusual solar activity; they may have an influence on climate, but the mechanism of interaction is not known.

Suture A join or weld left from a former collision of two continents.

Symbiosis Intimate living together of two organisms, usually within the body of one of them, in a relationship of mutual benefit.

Taxonomy The classification of organisms.

Tectonics The study of the movement and deformation of the earth's crust and mantle.

Thermocline The boundary between surface water and deep water in the ocean, usually at a depth of 100–200 m, and marked by a large and abrupt downward drop in temperature.

Transform fault A fault with horizontal movement connecting two segments of a mid-ocean ridge; more broadly, a plate boundary along which plates move horizontally past one another without either convergence or divergence.

Transgression A landward move of the shore produced by a sealevel rise or a sinking of the land. Erosion of the shore does not normally cause a significant transgression.

Unconformity A surface separating two beds of sedimentary rock and representing a gap in time during which sedimentation ceased or erosion removed the rock record.

Underthrusting A process pushing one rock body underneath another along an inclined fault plane. It is usually accompanied by intense deformation and some metamorphism and is common at the collision edge of two plates.

Upwelling Water rising to the surface from intermediate depth, usually as a result of surface water being driven away by wind or currents. Upwelling is the main agent recycling nutrients to the zone of primary production in the surface waters of the sea.

Weathering Decay of rocks at the surface of the earth under the influence of wind, water, plants, temperature, and all other processes that break them up into particles or dissolve them.

Wobble See precession.

Sources of illustrations

ON TITLE PAGES OF MAIN SECTIONS:

Itinerary
W. F. A. Zimmerman, *De Wonderen der Voorwereld,* p. 1, D. Bolle, Rotterdam, 1892.

Foundations
Charles Lyell, *Lectures on Geology Delivered at the Broadway Tabernacle,* frontispiece, Greeley and McElrath, New York, 1833.

Climate
M. Grouner, *Histoire naturelle des glaciers de Suisse,* plate X, Panzkouche, Paris, 1777.

Drifting continents
Thomas Sutcliff, *The earthquake of Juan Fernandez as it occurred in the year 1835,* plate 2, Manchester.

Ancient oceans
Louis Figuier, *La terre avant le déluge,* p. 430, Hachette, Paris, 1872.

Four-billion-year childhood
Amos Eaton, *Geological Textbook for Aiding the Study of North American Geology,* 2nd ed., p. 55, Websters and Skinners, Albany, 1832.

Life, time
G. F. Richardson, *An Introduction to Geology,* revised by Thomas Wright, frontispiece, H.G. Bohn, London, 1855.

Epilogue
B. Faujas Saint Fond, *Natuurlijke Historie van den St. Pietersberg bij Maastricht,* colophon, Johannes Allart, Amsterdam, 1802.

TEXT FIGURES

Most illustrations are based on concepts and data in the common domain, but in the construction of the following I have made use of specific data or ideas.

Figure 3.1. Data from H. H. Lamb, in H. Flohn (ed.), *World Survey of Climatology,* pp. 173–249, Elsevier, New York, 1969, and from R. A. Bryson, *Science* 184:753–760, 1974.

Figure 3.2. Adapted from J. and K. P. Imbrie, *Ice Ages, Solving the Mystery,* Fig. 43, Enslow Publishers, Garden City, N.J., 1979.

Figure 3.4. Based on an anonymous article in *Mosaic,* pp. 35–41, July-August, 1977.

Figure 4.1. Data from CLIMAP, in *Science* 196:1131–1137, 1976, and from G. M. Peterson et al., in *Quaternary Research* 12:47–82, 1979.

Figure 4.3. Adapted from A. McIntyre and N. G. Kipp, in *Memoirs of the Geological Society of America* 145:59, 1976.

Figure 4.4. Data from G. M. Woillard and W. G. Mook, in *Science* 212:159, 1982, and from Fig. 48 of Imbrie and Imbrie (see Figure 3.2).

Figure 4.6. Adapted from J. R. Curray, *Recent Sediments of the Northwestern Gulf of Mexico,* p. 260, American Association of Petroleum Geologists, 1960, and T. H. van Andel et al., in *American Journal of Science* 265:738, 1967.

Figure 5.1. Based on data from C. Emiliani and N. J. Shackleton, in *Science* 183:511–14, 1974, and from N. J. Shackleton and N. D. Opdyke, in *Memoirs of the Geological Society of America* 145:449–64, 1976.

Figure 5.3. Adapted from an idea of W. Ruddiman and A. McIntyre.

Figure 6.1. Based on data of M. W. McElhinny *Palaeomagnetism and Plate Tectonics,* Cambridge University Press, 1973.

Figure 6.9. Adapted from *The Age of the Ocean Basins,* a map by W. C. Pitman III et al., Geological Society of America, 1974.

Figure 6.10. Data from M. Barazangi and J. Dorman, in *Bulletin of the Seismological Society of America* 59:369–80, 1969.

Figure 7.2. Compiled from many publications by John C. Crowell.

Figure 7.3. See Figure 7.2.

Figure 7.4. Based on a map of C. Scotese et al., in *Journal of Geology* 87:217–78, 1979, and one by A. M. Ziegler et al., in *Annual Reviews of Earth and Planetary Sciences* 7:473–502, 1979.

Figure 7.5. Map base from A. G. Smith and J. C. Briden, *Mesozoic and Cenozoic Palaeocontinental Maps,* Cambridge University Press, 1977.

Figure 7.7. Based on various publications by Kevin Burke et al.

Figure 8.1. Adapted from Kevin Burke et al., in *Tectonophysics* 40:69–99, 1977.

Figure 8.5. Maps based on Scotese (see Figure 7.4).

Figure 8.6. Adapted from P. Molnar and P. Tapponnier, in *Scientific American* 236:30–41, 1976.

Figure 8.7. From an idea of Peter Bird, in *Journal of Geophysical Research* 83:4975–88, 1978.

Figure 8.8. Adapted from an anonymous article in *Mosaic,* p. 33, March-April, 1981.

Figure 8.9. Adapted from data of W. Schwan in *Bulletin of the American Association of Petroleum Geologists* 64:359–373, 1980.

Figure 10.1. From A. Hallam, in *Nature* 232:180, 1971, updated from A. Hallam, in *Nature* 269:771, 1977.

Figure 10.2. Base map from Smith and Briden (see Figure 7.5).

Figure 10.3. The seismic reflection record is from the shelf off Walvis Bay, southwestern Africa.

Figure 10.5. Simplified and adapted from P. R. Vail and R.M. Mitchum, in *Memoirs of the American Association of Petroleum Geologists* 29:470–1, 1979.

Figure 10.6. Adapted from W. C. Pitman III, in *Bulletin of the Geological Society of America* 89:1389–1403, 1978.

Figure 11.7. Adapted from T. H. van Andel, in J. Gray and A. J. Boucot (eds.), *Historical Biogeography, Plate Tectonics, and the Changing Environment,* p. 18, Oregon University Press, Corvallis, 1979.

Figure 11.8. Base maps from Smith and Briden (see Figure 7.5).

Figure 11.9. Data from S. M. Savin, in *Annual Reviews of Earth and Planetary Science* 5:319–55, 1977, and from N. J. Shackleton and J. P. Kennett, in *Initial Reports of the Deep-Sea Drilling Project* 29:801–7, 1975.

Figure 11.10. Adapted from S. L. Thompson and E. J. Barron, in *Journal of Geology* 89:143–68, 1981.

Figure 12.1. Adapted from A. G. Fischer and M. A. Arthur, in *Special Publications of the Society of Economic Paleontologists and Mineralogists* 25:32, 1977.

Figure 12.2. From data in G. Dietrich, K. Kalle, W. Kraus, and G. Siedler, *Allgemeine Meereskunde,* 3rd ed., pp. 469–74, Borntraeger, Stuttgart, 1975.

Figure 12.3. Based on Fischer and Arthur (see Figure 12.1).

Figure 12.4. Base maps from Scotese (see Figure 7.4).

Figure 13.3. Based on a hypothesis of R. B. Hargraves, in *Science* 193:363–71, 1976.

Figure 13.4. Data from A. B. Ronov, in *Geochemistry* 8:715–43, 1964, and from R. M. Garrels and F. T. Mackenzie, *Evolution of Sedimentary Rocks,* Fig. 10.2, Norton, New York, 1971.

Figure 13.5. Based on data from A. M. Goodwin, in *Science* 213:55–61, 1981.

Figure 13.6. From a landsat photograph in D. I. Groves et al., in *Scientific American* 245:66, 1981.

Figure 14.4. Adapted from Preston Cloud, in *Paleobiology* 2:179, 1976.

Figure 14.5. Drawn on a base map of P. Morel and E. Irving, in *Journal of Geology* 86:535–62, 1981.

Figure 15.3. Adapted from a diagram by R. Woese, in *Scientific American* 244:163–92, 1981.

Figure 15.4. Data from an anonymous article, in *Mosaic* 9(2):5, 1979, and articles by G. Vidal, in *Scientific American* 250:48–58, 1984, and J. W. Schopf, in *Scientific American* 239:84–102, 1978.

Figure 15.5. Data for the banded iron formation from M. J. Cole et al., in *Journal of Geology* 89:169–184, 1981; the oxygen curve is from Cloud (see Figure 14.4).

Figures 17.4 and 17.5. Adapted from drawings in *Mosaic,* p. 21, March-April, 1979.

Figure 18.3. Based on data of W. C. Fallaw et al., in *Geology* 7:398–400, 1979, and *Journal of Geology,* 88:723–9, 1980.

Figure 18.4. Based on a drawing of B. Kurtèn, in *Scientific American* 220:54, 1969, with a base map from Smith and Briden (see Figure 7.5).

Figure 18.5. Based on a drawing of A. Hallam, in *Scientific American* 227:56–66, 1972, using a more modern base map.

Figure 19.1. Adapted from D. M. Raup and J. J. Sepkoski, in *Science* 215:1502, 1982.

Figure 19.2. Compiled from graphs of K. S. Thompson, in *Nature* 261:578–80, 1976.

Figure 19.3. Modified, with new data on continental configurations from Scotese (see Figure 7.4), from a concept of J. W. Valentine and E. M. Moores, in *Journal of Geology* 80:167–84, 1972.

Figures 19.4 and 19.5. Based on diagrams of T. J. M. Schopf, in Gray and Boucot (see Figure 11.7).

Figure 19.6. Data from H. Tappan and A. R. Loeblich, in *Earth Science Reviews* 9:207–40, 1973.

Index

Index

Index